CONSERVING BIODIVERSITY IN ARID REGIONS

Best Practices in Developing Nations

Cover Photography (as pictured clockwise)

1. "Village in arid mountains of Oman" by John Lemons
2. "Llama pack train in Peru" by Eurgenio Araza-Llona
3. "People in the Gobi, Mongolia" by Ch. Dugarjav
4. "Research workers in Kenya," by Nathan Gichuki

CONSERVING BIODIVERSITY IN ARID REGIONS

Best Practices in Developing Nations

edited by

John Lemons
University of New England
Biddeford, ME, U.S.A..

Reginald Victor
Centre for Environmental Studies and Research
Sultan Qaboos University
Sultanate of Oman

Daniel Schaffer
Third World Academy of Sciences
Abdus Salam International Centre for Theoretical Physics
Trieste, Italy

SPRINGER SCIENCE+BUSINESS MEDIA, LLC

ISBN 978-1-4613-5045-3 ISBN 978-1-4615-0375-0 (eBook)
DOI 10.1007/978-1-4615-0375-0

Library of Congress Cataloging-in-Publication

A C.I.P. Catalogue record for this book is available
from the Library of Congress.

Conserving Biodiversity in Arid Regions: best practices in developing nations
Edited by: John Lemons, Reginald Victor, Daniel Schaffer
Copyright 2003

Printed on acid-free paper.

Contents

ABOUT THE EDITORS

Dr. John Lemons is professor of biology and environmental science in the Department of Environmental Studies at the University of New England, Biddeford, Maine. Dr. Lemons has authored over 100 peer–reviewed articles and has produced seven scholarly books focusing on issues of science and sustainable development, ecological integrity, and scientific uncertainty and environmental problem solving. He also served as editor–in–chief of *The Environmental Professional*, the official journal of the National Association of Environmental Professionals (NAEP) from 1990 through 1995. Dr. Lemons is an elected Fellow of the American Association for the Advancement of Science for his contributions to science and public policy.

Dr. Reginald Victor is professor of biology and Director of the Centre for Environmental Studies and Research (CESAR) at Sultan Qaboos University, Muscat, Sultanate of Oman. Dr. Victor has over 80 international publications in areas of freshwater ecology and environmental issues concerning sustainability and management in arid lands. Dr. Victor is a member of several professional organizations in the fields of limnology and environmental science. He advises the Ministry of Regional Municipalities, Environment and Water Resources (Oman) on environmental issues. He was chair of the Terrestrial and Freshwater Biodiversity Committee (Oman) and a member of the Final Drafting Committee for National Biodiversity Strategy and Action Plan (NBSAP), Oman.

Dr. Daniel Schaffer is the public information officer for the Third World Academy of Sciences and the Abdus Salam International Centre for Theoretical Physics, Trieste, Italy. He has written on science and technology issues both in the developed and developing world for more than 15 years and was the editor–in–chief of *Forum*, a science policy journal published jointly by the University of Tennessee, Tennessee Valley Authority and Oak Ridge National Laboratory. He has published scholarly books with Johns Hopkins University Press, Temple University Press and the University of Tennessee Press, and has written and produced television documentaries on science and public health issues that have aired on public television stations across the United States.

PREFACE

On the eve of the World Summit for Sustainable Development (WSSD), held in autumn 2002 in Johannesburg, South Africa, United Nations Secretary General Kofi Annan recommended five specific areas as focal points of discussion for the global forum: Water, energy, health, agriculture and biodiversity. In his address, "Towards a Sustainable Future," delivered just four months before the WSSD, Secretary General Annan contended that concrete progress in each of these areas, often referred to by their acronym WEHAB, would be key to improving the quality of life not only in the developing world but across the globe.

For most people, I think it is fair to say that the inclusion of biodiversity in a list that focuses on basic human needs may not be self–evident. Water, energy, health and agriculture, yes. But why biodiversity?

The truth is that biodiversity is just as critical to global well–being as water, energy, agriculture and health. This is because biodiversity both drives and shapes nature's intricate and dynamic structure in an enduring form and force that enables both current and future generations to enjoy its bounty.

Biodiversity provides the basis for the treasure trove of natural products and services that allow us to live our lives and that will enable our children to live theirs. It is the direct source of employment and sustenance for hundreds of millions of people, especially in the developing world. It also serves as a "shock absorber" for environmental assaults and provides a firm backdrop for nature's aesthetic diversity that is a source of pleasure, recreation, inspiration and indeed spirituality for all of us. In the words of internationally renowned ecologist Sir Ghillean T. Prance, biodiversity is the "green glue" that holds nature together.

The case studies contained in this volume are part of a project that began with a series of discussions between officials from the Third World Network of Scientific Organizations (TWNSO) and the Global Environment Facility (GEF) in the late 1990s. The dissemination of the case studies is designed to help give biodiversity the recognition that it deserves. But more than that, it is designed to examine numerous examples of efforts to wisely use and conserve biodiversity based on the knowledge and skills of people, especially people working at the local and regional levels whose labours have often been neglected and ignored.

The project is unique in two aspects. First, the case studies are dedicated exclusively to activities in the developing world. Second, the studies examine experiences in arid and semiarid zones. Ecological degradation in Mongolia's Baikal basin, management of vicuña populations in Peru, coastal protection in Oman, and forest conservation in Burkina Faso are just four examples of topics covered in the 35 case studies presented in this book.

We don't often think of drylands as being rich in biodiversity. Moreover, even when recognizing the rich biodiversity that exists in arid and semiarid areas, we are likely to undervalue the critical role that such diversity plays in the lives of the people who live there. The fact is that these regions are home to more than 1 billion people, ranging from the Bedouins who inhabit the desert sands of the Arabian peninsula to the highlanders residing in the cold desert glacial mountains of Mongolia. All told, drylands cover 40 percent of the globe and a diversity of land forms and ecosystems that include deserts, savannas, grasslands, shrub lands and forests.

The full range of organisms found in such environments face common environmental stresses marked by long stretches of low rainfall and periods of drought sometimes intermixed with brief spurts of heavy rain. Despite such adverse conditions, drylands are indeed remarkably rich in biodiversity. Such diversity, however, has been placed at risk due to population growth, overgrazing, the spread of invasive species, excessive plant harvesting, deforestation and global climate change. Because people living in such dryland regions are among the world's poorest people and thus often find themselves dependent on indigenous resources for their daily survival, it is important to develop strategies that do not simply preserve these resources but encourage their wise and productive use.

Like coastal zones and forests, dryland areas cannot be turned into "hands–off" national parks and nature reserves. People must be part of the equation. This means we must develop strategies that enable the rich biodiversity found in dryland regions to be used wisely by present generations without pursuing development policies and land–use practices that degrade or diminish the resource for future generations. In other words, we must devise strategies that ensure the sustainability of dryland biodiversity.

Such efforts require programmes that promote local participation in the decision making process; support a continual exchange of information and ideas; raise public awareness; and nurture partnerships among scientists, governmental officials, and local people that encourage them to learn from each other's expertise and experience.

These are the principles that have driven this project, "Promoting Best Practices for Conservation and Sustainable Use of Biodiversity of Global Significance in Arid and Semi–Arid Zones." Our journey has taken us to Ulaanbataar, Mongolia, where we held a workshop attended by more than 30 people largely from central and northern Asia. It brought us to Santiago,

Chile, where we organized a workshop that attracted more than 60 scientists largely from Latin America; it led us to Muscat, Sultanate of Oman, for a workshop in which 45 scientists largely from Africa and the Middle East participated; and it guided us to Cairo, Egypt, for an international conference attended by more than 75 scientists from throughout the world. In Egypt, participants decided to launch an institutional network for the discussion of issues related to dryland biodiversity that they hope will continue to function well beyond the grant project itself. In all, more than 100 case study and other experiences have been presented and discussed at these events. I would like to thank both the GEF and the United Nations Environment Programme for supporting our project. Both institutions have been indispensable in our efforts. I hope that what we have accomplished is worthy of the "investment."

Our hope is that this book, together with the project's other activities including the project's network of participating institutions, website (www.gefbiodiversity.org), and a popularized version of the case studies that will also be published, will help raise public awareness concerning this critical resource issue. Specifically, the case studies are designed to provide a series of successful experiences or "blueprints" that may serve as models to help others develop their own strategies for the conservation and wise use of biodiversity in dryland regions. We also hope that the camaraderie generated by this experience will lead to a continual exchange of information and ideas and ultimately to the development of joint projects that can build upon the knowledge and know–how that project participants have acquired both individually and collectively.

This project would not have been possible without the help of our advisory board: Leonard Berry, Florida Atlantic University, USA; Donald Brown, Pennsylvania Consortium for Interdisciplinary Environmental Policy, USA; Hassan Hassan, World Bank, Sudan; Gloria Montenegro, Pontifical Catholic University, Chile; Sir Ghillean T. Prance, The Eden Project, UK; and Hamid Zakri, United Nations University, Japan. On behalf of the participants and staff I would like to thank them for the contributions that they have made in helping us devise and implement an effective strategy for fulfilling the project's goal, including the publication of this book.

I would also like to thank John Lemons, Department of Environmental Studies, University of New England, Maine, USA, who has served as the project's principal consultant. John has guided the day–to–day activities of this project with skill and enthusiasm, keeping both staff and participants informed of our problems and progress. I would also like to extend our thanks to Helen Martin, a member of the TWNSO secretariat, who has provided expert administrative assistance enabling the project to stay on track despite the complications caused by its global reach and wide–ranging activities.

Like tropical rainforests and coastal zones, drylands are enduring elements of the global landscape. These environments can neither be dismissed nor radically transformed. Instead they must be wisely managed. To achieve this goal, we must draw on the indigenous knowledge and proven expertise of the people who live there; identify their best practices; increase South–South cooperation; and assist efforts to oversee and sustainably use the biodiversity resources that are an integral part of these ecosystems. These have been the prevailing goals of the project from its inception – goals that I hope will be advanced by the publication of this book and the other related activities that TWNSO and the participants of this project have been engaged in over the past several years.

Mohamed H.A. Hassan
Secretary General
Third World Network of Scientific Organizations (TWNSO)
Trieste, Italy
April 2003

About TWNSO and TWAS

TWNSO, founded in 1988, is a network of nearly 160 scientific organizations in the South, including ministers of science and technology, research councils and academies of science. TWNSO's mission is to help build political and scientific leadership in the South for science–based economic development and to promote sustainable development through South–South and South–North partnerships in science and technology. Among the important projects designed by TWNSO to fulfill this mission is to establish networks of centres of excellence in such critical areas as fresh water management, medicinal and food plants, renewable energy and dryland biodiversity. The Third World Academy of Science (TWAS), based in Trieste, Italy, provides the secretariat for TWNSO and cosponsors a number of its activities.

The Third World Academy of Sciences (TWAS) is an international autonomous organization, headquartered in Trieste, Italy. TWAS was founded in 1983 and operates under the administrative umbrella of the United Nations Educational and Scientific Organization (UNESCO). TWAS membership currently consists of 666 members, including 16 Nobel Laureates. More than 80 percent of the TWAS's members live and work in the developing world. TWAS pursues a wide–range of programmes designed to build scientific capacity in the developing world, including the sponsorship of awards and prizes, research grants, fellowships and associateships visits, and meetings. The Italian government provides most of the operational funds. In addition to hosting TWNSO, TWAS also oversees the secretariats of the Third World Organization of Women in Science (TWOWS) and the InterAcademy Panel on International Issues (IAP).

Chapter 1

CASE STUDIES ON CONSERVING AND SUSTAINABLY USING BIODIVERSITY IN ARID AND SEMIARID REGIONS OF SOUTHERN NATIONS

John Lemons[1] and Reginald Victor[2]
[1]Department of Environmental Studies, University of New England, Biddeford, ME 04005, USA, email jlemons@une.edu; [2]Reginald Victor, Department of Biology and Centre for Environmental Studies and Research, Sultan Qaboos University, PC 123, Sultanate of Oman

1. INTRODUCTION

The book is a compilation of case studies focusing on the conservation and sustainable use of biodiversity of global significance in arid and semiarid regions of Southern nations. The case studies were developed as part of the GEF/UNEP funded project "Promoting Best Practices for Conservation and Sustainable Use of Biodiversity of Global Significance in Arid and Semiarid Zones" (project number GF/1300–99–03) implemented by the Third World Network of Scientific Organizations (TWNSO) in Trieste, Italy.

An overall goal of the project was to more widely disseminate the lessons learnt and other findings from already existing projects and relevant work conducted by institutions of excellence on the conservation and sustainable use of biodiversity in arid and semiarid areas. We hope that this dissemination will accomplish the following. One, increase the availability and access to information on best practices for the conservation and sustainable use of biodiversity. Two, increase participation of local people in decision making about the use and management of fragile ecosystems. Three, increase awareness of the values of the biodiversity of global significance in arid and semiarid ecosystems in accordance with the

Convention on Biological Diversity (CBD). Four, increase coordination between institutions working towards the conservation and sustainable use of biodiversity, resulting in turn, in more effective programming of scarce financial resources and lesser duplication of activities. And five, increase partnerships of institutions in the South and their capacity to develop and implement successful programs to protect biodiversity.

Here, we provide some contextual issues pertaining to conceptual clarity about biodiversity and sustainability; prospects and problems of case studies; examples of lessons learnt from the case studies; and recommendations from the case studies.

2. CONCEPTUAL CLARITY OF KEY TERMS AND CONCEPTS

In literature and practice, concepts such as "biodiversity of global significance," "sustainable development" and/or "sustainability" and even "biodiversity" itself are "fuzzy." There are no commonly agreed–upon or a priori definitions that are sufficiently prescriptive or fully useable in scientific, public policy, or management. In this sense, the definitions are qualitative and stipulative and do not provide a firm foundation for precise decision making norms in science, public policy, or management. This might not be particularly troublesome because "fuzzy" concepts can provide convenient shorthand for discussing and analyzing complex phenomena and also can stimulate further discussion. However, if definitions are too fuzzy or if assumptions about them have not been sufficiently tested or critically analyzed this can limit problem formulation and solution and/or create conflicting formulations and solutions.

To our knowledge, the literature on biodiversity and/or sustainability contains few if any precise definitions for "biodiversity of global significance" or critical analyses of definitions. Proposed definitions of the term "sustainable development" include "the continued development as that which meets the needs of the present without comprising the ability of future generations to meet their own needs" (WCED 1987); "the continued satisfaction of basic human needs such as food, water, and shelter as well as higher–level social and cultural necessities such as security, freedom, education, employment, and recreation" (Johnson 1993); meeting the "...needs of the present as long as resources are renewed or, in other words, does not compromise the development of future generations" (Johnson 1993); to meanings that are so vague as to be ill–defined (Shearman 1990). Increasingly, the meanings of sustainable development have emphasized social and economic development (IISD 2003). Despite the fact that most proposed definitions of sustainable development are centered on human needs, there is increasing support to protect biodiversity for intrinsic reasons (see e.g., Regan 1983, Hargrove 1992, Oates 1999).

Goodland and Daly (1995) have attempted to clarify the meaning(s) of sustainable development by distinguishing environmental sustainability (ES) from social sustainability (SS) and economic sustainability (ECS). Environmental sustainability seeks to improve human welfare by protecting resources used for human needs and ensuring that people do not cause long–term or irreversible damage to the environment. Social sustainability seeks to improve peoples' social systems and necessary infrastructure in ways that empower self and local control and increasing social justice and equity. Economic sustainability seeks to promote the maintenance of capital, human investments, and income for economic well–being. In Goodland's and Daly's view, if people do not clearly define the kind of sustainability they are concerned about and/or do not distinguish these types of sustainability from each other the definitions of sustainable development or sustainability become so vague that they are not capable of being operationalized.

Robinson (1993) provides examples where ES, SS, and ECS are in conflict. Many population levels at which a species can be harvested sustainably, and the extent to which a species can be harvested depends in large part on whether it exhibits density–dependent compensation by increasing its rate of growth. Species that exhibit density–dependent compensation tend to be found in earlier successional and less biologically diverse ecosystems. Consequently, it becomes more feasible for humans to utilize species from less mature ecosystems because their production can be maximized despite the fact that to do so requires the maintenance of earlier successional ecosystems containing less biodiversity. Alternatively, it can be problematic for indigenous peoples to utilize species from mature ecosystems for economic use because many such ecosystems do not appear to contain species with high enough densities and rates of population growth to support more than relatively low levels of utilization. Human use of species also is problematic for the overall conservation of biodiversity because harvest of one or more species can have significant and unpredictable ecological ramifications throughout the community. For example, managing the populations of some species for sustained harvest can lead to a shift in relative abundance of coexisting species, the extent of which will depend upon the tightness of coupling of harvested species to others in the food web and cycles of nutrients. Consequently, although species can be used to meet the goals of social and economic sustainability, the fact that they can be used does not say anything about the very same use causing the loss of species and/or biodiversity.

Conflicts between ES, SS, and ECS also exist at the community or ecosystem level (see e.g., Lemons and Brown 1995, Westra and Lemons 1995). Theoretically, communities or ecosystems can be sustained at different levels of intensity of managed use. However, ecosystems managed at more intensive levels will be less biologically diverse than those managed at less intensive levels, although the former might be able to provide for peoples' needs. Conflicts between the different types of sustainability also derive from the fact that ecosystem services are used at the local, regional, and global levels. Conflicts may arise among the different types of sustainability because at the local or regional level communities or ecosystems might be used to meet the local needs for SS and ECS, whereas at the global level communities or ecosystems might be required to be preserved or used at lower levels of use in order to contribute to the maintenance of global ecosystems, or ES. Conflicts between ES, SS, and ECS also can exist because the time scales for their assessments can be different. For example, the needs of impoverished local people may require rapid improvements in their living conditions, and business decisions need to reflect certain rates of return for investments even under conditions of sustainability. The time scales for assessments for improving protection of biodiversity or for using biological resources often is slower than those for making decisions about SS or ECS because assessing the damage to biodiversity typically requires relatively long time periods. The fact that these different time scales exist has profound implications for our ability to correctly identify what constitutes "best practices for the conservation and sustainable use of biodiversity."

In the case studies presented in this volume, few authors provide clear or precise definitions of "biodiversity of global significance," "sustainable development," "sustainability," or even "biodiversity" itself. Further, most case study authors do not clearly distinguish ES, SS, or ECS from one another, or comprehensively discuss known or potential conflicts between them. This reflects the aforementioned lack of clear or precise definitions found in literature or practice, as well as problems stemming from the complexity of assessing the different forms of sustainability and their conflicts. The absence of conceptual clarity for key terms and concepts should not deter scientists, decision makers and environmental managers from implementing practices that are believed to be the best for conservation and sustainable use of biodiversity of global significance in arid and semiarid regions. Each case study presented here uses terms and concepts with particular reference to its context (i.e., local or regional problems). We are confident that each reader will strive to find implicit meanings of the terms and concepts in these case studies to evaluate their validity and usefulness.

3. PROSPECTS AND PROBLEMS OF CASE STUDY APPROACHES

What is a case study? We suggest the following: A case study is a detailed investigation of a specific problem or event where people reason deductively from general principles to reach the solution of that particular problem. One reason to use case studies is to help stakeholders with different perspectives, interests, and needs to understand and solve the problem. A second reason to use case studies is because lessons learnt might be useful beyond the immediate or more local or site–specific problems. If this could be accomplished it would help in the development of projects which might serve as models for the bilateral and multilateral funding communities. A third reason to use case studies is because, in part, they have been developed in response to the limitations of methods and techniques of ecological and biodiversity science to yield scientifically robust predictions based on hypothetic deductive hypotheses and testing (except for relatively simple systems). In other words, in most instances ecological and biodiversity theories are not generalizable and operationalizable (see, e.g., Shrader–Frechette and McCoy 1993). Finally, a fourth reason to use case studies is because of their heuristic value. In this sense, case studies focus on human judgments about practical environmental policy and management.

Strengths of the case study approach include enabling scientists and decision makers to: gain a measure of practical control over environmental problems; make rough generalizations about complex problems; and learn about problem solving from case studies conducted on similar but different cases. Case study approaches also are applicable to unique situations not amenable to statistical tests and hypothesis testing, and they provide an organized framework for developing alternative models and explanatory accounts of complex environmental phenomena. Case studies also allow for an ability to deal with imperfect evidence even though controls and manipulations and experimental tests are not possible; they provide a way to see a problem like other problems, and they offer a way to gather information in order to formulate hypotheses.

Case studies also have weaknesses. These include the fact that their analysis of examples does not follow any algorithms and therefore there might be inadvertent bias or different ways of framing a problem by the practitioner; they provide little basis for robust scientific predictions; they can suffer from the fact that they can only be used to evaluate those interpretations the case study researchers presuppose and that such presuppositions typically are not testable; they do not follow pure hypothetic–deductive inference; and most real life decisions are made in dynamic contexts wherein many case studies describe static situations.

Despite the drawbacks, case study approaches are suitable for issues on biodiversity conservation and sustainable development because both these issues are globally diverse and do not allow direct comparisons using conventional scientific and ecological methodologies. However, we recommend that to be useful, case studies must sufficiently describe and analyze the threats to biodiversity and the causes of those threats. This does not mean that scientists are required to demonstrate proof about threats based on data with a high degree of statistical confidence, but rather have rational reasons for their conclusions. However, they should make explicit the assumptions their methods and techniques are based on, the sources of uncertainty, and the implications of uncertainty to their scientific reasoning and conclusions. A step–by–step description of how their information and data leads to conclusions should be attempted. Further, public policy makers and decision makers must not assume that scientific information is more certain than is warranted (see, e.g., Lemons 1996).

The usefulness of case studies also requires knowledge about the sources of threats to biodiversity and which governments at which scales have the authority to do something or are failing to do something or have little power to implement specific recommendations to avoid a threat. For example, if the threat to biodiversity is land use encroaching upon valued biodiversity, it will be helpful if case studies identify and analyze the threats caused by land use. The threats could stem from the lack of appropriate land use plans or lack of the legal authority of a regional government to develop such plans, or from the reluctance of a government to restrict people who are malnourished from using local resources. Where causation of threats is not well understood, case studies might contain recommendations on the need to do a strategic analysis of the causes of threats. In other words, it is difficult to make meaningful recommendations about how to protect biodiversity if one cannot ascertain which institutions have the responsibilities to implement a plan.

4. EXAMPLES OF LESSONS LEARNT FROM THE CASE STUDIES

The case studies in this book contain several types of lessons learnt and recommendations. One, is improving science used in biodiversity research, public policy, and management. Two, is making connections between local, national, and global biodiversity efforts. Three, is building institutional capacity for research and protection of biodiversity in drylands and for sharing access and benefits of biodiversity resources. Four, is clarifying practical strategies to implement biodiversity protection in drylands. And five, is effectively using information obtained from this project to address GEF and other critical donor issues and positively impact their future activities and programs. Many case studies are highly concrete and are based on programs and practices that have been successfully implemented; others focus on recommendations to protect biodiversity based on scientific research or other experiences.

The examples of lessons learnt and recommendations from case studies provided below are grouped according to their primary focus, i.e., scientific, public policy and management, participation of local people in decision making, and partnerships and capacity. Despite this grouping, each case study transcends a narrowly defined focus as required to address the interdisciplinary nature of biodiversity and sustainability problems.

4.1 Case Studies Focusing on Science

S. Donaldson et al. ("Conservation Farming With Biodiversity In South Africa: A Preliminary Evaluation Of Ecosystem Goods And Services In A Semi–Arid Landscape In The Bokkeveld Plateau") describe the processes controlling habitat and ecosystem integrity essential to prevent biodiversity loss in agricultural landscapes. Their case study focuses on a study site in a semiarid region on the Bokkeveld Plateau (Northern Cape province) in South Africa. The data are presented as a preliminary model where land use options influence both biodiversity and the provision of ecosystem goods and services. Included are preliminary data on the risk of extinction for plant and animal species under different forms of land use, as well as ecosystem goods and services related to water infiltration, soil health, carbon sequestration, and production. The ecological components of the project will be integrated with social and economic studies to develop an ecological–social–economic model for the region that can be used to guide land use planning in the region.

P.D. Gunin and S.N. Bazha identify the most serious risks of ecosystem degradation to Mongolia's ecosystems and their potential impacts on the Russian part of the Baikal basin ("Ecological Assessment Of Degradation Processes In The Ecosystems Of The Mongolian Part Of Baikal Basin"). This case study had three objectives: to conduct an inventory of the ecosystems of the southern part of the Lake Baikal basin and classify them according to economic utilization and modification by humans; to analyze processes affecting the Lake Baikal basin's ecosystems and components such as vegetation, soil, topographical relief, etc.; and to assess integrated zoning of the southern part of the basin to help mitigate environmental degradation. The authors describe the results of an inventory of the ecosystems of the southern (Mongolian) part of the Lake Baikal basin at the level of mesoecosystems and classify the ecosystems according to their economic use and level of modification by humans; they also describe the spatial distribution of ecosystem degradation processes and the levels of their impacts on the biota of both the Lake Baikal territory within Mongolia as well as in contiguous areas in Russia. This information can be used in the monitoring of anthropogenic pressures on the ecosystems of the Lake Baikal basin. Gunin and Bazha provide a strong scientific basis for recommendations to protect the Baikal basin's biodiversity and environment. These include: ensuring that social and economic public polices are consistent with national and international biodiversity laws and the area's status as a World Heritage Site; establishing a network of strictly protected areas; and improving monitoring of the dynamics of aquatic and solid runoff in bottom sediments of the area's important rivers.

Two case studies are from China. One is "Plant Production And Diversity At Desertification Stages In Horqin Sand Grassland Region, China" by C. Xueli and Z. Halin illustrating the functions and effects of biodiversity in the region so decision makers will have a better theoretical and practical scientific foundation for decisions to restore and rehabilitate vegetation. The relationships between plant productivity and diversity depend on the specific structural or functional indices chosen at different desertification stages. Ecologists and natural resource decision makers not always recognized by ecologists and natural resource decision makers. For example, the desertification process distinctly affects the numerical values of the diversity indices calculated for sandy grassland vegetation. The various stages of desertification affect species richness change differently by influencing the species and life form composition, the frequency and abundance of different functional groups, and species and functional group productivity. Some of the study's results conflict with others published in the open literature, and this points to the need to refine studies on the relationships between plant productivity and diversity at different desertification stages so that scientific capability can better inform decision makers to protect arid and semiarid grassland biodiversity.

In the second case study from China, L. Xin–Rong et al. (Plant Diversity In The Process Of Succession Of Artificial Vegetation Types And Environment In An Arid Desert Region Of China) provide a theoretical basis for further understanding of ecological mechanisms of reversing desertification trends and restoring biodiversity in desert regions in China. Their study documents plant diversity changes in desertified areas where attempts have been made to restore vegetation. In the areas studied, after more than 40 years of succession the composition of restored vegetation species tends to become dynamically balanced; the plant diversity increases with the succession of plant communities; and the plant diversity index of the restored vegetation reaches is higher compared to the corresponding figures of the restored vegetation planted 10 years ago. Beta diversity measurements show that the succession of restored vegetation experienced two relatively rapid stages of species turnover, as when, e.g., restored shrub species became less abundant while the percentage of annual herbaceous plants increased as succession progressed. These finding have implications to management attempts to restore vegetation because they contribute to the theoretical understanding of vegetation restoration in desertified areas.

The case study "Theoretical Models Of Regeneration For Medicinal Plants: An Example Of The Use Of Science In Promoting Sustainable Wild–Harvesting" (G. Montenegro et al.) demonstrated that by working closely with people selling medicinal plants in local markets in Chile, researchers were able to study the harvest areas to calculate the biomass produced for market. They were then able to determine the frequency of plant collection and shared that information with their local partners. The researchers went on to develop a model for regeneration based on the location and dispersion of the plants' renewal buds. Plant growth both in the field and in clipping pots was measured to evaluate post harvest regeneration rates and plant response to biomass extractions. Theoretical models led to predictions of potential plant regeneration in different ecological zones on the Andes coast, and data sheets were produced for each species. Subsequent cultivation and reforestation programs have followed recommendations from the scientific studies described in this case study.

J. Wheeler et al. (Genetic Diversity And Management Implications For Vicuña Populations In Peru) demonstrate how advances in population genetics research are an important tool for monitoring the impacts of programs to preserve the vicuña. Results of the research indicate that individual Peruvian vicuña populations are characterized by relatively low levels of genetic diversity and that high levels of genetic differentiation exist between these populations. Such patterns are commonly observed in threatened species with formerly large ranges which have become isolated from each other, and in species which have suffered drastic demographic contraction in recent generations. These patterns may be becoming more predominant in populations of the Peruvian vicuña and therefore should be taken into account in future conservation strategies designed to minimize further loss of genetic diversity. In Peru, four demographically distinct vicuña population groups have been identified which should form separate management units. Preservation of this vicuña genetic biodiversity is becoming critical because Peru's vicuña management programs have increased the overall numbers of animals but this, in turn, has caused greater control over the species through construction of fences, intensive rearing and growing selection. Such approaches are inherently not sustainable and represent a serious threat to survival of the vicuña. Population genetics research represents an important tool for monitoring the impact of these activities and designing best management practices.

P. Lima et al. (Choice Of Species For Recovering A Degraded Mining Area In The Semiarid Zone Of Brasil) present the results of a project designed to analyze the processes of species and ecosystem recovery in an area degraded by copper mining and the prospects for choosing multipurpose trees for the recovery. The first phase of the project consisted of soil analysis, characterization of climate and vegetation, and descriptions of local and commercial farming systems; the second phase consisted of experimental planting of possible species for recovery; and the third phase validated innovative technological alternatives for rehabilitation of the degraded area. Flora in the degraded mining area and surrounding unaffected areas were identified and characterized as to their species, structure, frequency, abundance, dominance, value index of importance, vulnerability to mining operations, and potential for use in rehabilitation.

J. Araya–Valenzuela and R. Espero–Guasp (Use Of Creeping Fog Water As A Non–Traditional Water Resource In Chile) describe experiments utilizing creeping fog water as a non–traditional water resource in Chile's high deserts. They describes creeping fog formation and characteristics; factors influencing collection of fog; different fog collector designs and their technical advantages and disadvantages; and recommendations on which fog collectors are most efficient and economical in different environments and situations. Results of the experiments help in assessing the availability of water for species and ecosystem recovery, help in establishing new areas with endemic species cover, and help in establishing small communities around stable water (fog collected) resources or in supplying some existing communities with such water.

Soil criteria for selecting suitable transplanting sites for mangrove trees (*Avicennia marina*) in the Sultanate of Oman were studied and assessed by P. Cookson and T. Shoju ("Site Selection Criteria For Mangrove Afforestation Projects In Oman"). The extension and conservation of mangrove forests has recently been declared by Sultanate of Oman's Ministry of Regional Municipalities, Environment and Water Resources to be of national importance to maintaining biodiversity in the country. Consequently, a long–term afforestation project of some twenty coastal sites has been proposed, in addition to conservation measures for the over 1000 ha of existing mangrove forests in Oman. Commencing in 2001, mangrove seedlings have been raised in constructed nurseries and transplanted to a number of sites. Transplanting activities are already into their second year and seedling survival rates at different sites have ranged from 0.1 to over 80 percent. In this study, salinity and soil physical properties at four transplanting sites with a range of seedling survival rates were compared. Seedling survival rates were higher in soils with ratios of fine to very fine sand particles higher than unity. As the proportion of fine to very fine sand fell below one, the saturated hydraulic conductivity of sand decreased leading to anaerobic soil conditions in the seedling root zone. The study concluded that anaerobic soil conditions were closely associated with seedling death. Based on the study's conclusions, recommendations were made for some soil criteria for selecting suitable transplanting sites when transplanting mangrove seedlings.

M.D. Robinson also focuses on trees in the Sultanate of Oman in his case study "The Importance Of Native Trees In Sustaining Biodiversity In Arid Lands" by considering how trees develop microclimates, their interactions with other plants and animals, and their roles in sustaining below–ground soil fauna and flora. This latter aspect is given special emphasis because of its central importance in ecosystem function combined with the fact that soil biota has received far less attention by ecologists, managers and planners than the more visible species above the ground. Robinson ends his case study with a review of the possible benefits arid ecosystems can offer societies.

4.2 Case Studies Focusing on Public Policy and Management

The problems and prospects of domesticating indigenous trees in drylands are assessed by R.R.B. Leakey's in "The Domestication Of Indigenous Trees As The Basis Of Sustainable Land Use In Africa" for the purpose of developing national strategies for sustainable land use. The sustainability of land use in the tropics has typically been lost when species–rich natural vegetation has been cleared to make way for monocultures of improved staple food crops grown intensively with high inputs of agrochemicals. These intensive farming systems have helped to feed growing human populations, but at an environmental cost that cannot be sustained into the future. As Leaky suggests, an acceptable alternative that both feeds the people and restores some of the diversity found in natural vegetation is the development of lower input agroecosystems that combine the cultivation of "Green Revolution" staple food crops with a number of the indigenous food–producing tree species that can restore ecosystem function. This return towards more traditional land use practices can be enhanced by the judicious development of high quality, high yielding cultivars of the trees through the adaptation of standard horticultural practices and planting of these cultivars in agroforests. In dryland Africa, there are many traditionally important tree species producing marketable non–timber forest products, which have potential to be domesticated in this way. The techniques, methodologies and strategies exist and are being implemented in small ways. There is the potential to expand these programs to a scale where they can begin to have meaningful impact on land use and create new and more biologically diverse functioning agroecosystems which also support and enhance the livelihoods of local people.

In his case study "Strategies For In Situ Conservation Of Crop Genetic Resources In Dryland Areas Of Africa," M. Grum describes a program designed to reduce the loss of genetic biodiversity caused by drought and desertification in the dryland ecosystems of Africa. The aim is to mitigate the impact of temporary drought–induced conditions through the development of comprehensive community–based strategies for monitoring, analyzing, assessing and addressing the desertification phenomenon. The project attempts to understand the complex role that genetic diversity plays in coping strategies of farmers. This research focuses on a description of the crop genetic diversity, its extent and distribution, the socioeconomic situation of farmers, environmental factors, and on opportunities for stakeholder participation in decision making. The project also examines the positive and negative dynamics influencing the status of genetic resources and loss, including traditional management practices for selection, conservation and multiplication of the genetic resources. Finally, the project is testing models of community–based activities, such as community gene banks, improved seed storage systems, seed diversity fairs, farmer field schools, and on–farm seed production systems.

In "Aquatic Biodiversity In Arid And Semiarid Zones Of Asia And Its Linkages With Water Resource Management," B. Gopal points out that many scientific studies of arid lands biodiversity and its management focus on terrestrial species and ecosystems and overlook the aquatic habitats. Management which focuses on aquatic biodiversity in arid and semiarid zones often fails to take into consideration the unpredictable variability in precipitation and its relationships with native biota and consequently results in loss of biodiversity. Whereas natural aquatic habitats are being lost due to excessive withdrawal of water for various purposes, overexploitation of natural resources and changes in land use, numerous aquatic habitats have been created by transporting water through extensive networks of canals and water storage reservoirs. These water bodies are causing salinization, loss of biodiversity and a plethora of socioeconomic problems; loss of biodiversity from such practices is not sufficiently recognized by natural resource decision makers and managers. Pollution of water from industrial wastes, increasing salinity in coastal areas, and introduction of exotic species are other major problems affecting native biodiversity. This case study highlights the importance of public policy makers and decision makers to take into account understudied characteristics of aquatic habitats and their biodiversity in Asia's arid and semiarid regions.

Two other case studies from India focus on minimizing conflicts between local people and public policy and management of biodiversity. K. Chandrasekhar et al. ("Traditional Management Of Biodiversity In India's Cold Desert") describe areas of agreements and/or conflicts between local peoples' traditional use of biodiversity and government public polices and management of biodiversity in cold deserts in India. Their study shows that local peoples' concepts of the values of biodiversity have been key organizing principles in their traditional and sustainable landscape management and practices. However, the importance of local peoples' practices is diminishing largely because of government conservation policies and programs which fail to take into account the traditional practices. Specifically, policy and management driven changes establishing government rights in areas traditionally used by local communities have not enhanced either the economic values of biodiversity and/or its protection, for example, in the transfer of decision making powers from the village communities to government institutions and in government incentives to increase the yields of food crops. According to this study, the reason is because government policies and management practices fail to take into account local peoples' wide–ranging knowledge and uses of biodiversity. Accordingly, Chandrasekhar et al. conclude that the goals of biodiversity conservation and its sustainable utilization can be better achieved if conservation and development policies build on strengths and weaknesses of local peoples' traditional knowledge and institutions.

The main lessons learnt from M. Chauhan's "Conserving Biodiversity In Arid Regions: Experiences With Protected Areas In India" is that protected area management in the arid regions of India is not necessarily enhanced by simply leaving them alone. An understanding of the science behind the resilience of the protected areas in the face of occasional human disturbance and the fact that the areas have thrived in continuous interaction with human and domestic animal populations is required. The continuation of ecological linkages of protected areas with surrounding areas and their biotic communities may be necessary for ecosystem health.

Recommendations contained in the case study "Conservation And Sustainable Use Of Globally Significant Biodiversity In The Trans Altai Gobi Desert In Mongolia" (Ch. Dugarjav and B. Tsetseg) have laid a foundation for the further development of conservation activities in this region. The Trans Altai Gobi Desert, which includes the UNESCO Great Gobi Strictly Protected Area (SPA) Biosphere Reserve (largest protected area in Asia and the fifth largest in the world), is one of the world's most extreme arid and biodiversity unique deserts. This desert has remained relatively intact ecologically because small numbers of nomadic people and their animals have historically used it in traditional ways. However, intensification of desertification and threats to biodiversity are increasing due to greater industrialization and collectivization of animal herds; economical difficulties encountered by Mongolia during its transition to a free market economy; increasing poverty; and increasing demand for profit making by industries. Based on analyses of renewal of biodiversity legislation, establishment of the Great Gobi SPA, development of international cooperation and funding to support biodiversity protection, and the active involvement of scientists and broad application of research, lessons learnt from this case study can provide a foundation for the further development of conservation activities.

C. Richard discusses how Tibetan pastoralists face a number of natural, socioeconomic, organizational and policy challenges, especially "one size fits all" policies and development programs that have promoted intensification and private land tenure models that are more appropriate to moister lowland regions ("Co–Management Processes To Maintain Livestock Mobility And Biodiversity In The Diverse Alpine Rangelands Of The Tibetan Plateau: Rationale And Practice"). According to Richard, "indigenous common property regimes" (CPR's) have been shown to be an effective means to manage and protect common pool resources, especially among pastoral communities in dryland regions where survival and the maintenance of a healthy rangeland ecosystem is dependent upon collective action to maintain livestock mobility. Richard describes a set of hypothetical rangeland tenure models along an ecological gradient from the Tibetan Plateau in China to highlight the distinction between local autonomous control of pasture, top–down imposition of policy, a co–management model, and the influence that environment plays to enable tenure and management options. In today's complex and rapidly changing pastoral landscapes, participatory processes will not be sustained without a supportive external environment that protects the rights of users, facilitates conflict resolution, and promotes timely financial, marketing and technical inputs. A conceptual framework and set of strategies that foster a more collaborative organizational and policy environment within which pastoral communities can more effectively influence the course of their own development are proposed, based on the experiences of the International Centre for Integrated Mountain Development's (ICIMOD) Regional Rangeland Programme on the Tibetan Plateau.

Two case studies focus on the management of vicuña and other species in Bolivia, Chile, and Peru. C. Bonacic and J. Gimpal ("Sustainable Use Of The Vicuña [Vicugna Vicugna]: A Critical Analysis And The MACS Project) describe the current status of the vicuña's sustainable use status in the region. In their view, the vicuña conservation program in South America is entering a critical stage, and the direction that the program takes could conflict with the original objectives of the Vicuña Convention (i.e., to use the species in the wild for local communities' benefits). The trend for exploiting in captivity by fencing wild vicuña or captive breeding programs is not compatible with the initial criteria of sustainable use. Alternatively, a sustainable use program based on capture and release of small groups of vicuña with minimal interference on the natural populations should be considered the method of choice, to which other methods should be compared. Some of the main aspects to compare between methods of management are: the impact of capture, handling and shearing on different systems of use; the comparative studies of carrying capacity in different regions of the altiplano and livestock competition; and the social and potentially evolutionary consequences of captive programs.

B. Peredo ("Sustainable Use Of Andean Wildlife And Local Development Of Rural Communities In Dry Areas Of Latin America: A Commentary") discusses how population recovery programs based on sustainable use have increased the number of vicuña but threats to other Andean species such as the quirquincho and Andean Ostrich remain; the numbers of the latter two species may be declining due to unrestricted use. Studies and plans are being developed to improve the quality of vicuña fiber and distribute the benefits of the animal's sustainable use as well as to enable the sustainable use of the quirquincho and the Andean Ostrich, respectively, in traditional cultural activities. In this way, local communities are being provided with alternatives that will contribute to the conservation of these species as well as to the biodiversity of the Andean region of Bolivia.

In "Conflicts And Dilemmas Between Poverty And Biodiversity In The Semi–Arid Serido Of Northeast Brazil," E. Beaugrand examines the prospects and problems of reducing conflicts between long–term goals of sustainable development and short–term economic goals. She analyzes the current "Plan for Sustainable Development of the Serido Region" of northeastern Brazil, particularly with respect to the ceramics sector which is one of the region's most important social and economic sectors. Many government and nongovernmental institutions and stakeholders representing 28 municipalities in the region were involved in development and implementation of the plans. The ceramics industry is the primary source of jobs and income for local people in the region; few other opportunities exist for the livelihood of local people. However, the practices of the industry have significantly degraded both terrestrial and aquatic resources of the region. Based on an analysis of the plan's environmental, scientific, economic, social, political, and administrative indicators of sustainable development and their efficacy in achieving sustainable development goals, the study reveals a wide participation of local people in its development and implementation. Based on analysis of the indicators, conditions of poverty and lack of social and economic opportunities for local people of the region create pressures for public policy and management of the region to focus on short–term employment and economic opportunities without undertaking necessary precautions to protect the long–term sustainability of the region's environment.

The case study "In Situ Conservation Of On Farm Crop Biodiversity In Morocco: A Case Study," by F. Nassif and A. Birouk focuses on strengthening the scientific basis of in situ conservation of agricultural biodiversity. This in situ on–farm conservation project in Morocco involves nine countries with participating farmers, scientists, development workers and many others. The most important lesson learnt is that the in situ conservation of on farm agrobiodiversity is a very complex and multifaceted process. Scientific lessons learnt include: the need to establish multidisciplinary teams which include not only agricultural and genetic scientists but social scientists as well; the need to include a balance of gender perspectives in the study of in situ conservation of on–farm crop genetic diversity; the need to understand that landraces of crops are used by farmers because they represent the best options from the farmers' points of view. Public policy and management lessons learnt include: the most important gap between Morocco's official policy and implemented measures towards conservation and sustainable use of crop biodiversity stems from the absence of a national strategy plan on biodiversity as well as from the absence of a national gene bank. The main lessons learnt to assist local populations in utilizing and managing biodiversity are: that it takes time and effort on the part of researchers to understand and respect local knowledge systems and management practices and farmers' points of view; and, that there is a need to empower the local population by recognizing the key contributions of farmers in the management of crop biodiversity.

R. Victor ("Biodiversity Conservation And Sustainable Development: A Case Study Of Oman's National Biodiversity Strategy And Action Plan") discusses how concepts of biodiversity and sustainable development have evolved over the years but nevertheless remain vague and ambiguous; some concepts emphasize development while other emphasize biodiversity protection. This has led to an inability to explain the link between biodiversity and sustainable development in unambiguous terms. Examples are the national biodiversity strategies and action plans produced by the nations signatory to the CBD. The worst–case scenarios in some of these strategies and action plans indicate very remote, if at all any, links between biodiversity and sustainable development. Victor describes a study of a fragile mountain ecosystem in the Sultanate of Oman with reference to the country's National Biodiversity Strategy and Action Plan– Oman 2001. He provides a critical overview of biodiversity conservation and sustainable development in order to compare the past and present day thinking on biodiversity conservation and sustainable development and recommend a balanced perspective while formulating important instruments such as national strategies and action plans. Victor concludes with recommendations on methodologies to use to identifying the state of threatened habitats and best practices to mitigate the threats.

In "The Globally Threatened Corncrake *Crex Crex* (Egypt)" A. Grieve and W. Salama focus on the corncrake (*Crex crex*) which is a globally threatened and vulnerable species due to a long–term and steep decline of its breeding numbers and range. They describe a project in Egypt comprising the following activities: to carry out field surveys to calculate the number of corncrakes trapped in autumn on the Mediterranean coast of Egypt; to try to reduce the number of corncrakes killed in Egypt during the autumn migration period by talking to hunters, local people, decision makers and distributing information to support the releasing of corncrakes; to carry out a study program on migrant corncrakes to provide biometric data from trapped birds and feather samples for DNA analysis; to provide training for personnel of the Egyptian Environmental Affairs Agency in survey techniques, public awareness campaigns and management of protected areas for migrant corncrake; and to assist with the development of a visitor center and educational program at Zaranik Protected Area, north Sinai, Egypt, to raise public awareness for the corncrake.

4.3 Case Studies Focusing on the Participation of Local People in Decision Making

The case study "The Value Of Local And Indigenous Knowledge For The Development Of Information Systems For Conservation Management" by K. Kellner and O.J.H. Bosch provides an overview of the need to develop and disseminate more effective tools and methods to obtain information for conservation and sustainability programs, especially where the use of local indigenous knowledge is important or essential. Kellner and Bosch discuss how the development of inventories and information systems where indigenous and scientific knowledge are incorporated into single expert knowledge systems will not only help in the connection of science and community action but also in the awareness, education, training and capacity building of agriculturalists and conservationists. Accordingly, this will provide support for land users and managers in their future decision making and will also enhance the adoption of the management options by the end–users. Kellner and Bosch go on to discuss how important it is that information systems also include the verification and validation of local knowledge which can be based on literature, statistical analysis and scientific experimentation, as well as sampling theory and logical analysis of data collected by different methods. Finally, they discuss the difficulties of sustaining long–term funding in arid and semiarid environments and how this often leads to discontinuity in the research process. It is therefore essential to develop mechanisms whereby knowledge–building processes such as community–based research, adaptive management, monitoring, feedback and community dialogue can be institutionalized or embedded in the community before funding is ceased. In this way, knowledge building becomes an ongoing process and the information system in which the knowledge is captured evolves as more information becomes available through either research or management.

M.B.K. Darkoh ("Agriculture And Biodiversity Conservation In Africa Through Indigenous Knowledge") explores some of the important relationships between agriculture and biodiversity. Often, indigenous or local farmer knowledge about production systems has been largely overlooked despite the fact that indigenous people and farmers are as a matter–of–fact partners to conserve and manage biodiversity whether for nature reserves, or to improve crops and livestock yields. Darkoh discusses how local knowledge systems, traditions, institutions and environmental conditions are fundamental to biodiversity conservation and management. As a matter of policy, he advocates that the promotion of agricultural programs and biodiversity conservation projects should ensure that indigenous knowledge is incorporated in their design and implementation. A blend of modern science and indigenous knowledge will be required to face the challenges of increasing agricultural production and managing the environment on a sustainable basis in the decades ahead in Africa. It is also important to ensure that the relevant local communities are given appropriate control over and access to land and other resources as well as management responsibility for the natural areas upon which their continued prosperity depends. Furthermore, development assistance to agriculture and biodiversity conservation and land management will be most beneficial when it attempts to enhance existing agricultural systems.

The purpose of N. Gichuki's and J.M. Macharia's case study "Participation Of Local Communities In The Management Of Wetlands In Magadi Area, Kenya" is to show how wetlands in the arid zone of Magadi, Kenya, provide essential resources that sustain local economies and livelihoods of people. They also discuss how the full range of wetland benefits is not well understood and appreciated. Approaches to management have been focused on a few resources (e.g., water, land for agriculture or minerals). This narrow approach has led to unsustainable exploitation of wetland resources, thereby threatening ecological integrity of these ecosystems. Arid zone wetlands, however, have immense potential for supporting sustainable development and fighting poverty. Gichuki and Macharia discuss how local communities should be empowered to participate in shaping their own destiny through management of environment and resources according to their need and visions of the future in ways that will produce sustainable outputs, integrate environmental concerns and livelihood issues and establish mechanisms for long–term management of environment and natural resources.

L.G. Ouedraogo et al. demonstrate using traditional knowledge of crop seed storage to conserve forest genetic resources in Burkino Faso ("A Participatory Approach For Conservation Of Forest Genetic Resources"). In Burkina Faso as well as in the other Sahelian countries people rely on trees and shrubs for their daily life. Despite the importance of the forest resources to the livelihood of rural people, Burkina Faso loses approximately 32,000 ha of forest annually. In order to protect forest genetic resources and improve the well–being of local populations, the National Forest Seed Center of Burkina Faso developed and implemented methods to utilize local peoples' knowledge of crop seed storage and improve on this knowledge and ability to better protect forest genetic resources. All of the management areas were located in the Sudanian and Sahelian zones of Burkina Faso and part of the semiarid zones of West Africa. Based on local people's traditional knowledge of crop seed storage and subsequent experimentation, Ouedraogo et al. describe innovative methods for characterizing and mapping of forest areas for management; selecting the most important tree species whose growth and production should be increased; improving handling and storage of forest seeds; determining practical and reliable methods to maintain forest seed viability and direct sowing efficiency; identifying simple traditional forest seed storage which allow protection of the seeds and control of parasite attacks during storage; and evaluating direct sowing impact on forest regeneration. Finally, the study shows how involvement of local people in developing strategies to improve forest seed storage can generate jobs and income in rural areas while at the same time contributing to forest biodiversity protection.

4.4 Case Studies Focusing on Partnerships and Capacity Building

"Best Practices In The World's Oldest Desert" by M.K. Seely and J.R. Henschel describes informal public and private NGO organization partnerships and their contributions to long–term research and protection of biodiversity in the Namib Desert, Namibia. The public sector manages established national parks and more recently in partnership with NGOs is promoting community based natural resource management on farmlands, while the private sector is involved in tourism that provides the foreign exchange income motivating the public sector to retain interest in biodiversity protection. Long–term ecological research in support of best practices covers a wide range of basic and applied research. For example, lessons learned concerning environmental variability include the role of ephemeral rivers that cross desert areas while supporting a riparian forest with its associated fauna and farming opportunities and recharging underground aquifers upon which urban coastal development depends. Several research projects undertaken with the indigenous community focus on developing markets for fruit products, on community based tourism and on harvesting fog. Public policy is driven by changing global perceptions of biodiversity values and this in turn fuels international tourism. The tourist value of the Namib Desert has been enhanced by research and dissemination of the results of biodiversity research. As vast protected mining areas are opened up to alternative use, their status as a biodiversity hotspots influences public policy and future use. Seely's and Henschel's case study focuses on the Desert Research Foundation of Namibia and on the roles of other institutions participating in the government's biodiversity task force.

G. Prance ("Plants Of Northeastern Brazil: A Programme In Sustainable Use Of Plant Resources") describes an eight state program run by a consortium of governmental and nongovernmental organizations in the semiarid region of northeast Brazil which helps local people to use plants resources sustainably. The program "Plantas do Nordeste" or plants of northeastern Brazil is run by a consortium of Brazilian governmental and NGOs in the region in cooperation with the Royal Botanic Gardens, Kew, England, and led by the Brazilian NGO, Associação Plantas do Nordeste, based in Recife in the state of Pernambuco. The goal of the project is to promote sustainable use of the plant resources of the eight state arid regions of northeastern Brazil. Activities are divided into three subprograms: biodiversity, economic botany and information. The biodiversity subprogram works on the basic survey of plants and vegetation types of the region to facilitate their identification and use. The economic botany subprogram seeks to promote the sustainable use of regional plants and has had projects on medicinal plants, fodder plants, and fuel wood. The information subprogram has set up a Plant Information Centre in the Botany Department of the Federal University of Pernambuco in Recife that collects and disseminates the information resulting from the other subprograms, so that it may be used effectively by those aiming to improve the region's environment. More recently the Program has focused on integrated projects that combine the elements of the different subprograms.

In their case study ("Latin American Plant Sciences Network: A Higher Education Program For The Development Of Plant Sciences And Conservation Of Biodiversity In Latin America") S. Maldonado et al. describe a program for increasing the number of qualified botanists capable of protecting the biodiversity of native flora in the region. Many arid and semiarid areas in Latin America are losing native species and ecosystem services at an unprecedented rate due to factors such as ecosystem conversion, inefficient agricultural and grazing practices, and exploitation of timber resources. Often, these factors are driven by poverty, economic marginalization and undervaluing of natural resources, and low participation of local people in decision making about conservation and use of natural resources. In turn, these factors are exacerbated by constraints to educational opportunities at all levels. Ironically, despite the large amount of biodiversity in the Neotropics there is a dearth of human scientific and technical capacity. This case study describes the development and design of the Latin America Plant Sciences Network, which is a consortium of 23 academically prestigious institutions in Argentina, Brazil, Chile, Costa Rica, Mexico, and Venezuela which have worked together collaboratively since 1988 to offer graduate level training to students from Latin America; to organize scientific meetings and workshops; and to undertake collaborative research projects. Through these activities, the consortium is helping to increase the number of well–trained people capable of protecting the biodiversity of the region's native flora.

Érica Speglich and Carlos Alfredo Joly discuss how a major problem confronting biodiversity scientists and decision makers is the fact that available information regarding biological resources is fragmented, dispersed, difficult to access and underused. As a consequence of the lack of an updated cartographic base, key information such as the location of sampling sites is usually inaccurate. In their chapter "The Brazilian Biodiversity Virtual Institute," Speglich and Joly describe the creation process for an institute and the strategies being applied for connecting research projects, researchers, research students, the data they produce as well as how to qualify people for working with biodiversity conservation. In particular, they pay close attention to the "lessons learned on creating partnerships and capacity building."

In his case study "Conservation Of Fig (*Ficus Carica* L.) And Pomegranate (*Punica Granatum* L.) Varieties In Tunisia," M. Mars shows that the conservation of local cultivars is improved by strengthening relationships between all partners (research institutions, development agencies, local organizations, local communities, authorities, growers). However, conservation of local genetic resources also requires new approaches and collaborative efforts, especially if the conservation is to be integrated in sustainable rural development programs. The project also has shown that it is necessary to emphasize more in situ preservation of local genetic resources because ex situ collections of perennial woody plants present many technical problems particularly in arid zones with scarce water resources.

L. Berry critically analyzes prospects and problems of improving institutional partnerships and capacity for biodiversity conservation in Latin America and the Caribbean ("Capacity Building To Sustainably Use Biodiversity In Dryland Regions Of Latin America And The Caribbean"). He discusses how these regions are well known for their high biodiversity, but the focus of attention is most often on the admittedly important world heritage of the tropical and mountain rainforests and wetlands of the regions. Nevertheless, there is an increased need to improve funding, the development of partnerships, and capacity building to better protect biodiversity in the arid and semiarid zones of the regions. Berry reviews investments in the conservation of biodiversity in the regions; the need for improved networking of scientists; the need for networks to focus more efficiently and comprehensively on particular arid and semiarid ecosystems; the need for development of networks which make greater use of ethno scientists; the need for more effective networks of scientists and public policy makers with educational institutions; the need for greater balance between regional, national, and global networks; and the need for institutional development to make greater use of practical but innovative communication methods and technologies to promote capacity building. Based on this review, Berry recommends more effective processes to build on and expand the already sound capacity of the region to conserve biodiversity.

H. Hassan ("A Commentary On Strategies And Incentives To Improve Biodiversity In Arid And Semi–Arid Zones") argues that policies and programs to sustain biodiversity in arid and semiarid zones have been misdirected because they have been designed to combat desertification. Alternatively, Hassan argues that solutions to improve biodiversity should aim at improving the peoples' livelihood in a sustainable manner while conserving biodiversity. Specifically, Hassan argues for a greater recognition of the underlying causes of biodiversity loss in arid and semiarid regions of Southern nations. This includes understanding that: the poor depend on biodiversity and also utilize it more than the more affluent; macroeconomic policies shape possibilities for resource management and biodiversity conservation of which the poor often are excluded; market failure often leads to unsustainable patterns of resource use and consumption; appropriate institutional and social settings are needed to provide conditions for achieving resource conservation and sustainable development; unclear and insecure land tenure systems can be major deterrents to long term sustainable use of resources; and lack of ownership and participation by local people, especially women, in decisions about resource use is highly problematic.

D.A. Brown ("Achieving Institutional Cooperation For Implementation Of Sustainable Development Plans And Strategies") identifies barriers to obtain better cooperation between institutions trying to achieve sustainable development. First, is the failure to fix responsibility for rigorous interdisciplinary strategic planning to solve sustainable development problems; this can stem from a lack of institutional focus on strategic planning, or from institutional fragmentation about the scope of responsibilities. Second, is the failure to understand or consider what is necessary for implementation of sustainable development plans and strategies. Often the plans or strategies are prepared by organizations that have no legal authority to implement them. Where cooperation is needed among many institutions to implement plans or strategies, they often are prepared without full agreement by all the persons who must implement them and this limits their efficacy. Third, is a failure to match the scope of scientific investigation with the scale of the problem. Sustainable development problems often transcend ecological boundaries and political jurisdictions. Yet, decision makers who authorize the development of plans and strategies only have authority over decisions within their jurisdiction. For this reason, plans and strategies that match the scale of problems often are not implemented. In order to achieve sustainable development it is necessary to understand and control human activities that will threaten the entire social and ecosystems of concern. Fourth, is a failure to build an information base which is practicable to use in implementation plans and strategies. Often the information collected lacks understanding of the need to use it at different spatial and temporal scales. For instance, many countries have developed their own methods and indicators that cannot be aggregated at global scales because of incompatibility with the United Nations' indicators. Fifth, is a failure to consider prior plans, decisions, or information in decision making. Many international bodies meet and discuss problems with insufficient understanding of the prior legal or institutional decision making that has considered the same or similar issues. For instance, for many years various international institutions have placed the conservation of freshwater on their agenda, including UNCSD, UNEP, WHO, UNDP, and the Global Water Forum. As each new institution has considered global water issues they have often begun deliberations without an understanding of prior decisions on the very same subjects under consideration.

5. SUMMARY OF RECOMMENDATIONS

Recommendations contained in some case studies were site or area specific; many others were applicable to similar problems in different locations. Most recommendations focused on the following issues.

Sound scientific, public policy, and social science research should be used to inform public policy and decisions makers. This research should draw on disciplinary expertise when needed but overall be interdisciplinary in its nature and focus. The research, where appropriate, should address linkages between biodiversity and related problems. Two such examples are climate change and desertification, and poverty reduction and capacity for biodiversity protection. The research should be long–term because most biodiversity research questions cannot be answered sufficiently with information from short–term studies. The research also should be designed and conducted on the appropriate spatial scales, e.g., local habitat, landscape, or regional. Accomplishing this kind of research requires not only an increase in overall funding, but greater recognition by funding agencies of the need to fund interdisciplinary research, long–term research, and research which is focused on appropriate spatial scales.

Local, national, and regional governmental policies and plans need to be developed and implemented based on the best scientific and other research available. This requires local, national, and regional governments to promulgate policies and plans based on interdisciplinary research; on the recognition that one problem often is linked to others; on long–term studies; and on appropriate spatial scales.

The participation of local people and other stakeholders in policy and decision making should be increased, including the roles of women.

In order for institutions to enhance the relevance and applicability of their work there must be more effective means developed to obtain the views of key personal working on high–quality national plans and strategies of the CBD as well as the CCC and CCD regarding priorities for implementing the conventions in areas that link with protection of biodiversity. A similar need exists to obtain more comprehensive information about priorities for biodiversity protection from GEF and other donor groups so that institutions can enhance the relevance of their work to the priorities.

Education at all levels on the values of biodiversity should be increased. This means reform of education at the primary, secondary, and university levels; the development of out–research and extension education for civil society; educating public policy and decision makers about the values of biodiversity and the capabilities and limitations of science; and educating scientists and other specialists to more effectively communicate their findings to members of the general public and nonspecialists.

Finally, coordination and collaboration between public and private stakeholders should be increased, including the formation of regional networks and greater interaction between institutions of Southern nations.

6. REFERENCES

Goodland, R., and Daly, H. 1995. Universal environmental sustainability and the principle of integrity. Pages 102–124 In: Westra, L., and Lemons, J. (Editors) Perspectives on Ecological Integrity. Kluwer Academic Publishers, Dordrecht, The Netherlands.

Hargrove, E. 1992. The Animal Rights/Environmental Ethics Debate. State University of New York Press, Albany, NY.

(IISD) 2003. A Summary Report from "Biodiversity After Johannesburg." Sustainable Development 81: 1–7.

Johnson, S.P. 1993. The Earth Summit. The United Nations Conference on Environment and Development (UNCED). Graham & Trotman/Martinus Nijhoff, London.

Lemons, J. (Editor). 1996. Scientific Uncertainty and Environmental Problem Solving. Blackwell Science, Cambridge, MA.

Lemons, J., and Brown, D.A. (Editors) 1995. Sustainable Development: Science, Ethics, and Public Policy. Kluwer Academic Publishers, Dordrecht, The Netherlands.

Oates, J.F. 1999. Myth and Reality in the Rain Forest. How Conservation Strategies Are Failing in West Africa. University of California Press, Berkeley. 310 pages.

Regan, T. 1983. The Case for Animal Rights. University of California Press, Berkeley, CA.

Robinson, J.G. 1993. The limits to caring: Sustainable living and the loss of biodiversity. Conservation Biology 7: 20–28.

Shearman, R. 1990. The meaning and ethics of sustainability. Environmental Management 14: 1–8.

Shrader–Frechette, K.S., and McCoy, E.D. 1993. Method in Ecology. Cambridge University Press, Cambridge.

(WCED) World Commission on Environment and Development. 1987. Our Common Future. Kluwer Academic Publishers, Dordrecht, The Netherlands.

Westra, L., and Lemons, J. (Editors). 1995. Perspectives on Ecosystem Integrity. Kluwer Academic Publishers, Dordrecht, The Netherlands.

Chapter 2

THE DOMESTICATION OF INDIGENOUS TREES AS THE BASIS OF SUSTAINABLE LAND USE IN AFRICA

R.R.B. Leakey
Agroforestry and Novel Crops Unit, School of Tropical Biology, James Cook University, P.O. Box 6811, Cairns, QLD 4870, Australia; email roger.leakey@icu.edu.au

1. INTRODUCTION

The need for sustainable land use, especially on agricultural land in the tropics, has been recognized for a long time and was highlighted by the Earth Summit in 1992. The problem of land degradation from agricultural and livestock pressures received particular prominence for dryland Africa in the UN Convention to Combat Desertification. The means to a practical solution to the problem of unsustainable farming practices has, however, been very elusive.

The sustainability of land use in the tropics has typically been lost when species–rich natural vegetation has been cleared to make way for monocultures of improved staple food crops grown intensively with high inputs of agrochemicals. These intensive farming systems have indeed helped to feed growing human populations, but at an environmental cost that cannot be tolerated into the future. Traditionally, in many parts of dryland Africa, shifting cultivation has been practiced. In the past this was sustainable, as certain trees were protected and retained within the farm, while a fallow period of sufficient length allowed the regeneration of natural vegetation, especially the trees and shrubs. In more recent times, however, trees have been removed from the landscape because they are a vital resource of fuelwood and because they are perceived to be getting in the way of modern agriculture. Additionally, pressures from growing numbers of humans and livestock have typically led to shorter and shorter periods of fallow and to the prevention of tree regeneration on farmland. This pressure on the land has, in some areas, been exacerbated by land tenure issues, which have discouraged farmers from protecting, let alone planting, trees. The lifestyle of nomadic herdsmen has also resulted in conflicts between sedentary farmers and the livestock of the nomads, and has been a further disincentive to tree establishment and management.

The importance of land degradation is only matched by the complexity of its causes. Degradation results from highly interactive processes involving numerous social, economic and biophysical factors, and consequential feedback loops that exacerbate the problem (e.g., between poverty, ecosystem degradation, loss of biodiversity, declining livelihoods). There are some suggestions that enhanced climatic variability and perhaps climate change are also a factor in the process. Consequently, there is not a simple solution to land degradation and any solution will have to be equally interactive and solve many individual problems simultaneously. There is therefore a need to take an innovative approach to integrated natural resource management, aimed at simultaneously restoring biological resources, alleviating poverty and reversing the degradation processes in agricultural land. This has been dubbed as the "Win–Win" scenario (World Bank 1992, Tengberg and Stocking 2000, Leakey 2001a/b), in which improved human welfare (sustainable rural livelihoods) and the causes of environmental degradation (including the loss of biological diversity) are tackled together.

2. AGROFORESTRY AND SUCCESSION

Agroforestry involves the integration of trees into farming practices and occurs in many forms (Nair 1993) and should be recognized as a land use option that provides both products and environmental services (ICRAF 1997), and has been proposed as one way of diversifying and intensifying agroecosystems in a way that generates income and is beneficial to the environment (Sanchez and Leakey 1997). The marketable products are: fruits, nuts, medicines, resins, fibres, timber, etc., while the relevant services of trees are those that increase the crop yields (nitrogen fixation, soil organic matter, nutrient cycling, soil conservation, etc.) create environmental resilience (niche diversification, food web complexity, carbon sequestration, reduced greenhouse gas emissions, hydrological control function, etc.) and provide for social needs (boundary delineation, shade, etc.). By integrating trees into farmland, Leakey (1996) suggested that agroforestry practices allow the establishment of successional phases in the development of productive and complex agroecosystem that can be similar to the succession of natural ecosystems. This is very different from the earlier concept in which agroforestry was seen more as an agronomic practice that was maintained more or less unchanged for long periods. Just as natural ecosystems progress from relatively simple early successional phases dominated by a few pioneer species to a species rich, multistrata tree–based mature phase (climax) phase, so agroecosystems can progress from a monocultural crop to a multistrata agroforest (Leakey 1999a). It is anticipated that this progression will be accompanied by an increase in biological diversity with each phase of the agroecological succession, as the planned biodiversity (the planted crops and agroforestry trees) is enriched by the unplanned biodiversity, made up of all those organisms, above and below ground, that have found niches to fill among the planted trees and crops. Trees producing different products can be used to create and fill different niches, thus the productive components of the agroecosystem are not tied to any one part of the agroforest.

Again, as in natural ecosystems, not all parts of agroecosystems will be in the same temporal phase, thus creating spatial variation in physical structure which creates a mosaic of patches in the landscape and making the system ecologically more stable and biologically and economically more diverse. In addition, further diversity on a landscape scale will be developed by farmers as they apply different management practices and implement different farming systems (e.g., food crops versus livestock). Farms too will differ due to the economic impacts of different farm sizes, the wealth of the farmer, access to market, the tenure systems, the availability and price of labour, other sources of income, gender of the head of household, etc. All of these sources of diversity are beneficial environmentally and socially, and have been termed "agrodiversity" (Netting and Stone 1996, Stocking 2000).

All of the above suggests that progress towards sustainable land use can be made if ways can be found to promote the development of the temporal, spatial and managerial diversity of agroecosystems which also meet the food security and economic needs of rural people.

3. NOVEL CROPS FOR INCOME GENERATION

Meeting the needs of subsistence farmers for food security and economic development has been a major focus of aid projects and the international research institutes of the CGIAR over the last 40–50 years. The "Green Revolution" made enormous strides towards increased food production (Tribe 1994), but it has also led to enhanced environmental degradation and reliance on high capital and production inputs. It did not benefit the poor as much as expected and has not managed to provide food for all (Conway 1997). New approaches are needed that provide a wider array of opportunities, with a stronger focus on indigenous species, rather than on a handful of exotic staple food crops (Leakey 2001a/b). There also has to be a much stronger emphasis on maintaining the natural environment and therefore on ecological agriculture (McNeeley and Scherr 2001).

Throughout the tropics, and dryland Africa is no exception, there are a great many species, whose products have been used by indigenous peoples for food, medicines and to meet their many other day–to–day needs (e.g., von Maydell 1990, Palgrave 1983). These products have been gathered from wild trees and have been important traditionally in food preparation (Lamien et al. 1996.), for food security and for marketing (Boffa et al. 1996). Using a participatory prioritization procedure, ICRAF and its national agroforestry partners around Africa have identified those species which have the greatest value for domestication as new tree crops (Franzel et al. 1996, Bonkoungou et al. 1998, Maghembe et al. 1998). For the Sahel, the top five species in order of priority, were identified as *Adansonia digitata*, *Vitellaria paradoxa*, *Parkia biglobosa*, *Tamarindus indica* and *Zizyphus mauritiana*. These are all fruit trees. In addition, some fodder species have also been identified for domestication (*Pterocarpus erinaceus*, *Bauhinia rufescens*, *Combretum aculeatum* and *Prosopis africana*). *C. aculeatum* also produces an edible nut. For southern Africa, indigenous fruit trees also predominated, with *Uapaca kirkiana*, *Sclerocarya birrea*, *Zizyphus mauritiana*, *Vangueria infausta*, *Azanza garckeana* being the top priorities (e.g., Leakey et al. 1999).

Although there is some information at the species level about the nutritional value of these fruits from proximate analyses, there is inadequate knowledge about the tree–to–tree intraspecific variation in these nutritional components (Leakey 1999b). This information is especially important if the cultivars arising from the domestication program are to maximize the health benefits to the consumers and meet the needs of the food industry. The latter could greatly influence the market opportunities, and hence the potential for tree domestication to provide an incentive to farmers to grow the trees within their farming systems. The cash generation potential of non–timber forest products from mature agroforests is crucial if farmers are to implement this model for sustainable development in Africa.

4. POLYCULTURE AGROECOSYSTEMS

The best examples of multistrata polycultural systems are in the wet tropics (Leakey 2001a). However, in arid and semi–arid lands the tree carrying capacity is considerably lower (optimal woody plant cover is around 15–20 percent, according to Breman and Kessler 1995) due to the limitations of water scarcity and nutrient availability and the consequently increased competition for water and nutrients between the trees and other crops. Nevertheless, there is no reason not to consider the principle that the development of a mature agroecosystem may provide the best opportunity for sustainable production. Indeed, the Sahelian Parklands have been a very extensive traditional mature agroecosystem over large tracts of Africa (Boffa, 1999). In Mali alone these Parklands cover 90 percent of the country with a population of 2.5 million people. The problem now is that it is starting to degenerate as traditional practices, which used to allow it to regenerate, are being eroded. What is needed to restore the Parklands?

Traditionally, the Parklands were regenerated during a fallow cycle, which allowed the regrowth of the indigenous trees, some of which were retained when the land was cleared for the next cycle of cropping. The farmers particularly retained the trees that produced edible fruits and fodder, such as *Vitellaria paradoxa* (Sheanut or Karité), *Adansonia digitata* (Baobab), *Parkia biglobosa* (Néré), *Faidherbia albida* (Gawo, Kad, Winterthorn), and *Pterocarpus spp.*, usually at a density of around 40–80 trees of up several species per hectare (Boffa 1999). Commonly, one or two species would dominate the Parklands in any given area, although there are 20–30 species commonly found in Parklands. This traditional practice has, however, suffered from the increasing pressure for farmland arising out of growing populations of people and their livestock, and the consequently shortened fallows have not provided the next generation of trees. Unfortunately, the obvious solution of just planting new trees within the farmland is not a solution because in many areas the livestock herds of nomadic herdsmen have the right to wander over the fields of the sedentary farmers during the dry season, and they eat the young trees. Additionally, in many areas the fields are only fenced temporarily during the wet growing season, using the stalks of the previous sorghum or millet crop, or thorny branches. These flimsy fences provide security for the young cereal crops, but are subsequently devoured by termites. A longer–term solution with more durable fences is often too expensive.

An alternative to fences could be provided by hedges (or living fences) of thorny, non–palatable tree or shrub species, some of which (e.g., *Zizyphus spp.*) produce edible and marketable fruits. Research to examine these options was initiated by ICRAF in 1990's (ICRAF 1996), and was successfully adopted by some farmers in Burkina Faso. In addition to providing protection to cereal fields they can also protect longer–lived crops, such as tree fodder banks of *Gliricidia sepium*, vegetables and young fruit trees. Potentially, hedges could allow the diversification of the farming system and the re–establishment of the scattered fruit trees, and so allow the regeneration and even the improvement of the Parklands. In the semi–arid area of Kenya, near Machakos, somewhat similar regeneration of degraded farmlands through tree planting has occurred, interestingly at the same time as a large increase in population pressure. This arose in response to the establishment of measures to reduce soil erosion (Tiffen et al. 1994). This has particularly occurred where farmers have secure land or tree tenure.

Incentives are needed to enhance this process. One approach would be to domesticate a wider range of the indigenous trees producing fruits and fodder, thereby improving the yields and quality of the products and making them more marketable locally, regionally and even perhaps internationally. These domesticated trees would then have to be integrated into the farmland in ways that create improved mature parkland agroecosystems. For example, a mosaic of food crop fields and fodder banks could be established between widely spaced fruit trees, within living hedges. These, and similar enclosed areas for vegetables and other perennial or valuable crops, would not be in direct competition with traditional staple food crops. The agrodiversity arising from all the different needs of households of different sizes, structures, wealth and labor availability would further ensure the establishment of a mosaic of land uses and cropping patterns, which would enlarge the environmental (and perhaps the economic and social) benefits.

The enclosure of farmland for field and tree cropping could allow the regeneration of the Parklands and the development of a mature agroecosystem. The tree crop component of these enclosed areas could be made more profitable by their domestication and the development of cultivars. This approach could, it seems, go a long way towards the establishment of land uses that will fulfil the needs of rural and urban populations for food and income, while maintaining much of the biological diversity of forests or rehabilitating degraded ecosystems. However, the extent to which unplanned biodiversity would develop within the different agroecosystems and how this improves agroecosystem function is not yet known or understood, especially in the drylands of Africa.

5. AGROECOSYSTEM FUNCTION

The understanding of the interactions between biological diversity and agroecosystem function is remarkably poor, although it is an area of increasing interest for research (Gliessman 1990, 1998; Altieri and Nicholls 1999). Generally, it is accepted that increasing the numbers of species increases structural complexity and niche formation, and so enhances ecosystem processes (nutrient cycling, production of different products, light requirements, etc.). However, there are likely to be scale–related benefits, as species vary in their role and importance in different food webs. Typically, predators require larger range than their prey, thus for the sustainable production the area may need to be big enough to support predators. The absence of top predators may create pest management issues for farmers, such as the increase in fruit or seed–eating birds, but at the same time, some top predators are unwelcome in farmland. In this context, a land use mosaic with corridors between forest patches may be the most appropriate means of acquiring the scale needed to achieve some level of ecological equilibrium. This raises the question at what point in diversification of farming systems the controls which prevent isolated outbreaks of pests and disease are in place. Not enough is known about the patch dynamics of such outbreaks, relative to species numbers and configurations, to steer agroforesters in the development of risk–averse strategies. This reaffirms the need to consider the ecological implications of agricultural landscape design (e.g., Leakey 1999a).

It is important to remember that many of the important ecological processes go on in the soil below ground. For example, the roots of plants are covered with symbiotic fungi and mycorrhizas which enhance nutrient cycling. Deforestation seriously depletes the inoculum of these species in the soil with serious consequences for plant (especially tree) regeneration (Wilson et al. 1992, Mason and Wilson 1994). Conversely, the planting of inoculated trees can help to restore the inoculum and improve soil processes. For example, at Olorgasailie in Kenya, an alkaline site with about 475mm rain y^{-1} arbuscular mycorrhizal spores declined rapidly with increasing distance from Acacia *tortilis* trees (from about 400 spores 50 g^{-1} soil close to the trunk to < 100 spores outside the tree canopy). Inoculating nursery plants with soil collected from close to the tree trunk resulted in larger plants at planting time with well–infected root systems and subsequently in substantially improved (up to 80 percent) establishment and survival (Wilson et al. 1991, Munro et al. 1999) and higher water use efficiency (Newton et al. 1996b). In addition, when interplanted with crops there was a trend of improving crop yield with time.

The role of trees in maintaining mycorrhizal inoculum has also been demonstrated by Mbuthia (1992) and McGreevy (1996), who found that maize grown in soil taken from close (0.5 m) to trees (*Senna siamea*) formed more mycorrhizas than when it was grown in soil collected at 2 m distance. Microsymbionts can thus enhance the ability of trees to become established on degraded soils and, in turn, these trees have an important role in maintaining populations of microsymbionts for short–lived crops.

The longevity and ability of trees to scavenge for nutrients over large areas contributes considerably to their other service functions. Deep roots can bring up nutrients otherwise inaccessible to crops from great depths. In Sahelian aquifers there is increasing evidence of a large reserve of nitrogen at depth in groundwaters in arid zones (e.g., Leakey et al. 1999). This nitrogen is thought to derive from nitrogen fixation, to have been concentrated by evaporation and to be leached to ground waters. Tree species have been found to be capable of rooting to depths of 30m or more and thus are capable of recycling the nitrogen to the soil surface by root and leaf litter turnover. Trees also scavenge laterally over a wide area for water and nutrients.

6. CONSTRAINTS TO DEVELOPING MATURE AGROECOSYSTEMS

6.1 Conflict Between Sedentary Farmers and Nomadic Herdsmen

The use of hedges to enclose land for the protection of crops and regenerating fruit and fodder trees could, if not carefully planned and executed, create another problem; a social conflict between nomads and sedentary farmers arising from the exclusion of the nomads' herds from their traditional grazing lands (Leakey et al. 1999). Clearly there is a need for negotiation and planned development to avoid this conflict and the development of agroforestry must not be allowed to exacerbate the existing tensions. Consequently, any development that includes changes to the land use by nomadic herdsmen must look at solutions to their needs for livestock grazing/fodder in the dry season. For example, this might involve corridors for livestock herds between blocks of enclosed farmland to allow access to watering holes, and the establishment of tree fodder banks at strategic places.

6.1.1 Property Rights

The incentive to plant tree crops can be adversely affected by land and tree tenure issues. These vary considerable between different areas depending on local cultural customs and land management legislation. In some areas of Sahelian Africa, for example, sedentary farmers only have rights to their land during the cropping season and hence in the dry season; however, livestock belonging to other farmers and to nomadic herdsmen can graze anywhere and young trees are vulnerable. These practices pose a severe constraint to implementing the sustainable land use options.

6.1.2 Degradation of Soil Microflora

Mycorrhizal fungi are an integral part of practically all plant communities and have a fundamental role in plant growth and in the functioning of ecosystems (Mason and Wilson 1994). Like herbaceous plants, the majority of tropical trees are mycorrhizal and benefit from inoculation in the nursery (Garbaye 1991). Without appropriate inoculum, plantations can fail. As mentioned earlier, deforested and degraded land has often lost much of its natural soil inoculum or had it replaced by other fungi which are less effective. This problem has been widely recognized by foresters establishing conifers in the tropics, but is not so commonly understood by foresters or farmers planting hardwoods, especially native species, as it is generally believed that all soils contain the appropriate inoculum for native species. This is not necessarily the case in either the humid tropics (Mason et al. 1992) or the dry tropics of Africa (Wilson et al. 1992).

6.1.3 Availability of Markets for Tree Products

The non–timber forest products from local tree species are often marketed locally to meet the traditional needs of the population. However, an expansion in production arising from large scale planting programs would cause a glut and the collapse of market prices. To ensure that the opportunities offered by the domestication of agroforestry trees for non–timber forest products are captured, it is essential that there is parallel development of market opportunities, entrepreneurial marketing, and processing activity in the villages, towns and cities. This is required to ensure that the market demand is increased and hence that the market prices of the products are at least maintained, if not increased. Typically, the markets for tree products are seasonal and hence it is important that the domestication process also tries to extend the productive season and that post–harvest technologies, especially processing, extends the shelf–life of the products. The continuity of supply and the maintenance of product quality are also critical components of an improved commercial process. Market development of this sort is also critical for the overall success of sustainable development aimed at poverty alleviation and the promotion of employment opportunities for landless people, especially in the growing urban centers of Africa. Thus, there is complementarity between the development needs of the farmers and those of the urban communities.

The domestication process also has to be integrated with the commercialization process (Leakey and Izac 1996), and it is important that the emerging trade and industry partners have an input into the identification of the traits for the genetic selection (Leakey 1999). For example if a nut is being domesticated for oil products, the industry may see an enhanced market niche from the product having a certain concentration of a particular fatty acid, amino acid, vitamin, etc. Without this input to the domestication process, this opportunity will be lost. It is also important to ensure that domestication does not lead to the point where outsiders with capital to invest come in and develop large–scale monocultural plantations for export markets (Leakey and Izac 1996).

The need to ensure that the beneficiaries of domestication of agroforestry trees are small-scale subsistence farmers is being studied in southern Africa, using the marula fruit (*Sclerocarya birrea*) as a case study (Shackleton et al. in press). This project explores the numerous industrial and community pathways associated with the commercialization process and seeks to identify the "winners and losers" arising from commercialization of marula products. The study is examining the commercialization benefits and costs in terms of the five sustainable livelihood capital entitlements of the producing communities (physical, social, human, natural and financial capital).

There are a number of examples of interesting and novel approaches to marketing forest products. Two contrasting, but positive, examples are illustrated by the development of the medicinal plant market in Durban, South Africa (Mander 1998) and the Daimler Chrysler car factory in Brazil, which is supporting the small–scale agroforestry production of raw materials for the manufacture of novel car components (Mitschein and Miranda 1998, Panik 1998).

7. CONCLUSIONS

In this review, I have indicated that there are many opportunities for more intensive and imaginative use of trees for both the production of non–timber forest products and the capture of their environmental services. I have recommended that the challenge for the project "Promoting Best Practices for Conservation and Sustainable Use of Biodiversity of Global Significance in Arid and Semi–Arid Zones" is to use agroforestry as a means to develop economically and socially acceptable land use systems that function like undisturbed ecosystems and maintain biodiversity.

There is a need for policy and development activities to promote the large–scale adoption of lower input agroecosystems that combine the cultivation of "Green Revolution" staple food crops with a number of the indigenous food–producing tree species that can restore agroecosystem function. This return towards more traditional land use practices can be enhanced by the judicious development of high quality, high yielding cultivars of the trees through the adaptation of standard horticultural practices and their growth in agroforests. This approach should be based on the participatory tree domestication activities developed by ICRAF and partners, so as to allow subsistence farmers to help themselves.

There are at least four needs that have to be met in order to reform land use practices. Firstly, there is the need to identify the species of commercial importance and relevance to local markets. Secondly, there is the need to train individuals from rural communities in the already developed techniques for creating cultivars of the chosen species. Thirdly, there is a need to develop an entrepreneurial mentality among the community, who have traditionally been subsistence farmers and hunter–gatherers. The growth of the urban market and the absence of jobs in the urban areas may provide the commercial incentive needed. Finally, there is a need to determine how best the cultivars may be integrated into an agroecosystem to develop mature agroforests modeled on mature "parkland" agroforests, which more closely mimic the tree and grass interactions of natural ecosystems, and provide annually harvested products for food, fodder, pharmaceutical and extractive markets.

All of this will require a better understanding by the international development community of the role of trees in farming systems, and further applied and adaptive research to domesticate trees. This research effort needs to be directed towards the creation of farming systems that enhance the market potential of indigenous plants and which meet the needs of the people. In pursuit of this approach in dryland Africa, it is particularly important to do so within a policy framework that avoids the creation of conflict between sedentary farmers and nomadic herdsmen.

8. REFERENCES

Altieri, M.A., and Nicholls, C.I. 1999. Biodiversity, ecosystem function and pest management in agricultural systems. Pages 69–84 In: Collins, W.W. and Qualset, C.O. (Editors) Biodiversity in Agroecosystems. CRC Press, Boca Raton, USA.

Boffa, J.M. 1999. Agroforestry Parklands in Sub–Saharan Africa, FAO Conservation Guide 34, FAO, Rome. 230 pages.

Boffa J.M., Yaméogo, P., Nikiéma, P., and Knudson, D.M. 1996. Shea nut (Vitellaria paradoxa) production and collection in agroforestry parklands in Burkina Faso. Pages 110–122 In: Leakey, R.R.B., Temu, A.B., Melnyk, M., and Vantomme, P. (Editors) Domestication and Commercialization of Non–Timber Forest Products in Agroforestry Systems, Non–Wood Forest Products No 9, FAO, Rome, Italy.

Bounkoungou, E.G., Djimde, M., Ayuk, E.T., Zoungrana, I, and Tchoundjeu, Z. 1998. Taking Stock of Agroforestry in the Sahel: Harvesting Results for the Future, End of Phase Report: 1989–96. ICRAF, PO Box 30677, Nairobi, Kenya. 58 pages.

Breman, H., and Kessler, J.J. 1995. Woody Plants in Agro–Ecosystems of Semi–Arid Regions: With an Emphasis on the Sahelian Countries. Advanced Series in Agricultural Sciences, 23. Springer–Verlag, Berlin, Germany. 340 pages.

Conway, G. 1997. The Doubly Green Revolution: Food for All in the 21st Century. Penguin Books Ltd, London. 335 pages.

Franzel, S., Jaenicke, H., and Janssen, W., 1996. Choosing the Right Trees: Setting Priorities for Multipurpose Tree Improvement. ISNAR Research Report 8: 1–87, The Hague, Netherlands.

Garbaye, J. 1991. Utilisation des mycorhizes en sylviculture. Pages 197–248 In: Strullu, D.G. (Editor) Les Mycorhizes des Arbes et des Plantes Cultivées. Technique et Documentation Lavoisier, Paris, France.

Gliessman, S.R. 1990. Agroecology: Researching the Ecological Basis for Sustainable Agriculture. Springer–Verlag, New York, USA. 380 pages.

Gliessman, S.R. 1998. Agroecology: Ecological Processes in Sustainable Agriculture, Ann Arbor Press, Michigan, USA. 357 pages.

ICRAF 1997. Annual Report 1996. ICRAF, Nairobi, Kenya. 340 pages.

Lamien, N., Sidibe, A., and Bayala, J. 1996. Use and commercialization of non–timber forest products in Western Burkina Faso. Pages 51–64 In: Leakey, R.R.B., Temu, A.B., Melnyk, M., and Vantomme, P. (Editors) Domestication and Commercialization of Non–timber Forest Products in Agroforestry Systems. Non–Wood Forest Products No 9, FAO, Rome, Italy.

Leakey, R.R.B. 1996. Definition of agroforestry revisited. Agroforestry Today 8: 5–7.

Leakey, R.R.B. 1999a. Agroforestry for biodiversity in farming systems. Pages 127–146 In: Collins, W.W., and Qualset, C.O. (Editors) Biodiversity in Agroecosystems. CRC Press, Boca Raton, USA.

Leakey, R.R.B. 1999b. Potential for novel food products from agroforestry trees: A review. Food Chemistry 66: 1–14.

Leakey, R.R.B. 2001a. Win:win landuse strategies for Africa: 1. Building on experience with agroforests in Asia and Latin America. International Forestry Review 3: 1–10.

Leakey, R.R.B. 2001b. Win:win landuse strategies for Africa: 2. Capturing economic and environmental benefits with multistrata agroforests. International Forestry Review 3: 11–18.

Leakey, R.R.B., and Izac, A.M.N. 1996. Linkages between domestication and commercialization of non–timber forest products: Implications for agroforestry. Pages 1–7 In: Leakey, R.R.B., Temu, A.B., Melnyk, M., and Vantomme, P. (Editors) Domestication and Commercialization of Non–timber Forest Products in Agroforestry Systems. Non–Wood Forest Products No 9, FAO, Rome, Italy.

Leakey, R.R.B., and Tomich, T.P. 1999. Domestication of tropical trees: From biology to economics and policy. Pages 319–338 In: Buck, L.E., Lassoie, J.P., and Fernandes, E.C.M. (Editors) Agroforestry in Sustainable Agricultural Systems. CRC Press/Lewis Publishers, New York, USA.

Leakey, R.R.B., Wilson, J., and Deans, J.D. 1999. Domestication of trees for agroforestry in drylands. Annals of the Arid Zone 38: 1–26.

Maghembe, J.A., Simons, A.J., Kwesiga, F., and Rarieya, M., 1998. Selecting Indigenous Trees for Domestication in Southern Africa: Priority Setting with Farmers in Malawi, Tanzania, Zambia and Zimbabwe. ICRAF, Nairobi, Kenya, 94 pages.

Mander, M. 1998. Marketing of Indigenous Medicinal Plants in South Africa: A Case Study in KwaZulu–Natal. Food and Agricultural Organisation of the United Nations. Forest Products Marketing Programme. Rome. 151 pages.

Mason, P.A., and Wilson, J. 1994. Harnessing symbiotic associations: Vesicular–arbuscular mycorrhizas. Pages 165–175 In: Leakey, R.R.B., and Newton, A.C. (Editors) Tropical Trees: the Potential for Domestication and the Rebuilding of Forest Resources. HMSO, London, UK.

Mason, P.A., Musoko, M.O., and Last, F.T. 1992. Short term changes in vesicular–arbuscular mycorrhizal spore populations in Terminalia plantations in Cameroon. Pages 261–267 In: Read, D.J., Lewis, D.H., Fitter, A.H., and Alexander, I.J. Mycorrhizas in Ecosystems: Proceedings of 3rd European Symposium on Mycorrhizas, CAB International, England.

Mbuthia, K.W. 1992. Mycorrhizal inoculum potential for agroforestry. The effect of trees: Case study Cassia siamea Lam. M.Sc. thesis, University of Aberdeen, UK

McGreevy, S. 1996. Indigenous arbuscular mycorrhizal fungi of a tropical agroforestry system and their association with the intercrop, Zea mays L. PhD thesis, University of Edinburgh. 223 pages.

McNeely J.A., and Scherr, S.J. 2001. Saving our Species: How the Earth's Biodiversity Depends on Progress in Agriculture. Future Harvest, Washington DC, USA.

Mitschein, T.A., and Miranda, P.S. 1998. POEMA: A proposal for sustainable development in Amazonia. Pages 329–366 In: Leihner, D.E., and Mitschein, T.A. (Editors) A Third Millenium for Humanity? The Search for Paths of Sustainable Development. Peter Lang, Frankfurt am Main, Germany.

Munro, R.C., Wilson, J., Jefwa, J., and Mbuthia, K.W. 1999. A low–cost method of mycorrhizal inoculation improves growth of Acacia tortilis seedlings in the nursery. Forest Ecology and Management 113: 51–56.

Nair, P.K. 1993. An Introduction to Agroforestry, Kluwer Academic Publishers, Dordrecht, The Netherlands.

Netting, R., and Stone, P. 1996. Agro–diversity on a farming frontier: Kofyar smallholders on the Benue Plains of central Nigeria. Africa 66: 52–77.

Newton, A.C., Dick, J.McP., and Heaton, T.H.E. 1996b. Stable carbon isotope composition ($*^{13}C$) of Acacia tortilis subsp. spirocarpa (A. Rich.) Brenan growing at three semi–arid sites in Kenya. Journal of Arid Environments 34: 325–330.

Palgrave, K.C., 1983. Trees of Southern Africa. Struik Publishers, Cape Town, South Africa. 959 pages.

Panik, F. 1998. The use of biodiversity and implications for industrial production. Pages 59–73 In: Leihner, D.E., and Mitschein, T.A. (Editors) A Third Millennium for Humanity? The Search for Paths of Sustainable Development. Peter Lang, Frankfurt, Germany.

Sanchez, P.A., and Leakey, R.R.B. 1997. Land use transformation in Africa: Three determinants for balancing food security with natural resource utilization. European Journal of Agronomy 7: 15–23.

Shackleton, S.E., Shackleton, C., Cunningham, A.B., Lombard, C., Sullivan, C., and Netshiluvhi, T., 2002. Knowledge on Sclerocarya birrea. subsp. caffra with emphasis on its importance as a non–timber forest product in South and southern Africa: A summary. Southern Africa Forestry Journal 194: 27–41.

Stocking, M. 2000. Biological diversity, land degradation and sustainable rural livelihoods. Acta Botanica Yunnanica Supplement XII: 4–17.

Tengberg, A., and Stocking, M. 2000. Land degradation, food security and biodiversity–examining an old problem in a new way. In: Sombatpanit, S. et al (Editors) Meeting the Challenges of Land Degradation in the 21st Century. Proceedings of the 2nd International Land Degradation Conference, Khon Kaen, Thailand. Delhi: Oxford & IBH Publishers.

Tiffen, M., Mortimore, M., and Gichuki, F. 1994. More People, Less Erosion: Environmental Recovery in Kenya. ACTS Press, Nairobi, Kenya and ODI, London, UK. 311 pages.

Tribe, D. 1994. Feeding and Greening the World: The Role of International Agricultural Research. CAB International, Wallingford, England. 274 pages.

von Maydell, H.J, 1990. Trees and Shrubs of the Sahel: Their Characteristics and Uses. GTZ, Eschborn, Germany. 525 pages.

Wilson, J., Munro, R.C., Ingleby, K., Mason, P.A., Jefwa, J., Muthoka, P.N., Dick, J.McP., and Leakey, R.R.B. 1991. Tree establishment in semi–arid lands of Kenya–Role of mycorrhizal inoculation and water–retaining polymer. Forest Ecology and Management 45: 153–163.

Wilson, J., Ingleby, K., Mason, P.A., Ibrahim, K., and Lawson, G.J., 1992. Long–term changes in vesicular–arbuscular mycorrhizal spore populations in Terminalia plantations in Côte d'Ivoire. Pages 268–275 In: Read, D.J., Lewis, D.H., Fitter, A.H., and Alexander, I.J. Mycorrhizas in Ecosystems: Proceedings of 3rd European Symposium on Mycorrhizas, CAB International, England.

World Bank, 1992. World Development Report 1992: Development and the Environment. New York: Oxford University Press.

Chapter 3

CONSERVATION FARMING WITH BIODIVERSITY IN SOUTH AFRICA: A PRELIMINARY EVALUATION OF ECOSYSTEM GOODS AND SERVICES IN THE BOKKEVELD PLATEAU

J.S. Donaldson, A. Mills, P. O'Farrell, S. Todd, A. Skowno, and I. Nanni
Conservation Farming Project, National Botanical Institute, Private Bag X7, Cape Town, South Africa; email Donaldson@nbict.nbi.ac.za

1. INTRODUCTION

Much of the worlds biological diversity exists outside of a formal protected reserve network, often on land that is managed for some form of agricultural production and the maintenance of rural livelihoods. Effective biodiversity conservation will therefore require innovative ways to merge the needs of farmers and rural communities with the need to conserve globally significant biodiversity. One of the key questions in this process is who actually benefits from biodiversity conservation. If farmers and rural communities do not benefit from biodiversity conservation, then society needs to provide incentives to achieve conservation objectives on agricultural lands. However, farmers and rural communities rely on healthy ecosystems to provide goods and services to sustain agricultural production (Pagiola et al. 1997) and to improve their quality of life, and the maintenance of healthy ecosystems may also provide benefits for biodiversity conservation. An important first step is to identify and promote conservation farming practices that can meet both these needs.

Arid lands make up nearly half of the earths land surface and therefore form an important focal area for identifying types of landuse that provide mutual benefits for farmers and biodiversity conservation. South Africa is a dry country, with 91 percent of the land classified as arid, semiarid, or dry subhumid (Hoffman et al. 1999). The country comprises less than 0.8 percent of the worlds total land area, but it is a globally significant area for biodiversity conservation, with ca. 8 percent of the world's vascular plants, and between 2 and 7 percent of the worlds amphibian, reptile, avian and mammalian species (Siegfried 1989). Approximately 84 percent of South Africa's land area is set aside for agricultural use (Macdonald 1989), and achieving conservation objectives on agricultural land is essential for an effective conservation strategy.

The objective of conservation farming is to reach and maintain a balance between the utilisation and conservation of natural assets. This principle of optimal resource use began to influence farming practice and agricultural policy in South Africa when the agricultural sector realised that declines in agricultural production, desert encroachment, and the high impact of disaster droughts were linked to the loss of key ecosystem services, specifically soil fertility, soil structure, and the resilience of natural systems. The development of conservation farming in South Africa indicates that agricultural practice has been continuously shaped by the need to secure additional ecosystem goods and services, including those linked to biodiversity (Donaldson 2002). The net benefits of conservation farming for agricultural production (through securing ecosystem services) and biodiversity conservation need to be properly assessed so that successful land use practices can be identified and promoted. These benefits are now being assessed in South Africa as part of a Global Environmental Facility (GEF) funded project on conservation farming, which is examining farming practices in four areas with globally significant levels of biodiversity.

This case study focuses on the Bokkeveld Plateau, an area of ca. 100 x 75 km in the semiarid northwestern part of South Africa (Figure 1). The plateau is situated at the junction of two globally recognised biodiversity hotspots, the Cape Floristic Region (CFR) and the Succulent Karoo (SK) and is best known for the high number of plant species. Collection records show that there are at least 1350 plant species on the Bokkeveld Plateau, of which 6 percent are local endemics (found only on the plateau) (Manning and Goldblatt 1997). Our study examined the ecosystem goods and services and biodiversity benefits associated with different forms of land use in the eastern part of the Bokkeveld Plateau. Materials and methods have been developed for use in dynamic ecological economic models using STELLA (HP Systems). The outline for such a model shows how production is affected by various ecosystem services (Figure 2). For the purposes of the model, services can be either positive (e.g., soil health) or negative (e.g., stock predation). Our discussion in subsequent sections deals only with factors impacting on production. However, the data we are collecting enables us to include other forms of ecosystem goods and services as well.

Figure 1. Map (top) of the western part of South Africa showing the Bookieveld Bokkeveld Plateau and (below) a more detailed map showing local ecoregions.

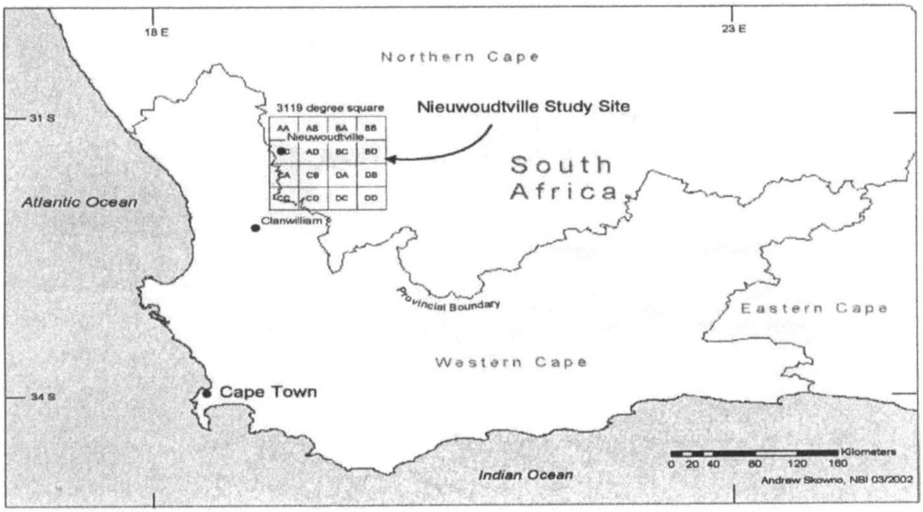

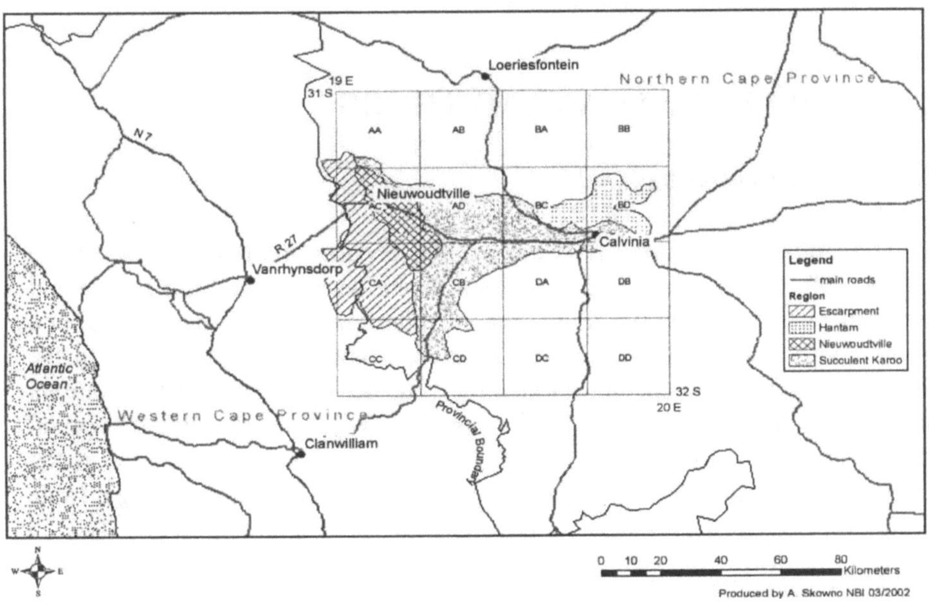

Figure 2. Outline of an ecosystem services model showing the relationship between agricultural activities, ecological variables, ecosystem services and ecosystem disservices (e.g., predation and contamination of wool)

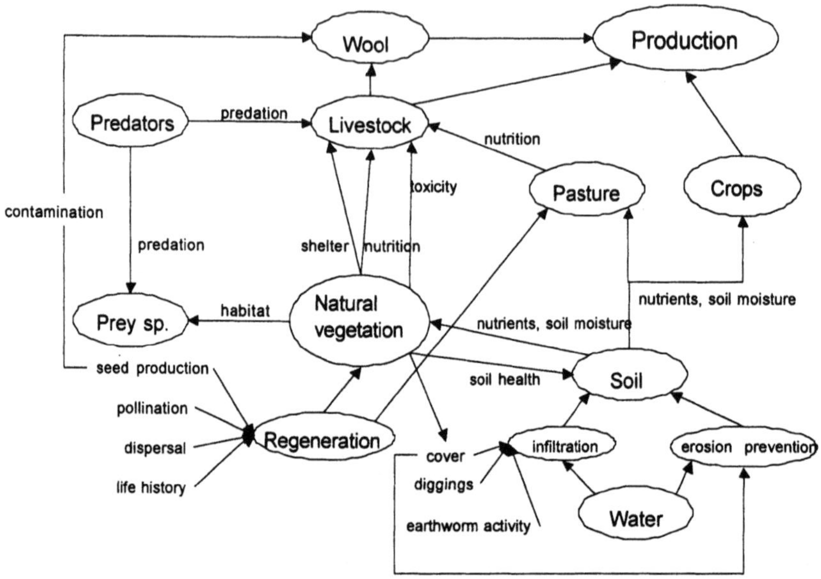

2. MATERIALS AND METHODS

2.1 Study Area

The Bokkeveld Plateau falls almost entirely into a single degree square (31° S, 19° E) (Figure 1) and is defined in the west by the high Bokkeveld escarpment that rises 600m above the surrounding plains (Knersvlakte) and in the south and north by slightly lower escarpments. The boundary in the east is less well defined, but an arbitrary cut off point beyond the Hantam mountains is justified by substantial changes in vegetation beyond that point. Highly variable soils and steep rainfall gradients are striking features of the Bokkeveld Plateau. The Bokkeveld escarpment in the west comprises Table Mountain Sandstone (TMS) with soils that are characteristically low in nutrients. Towards the east, the TMS is overlain by sedimentary deposits of Dwyka Tillite and Ecca shales. In the vicinity of Niewoudtville and the

Hantam mountains, tillite and shale soils have been further overlain by Karoo dolerite intrusions that have eroded to form heavy clay soils.

Annual rainfall varies from 550 mm on the edge of the Bokkeveld escarpment to below 250 mm in some areas towards the east. Along the Bokkeveld escarpment the rain falls mainly in the winter months (April–September) but summer thunderstorms influence rainfall patterns in the eastern part of the Bokkeveld Plateau.

This case study focuses mainly on the area around the town of Nieuwoudtville where cultivation of crops has had the highest impact on natural vegetation. The town is situated near the transition between TMS, Dwyka Tillite, and dolerite soils, and has an annual rainfall of 350 mm which occurs mostly in winter. Both Dwyka tillite and dolerite soils have been used mainly for intensive wheat production and sheep farming.

2.2 Biodiversity Surveys

Available information from previous surveys and collections of plant and animal diversity was obtained from published records and unpublished databases. Additional surveys were carried out for plants, insects, and reptiles (only plants are reported here). Plant diversity was sampled in different parts of the landscape (based on soils, geology, types, vegetation types, and landforms) using Whittaker–type plots (Stohlgren 1995) in order to partition plant diversity in relation to landscape structure and the likely impact of different types of land use. The effect of land use on plant diversity was also examined by using stratified random sampling where 100m x 100m plots were laid out in such a way that soil type, slope, aspect, and climate were kept as constant as possible and land use became the only significant variable.

2.3 Ecosystem Goods and Services

Ecosystem goods and services related to different types of land use were identified through interviews with farmers, comparison with studies of similar systems in other parts of the world, and the perceptions of scientists familiar with the Bokkeveld Plateau. Specific ecosystem goods and services were also measured under different forms of land use to quantify changes associated with land use. These included soil health, carbon sequestration, water infiltration, regeneration of habitat, maintenance of genetic resources, problem animal control, and pest control (methods for these assessments will be provided in other technical papers dealing with these issues). Soil electrical conductivity was measured using a 1:5 soil water solution. This gives an idea of the total soluble salt content of the soil, which is linked to fertility. Summary data are provided here to show how these aspects fit into models of landuse and ecosystem services.

3. RESULTS

3.1 Plant Diversity

Plant databases for the Bokkeveld Plateau indicate that plant diversity is not evenly spread throughout the plateau. Data on threatened species and on geophytes, which are among the best collected plant taxa in the region, show that the quarter degree square (QDS) incorporating the town of Nieuwoudtville has a disproportionately high number of plant species. Out of 185 species listed in the Red Data List (RDL), 98 species occur in the vicinity of Nieuwoudtville. The next most diverse QDS has only 40 RDB species. Similarly, the data for geophytes shows that the Nieuwoudtville QDS has 241 geophyte species out of a total of 309 geophytes for the entire Bokkeveld Plateau. The high diversity in the Nieuwoudtville QDS probably reflects some collecting bias but the plateau is traversed by several major roads (which tend to be well collected) and incorporates other areas that have been well sampled (e.g., the Hantamsberg mountains). As a result, additional sampling is unlikely to alter the status of Nieuwoudtville as a centre of diversity on the Bokkeveld Plateau.

The three main soil types in the vicinity of Nieuwoudtville give rise to four different vegetation types: fynbos vegetation on TMS, Renosterveld on Dwyka sediment derived soils (DSD) with relatively high rainfall (>300 mm), Karoo vegetation on DSD and Karoo dolerite derived soils (KDD) with low rainfall (< 300 mm), and a modified Karoo vegetation with low shrub cover and a high proportion of geophytes on KDD soils with relatively high rainfall (>300 mm). The vegetation types that have been most affected by agriculture are Renosterveld and the modified Karoo vegetation on KDD soils. In both cases, less than 20 percent of the original vegetation remains. Fynbos vegetation has a relatively high plant diversity but the nutrient poor soils have low agricultural value and more than 50 percent of the original vegetation is still intact. Similarly, the areas with <300 mm rainfall are generally too dry for cropping (except near rivers) and the vegetation is used primarily for grazing. Grazing impacts on plant diversity in these vegetation types, but the initial focus has been on the likely loss of species through cultivation.

Detailed plant surveys in Whittaker–type plots show that geophytes make up the majority of species (43 percent) in both Renosterveld and modified Karoo vegetation on KDD soils. Perennial herbs (14 percent), annuals (18 percent), and shrubs (16 percent) are important components of Renosterveld vegetation but shrubs are almost absent (1 percent) from modified Karoo vegetation where annuals make up 36 percent of the species. Plant diversity within Whittaker–type plots was higher in Renosterveld than on the dolerite derived soils across all spatial scales used in the Whittaker plots indicating that point diversity and turnover are higher in Renosterveld vegetation than on the adjacent dolerite soils. Exceptionally high numbers of plant species were recorded in Renosterveld vegetation with between 80 and 140 species per 50 x 20m plot (1000 m^2).

We are in the process of measuring the impact of land use on plant diversity on different soil types but the initial indication is that further transformation of Renosterveld will have the greatest impact on plant diversity. However, although the modified Karoo vegetation on KDD soils has lower overall diversity, it contains a high number of regional endemics (25 species) that now survive in a total area of a few thousand hectares.

3.2 Ecosystem Goods and Services

Ecosystem services were identified for all the major forms of land use in the Nieuwoudtville area of the Bokkeveld Plateau (Table 1).

Several ecosystem services are common to several forms of land use, such as maintenance of soil health, carbon sequestration, water infiltration and erosion control, and pest control. Others such as the provision of shade and shelter, maintenance and regeneration of habitat, problem animal control, resilience, and maintenance of genetic resources affect only a few types of land use. However, one type of land use can impact on the provision of ecosystem goods and services for another form of land use. For example, the maintenance and regeneration of habitat has little benefit for wheat farming, but the elimination of habitat means that this land has little value for anything other than cultivation. Similarly, wheat farming may not benefit from the regulation of river flows and groundwater, but any impact on this ecosystem service could influence ecotourism, residential water, and vegetable farming. The downstream costs of each land use therefore need to be considered in order to determine the overall costs and benefits of each land use. The connection between land use and ecosystem goods and services is outlined in Table 2.

The key element here is that different land uses and management practices affect ecosystem processes in different ways. As a result, they can influence the outcome in terms of ecosystem services. One of the important factors in the research program that is currently underway is to identify the important ecological processes and drivers so that they can be incorporated in ecological economic models.

We have analyzed some of the ecosystem services in greater detail. The analyses are not yet at a stage where we can identify the main forces influencing ecosystem goods or services or the interactions between land use and ecosystem services, but the preliminary data provide insights into the changes that may occur under different forms of land use.

Table 1. Ecosystem services related to the main forms of land use in the Nieuwoudtville area on the Bokkeveld lateau.

Land use	Ecosystem Services Required to Sustain Land Use
Cropping (cereals, including hay production)	Maintenance of soil health; Water infiltration and erosion control; Pest control; Absorption and breakdown of waste products; Maintenance of genetic resources
Crop and medic systems	Maintenance and restoration of soil health; Water infiltration and erosion control; Pest control; Absorption and breakdown of waste products
Grazing	Provision of shade and shelter; Maintenance of soil health; Water infiltration and erosion control; Maintenance and regeneration of habitat; Pest control; Problem animal control; Resilience; Maintenance of genetic resources
Vegetables	Pest control; Maintenance of soil health; Water infiltration and erosion control; Absorption and breakdown of waste products; Regulation of river flows and groundwater; Pollination
Rooibos tea	Pest control; Maintenance of soil health; Water infiltration and erosion control; Pollination and seed dispersal (ants)
Ecotourism	Maintenance and regeneration of habitat; Life fulfillment; Regulation of river flows and groundwater; Water infiltration and erosion control; Maintenance of genetic resources
Residential	Regulation of river flows and groundwater; Life fulfillment; Provision of shade and shelter; Firewood; Medicinal plants
Future options	Life fulfillment; Regeneration of habitat

Table 2. An illustration of the possible relationships between different types of land use and ecosystem goods and services. The lists are not complete and each column is independent of the others.

Land use →	Management →	Ecosystem processes →	Ecosystem services →	Drivers →	Goods/ Services →
Crops	Ploughing	Nutrient cycling	Soil fertility	Available nutrients	Wheat yield
Livestock	Herbicides	Microbial activity	Erosion control	Soil moisture	Wool production
Vegetables	Pesticides	Infiltration	Pest control	Pest damage	Tourist visits
Flowers	Fertilizers	Insect predation	Shelter/ Shade	Stock losses	Carbon credits
Tourism	Stock	Pollination dispersal	Resistance to invasion	Production	Biodiversity credits
Recreation	Fencing	Soil turnover	Carbon storage	Loss of biodiversity	
Residential	Reserves	Climate regulation	Pollination	Global warming	
		Predator-prey	Predator control		

1.3 Soil Health

Soil health is an important ecosystem service that underpins agricultural production and influences the regenerative capacity of cultivated and natural lands. The region has two main soil types: Karoo dolerite–derived (KDD) vertisols with reddish brown colours and Dwyka sediment–derived (DSD) duplex soils with characteristic yellowish brown to gray colours. The dolerite–derived soils (KDD) have a high shrink–swell capacity and are self–mulching due to the turnover that occurs when soils shrink and swell. Clay illuviation of the Dwyka soils (DSD) has resulted in the formation of dense and water impermeable B horizons, which may lead to water logging in the wet winter months. Plants may benefit from this water storage but it could also lead to anoxic conditions that impede root growth. The two soil types respond in different ways to land use.

1.4 Natural Vegetation

Soil heterogeneity is a characteristic feature of natural landscapes. Renosterveld vegetation on DSD soils in the Nieuwoudtville area comprises grass tussocks (*Merxmuellera stricta*), shrubs (mainly *Elytropappus rhinocerotis*), and open areas that may contain geophytes (bulbs and corms). Soils collected under *M. stricta* tussocks and *E. rhinocerotis* bushes had higher nutrient concentrations, higher organic carbon, lower modulus of rupture, and faster infiltration rates than soils in the open or under some forms of cultivation. The grass tussocks and bushes form islands of fertility within the landscape and play an important role in nutrient cycling and productivity within the ecosystem. The removal of these fertility islands can reduce soil heterogeneity and consequently soil health of the landscape.

The vegetation on KDD soils has fewer shrubs and is often dominated by geophytes. Soils collected under leaves of large bulb species (*Brunsvigia bosmaniae*; Amaryllidaceae), which create a form of ground cover, had higher EC, higher soluble K, Ca, Mg, and NO_3, lower extractable Mg, higher extractable P and Zn, and higher labile carbon than adjacent open areas. This can be attributed to higher mineralizing of organic matter under the leaves and shows that the vegetation influences soil structure and fertility.

1.5 Wheat Cropping

KDD soils under wheat had lower electroconductivity (EC), lower pH (water), lower total nitrogen, soluble K, Ca, Mg and SO_4 as well as lower extractable Mg and P in comparison to virgin soils. Farmers tend not to fertilise these soils due to their inherently high fertility status. Although these vertisols are clay and nutrient rich, and thus have some inherent resilience to nutrient removal via cropping, there appears to be a net export of some nutrients from these wheatlands. The depletion of phosphorus, in particular, is a factor that may limit sustained agricultural production as well as ecosystem recovery on these cultivated lands.

On DSD soils, there was a general pattern of higher nutrient concentrations in active wheat lands in comparison to areas with natural vegetation. (sampled between bushes). DSD soils under wheat had higher water soluble K and extractable P concentrations than virgin interbush (VIB) soils, probably as a result of fertilization of these lands. The rate of infiltration through simulated soil crusts and the modulus of rupture of DSD soils were unaffected by wheat farming.

1.6 Medic Pastures

Medicago pastures used for stock grazing are typically found on old wheatlands on DSD soils, either as a transition from wheat farming to sheep farming or as a form of crop rotation. Medicago pastures had higher pH (KCL) and higher soluble and extractable K, Mg, Ca and P concentrations than DSD soils under natural vegetation (between bushes). This is probably due to prior fertilisation of wheat lands as well as fertilisers added to Medicago pastures and greater nutrient cycling by livestock. Infiltration rates through simulated crusts were lower in Medicago pastures than VIB soils and thus Medicago pastures seem to have a negative effect on soil structure and the capacity of these soils to trap rainwater efficiently.

1.7 Carbon Sequestration

Under suitable land management, soil carbon is potentially a large and long–term sink for sequestrating atmospheric carbon (Post and Kwon 2000, Post et al. 2001). The reverse is also true, i.e., land management can result in the loss of carbon from the soil into the atmosphere. Soil carbon data collected on the Bokkeveld Plateau suggest that the removal of Merxmuellera stricta tussocks and *Elytropappus rhinocerotis* bushes from the renosterveld ecosystem on DSD soils can result in the loss of up to 11 tonnes of carbon per hectare in the top 10cm of soil. The amount of carbon stored in the topsoil appears dependent on the type and extent of plant cover. Indigenous grasses and bushes shade the soil and reduce carbon mineralisation, whereas cultivated lands such as wheatlands and Medicago pastures are likely to have warmer soil temperatures and greater rates of carbon mineralisation (Dalal and Chan 2001). On KDD soils, wheatlands had soil carbon stores 5tC/ha lower than natural dolerite lands and 26tC/ha lower than soil under bulb leaves in natural dolerite lands. The data collected to date shows that agricultural activities on the Bokkeveld Plateau that disturb the soil through tillage or exposure of the soil surface to sunlight, result in a considerable reduction of soil carbon.

1.8 Water Infiltration

Water infiltration is a major ecosystem service as it reduces erosion potential, and recharges ground water supplies on which farmers in this area are completely dependant.

Transects of 25 m were run in lands and renosterveld. Infiltration rates were recorded for 50 ml of water at meter intervals. Our preliminary findings show significantly faster infiltration in lands ($P < 0.01$). This is believed to result from a higher level of disturbance and greater earthworm activity on lands. The presence of earthworms was associated with faster infiltration rates in both veld types. Infiltration rates were also faster in porcupine diggings than in the adjacent undisturbed areas. Infiltration rates were found to be significantly slower on sheep paths than in the adjacent undisturbed areas.

1.9 Problems of Animal Control

To determine the relationship between land use and problem animals, livestock losses due to predators are being recorded for the period 1997 to the present. Field studies will compare vegetation and landscape structure (as defined by a list of habitat variables) under different land use or vegetation management. Surveys of the dominant prey species will be undertaken for individual farms to determine the relationship between habitat availability, prey availability and prevalence of problem animals. Potential prey include antelope, hyrax, hare or rabbit, birds, rodents, and reptiles. This can then be compared with the annual livestock losses experienced by the farmer.

1.10 Pollination

Pollination has been identified in other studies as a major ecosystem service. Although there are relatively few insect pollinated crops in the Nieuwoudtville area, *Medicago polymorpha* is one of the most important forage species for farmers in the Nieuwoudtville area and self–seeding is essential to sustain these pastures. We therefore set up exclosures to determine the importance of insect pollinators for seed set in *M. polymorpha* plants. Significant differences in seed set were recorded, with twice as much seed formed in those pods outside exclosures ($P < 0.001$). The affect of land use on the pollinators of *M. polymorpha* has yet to be established. Insect pollination is also an important service for the regeneration of habitat and this is also being investigated.

2. SUMMARY

South Africa is a dry country (91 percent of the country is classified as arid, semiarid, or dry subhumid) that is a globally significant region for biodiversity conservation. The country comprises less than 0.8 percent of the worlds total land area, but it contains ca. 8 percent of the worlds vascular flora, as well as between 27 percent of the worlds amphibian, reptile, avian and mammalian species. Despite a reserve network that covers ca. 6 percent of the land area, a significant proportion of South Africa's biological diversity exists only outside reserves, mainly in agricultural landscapes. Conserving biodiversity in agricultural landscapes is therefore critical for the conservation of biodiversity in this region. In some biodiversity hotspots in South Africa, farmers have experimented with alternative land use practices in order to secure ecosystem goods and services and to preserve the natural assets on their farms. The biodiversity benefits, as well as the provision of goods and services, are being assessed as part of a Global Environmental Facility project on conservation farming. This paper focuses on one study site in a semiarid region on the Bokkeveld Plateau (Northern Cape province) in South Africa. The data are presented as a preliminary model where land use options influence both biodiversity and the provision of ecosystem goods and services. We have included preliminary data on the risk of extinction for plant species under different forms of land use, as well as ecosystem goods and services related to water infiltration, soil health, carbon sequestration, and production. The ecological, social, and economic components of the project will be integrated to develop a model for the region that can be used to guide land use planning in the region.

Preliminary data of the sort we have presented here does not lend itself to detailed discussion. The main purpose of this paper is to outline an ongoing research program that will enable us to make more informed decisions about land use in an area with exceptional biodiversity. By quantifying the full extent of goods and services in this area, farmers and policy makers will have a better idea of the consequences of different forms of land use and different management actions.

3. ACKNOWLEDGEMENTS

This work forms part of the GEF Conservation Farming Project coordinated by the National Botanical Institute. The authors thank the Mazda Wildlife Fund for the use of a vehicle and the farmers of Nieuwoudtville for unrestricted access to their properties.

4. REFERENCES

Dalal, R.C., Chan K.Y. 2001, Soil organic matter in rainbed cropping systems of the Australian cereal belt. Australian Journal of Soil Research 39: 435–464.

Donaldson, J.S. 2002. Biodiversity and conservation farming in the agricultural sector. Pages 43–55 In: S.M. Pierce, R.M. Cowling, T. Sandwith, and K. MacKinnon (eds)

Mainstreaming biodiversity in development: Case studies from South Africa. The World Bank Environment Department, Washington D.C.

Hoffman, M.T., Todd, S., Ntshona, Z., and Turner, S. 1999. Land degradation in South Africa. Unpublished report for South African Department of Environmental Affairs and Tourism, Pretoria. Electronic copy: www.nbi.ac.za/landdeg.

Hoffman, T., and Ashwell, A. 2001. A Land Divided: Land Degradation in South Africa. University of Cape Town Press, Cape Town, South Africa.

Macdonald, I.A.W. 1989. Mans role in changing the face of southern Africa. Pages 51–78 In: B.J. Huntley (ed) Biotic Diversity in Southern Africa. Oxford University Press, Cape Town, South Africa.

Manning, J., and Goldblatt, P. 1997. Nieuwoudtville, Bokkeveld Plateau & Hantam. South African Wild Flower Guide 9. Botanical Society, South Africa. 204 pages.

Pagiola, S., Kellenberg, J., Vidaeus, L., and Srivastava, J. 1997. Mainstreaming biodiversity in agricultural development. World Bank Environment Paper 15. World Bank, Washington D.C.

Post, W.M, and Kwon, K.C. 2000, Soil carbon sequestration and land–use change: Processes and potential. Global Change Biology 6: 317–327.

Post, W.M., Izaurralde, R.C., Mann, L.K., and Bliss, N. 2001. Monitoring and verifying changes of organic carbon in soil. Climatic Change 51: 73–99.

Siegfried, W.R. 1989. Preservation of species in southern African nature reserves. Pages 186–201 In: B.J. Huntley (ed). Biotic Diversity in Southern Africa. Oxford University Press, Cape Town, South Africa.

Stohlgren, T.J, Falkner, M.B., and Schell, L.D. 1995. A modified–Whittaker nested vegetation sampling method. Vegetatio 117: 113–121.

Walker, B.H. 2000. Incorporating ecosystem services into the Australian economy: Investing in a new and sustainable future. http://www.foresttrends.org/whoweare/pdf/sydney/brianwalker.pdf

Chapter 4

STRATEGIES FOR IN SITU CONSERVATION OF CROP GENETIC RESOURCES IN DRYLAND AREAS OF AFRICA

M. Grum[1], V. Guerette[1], B. Kone[2], A. Sidibe[2], M. Kouressy[2], T. Huvio[3], A. Mafa[4], C. Mujaju[4], A. Bretaudeau[5], M. Vaxmann[6], M. Ouatara[7], I. Traore[8], B. Butaumocho[9], F. Zinanga[10] and O. Mamadou[11]

[1]International Plant Genetic Resource Institute, Nairobi, Kenya, m.grum@cgiar.org; [2]Institut d'Economie Rural, Bamako, Mali; [3]United Nations Food and Agriculture Organization, Rome, Italy; [4]Department of Research and Extension, Harare, Zimbabwe; [5]Institut Polytechnique Rural, Katibougou, Mali; [6]Cirad, Bamako, Mali; [7]USC–Canada, Douentza, Mali; [8]Fondation Pour le Developpement du Sahel, San, Mali; [9]Intermediate Technology Development Group, Mutare, Zimbabwe; [10]Commutech, Harare, Zimbabwe; [11]ACAS–GAO, Gao, Mali

1. INTRODUCTION

Many millions of farmers continue to depend on their traditional cultivars for food, fodder, and other economic, cultural and ecological activities (Brush 1991, Zimmerer and Douches 1991, Bellon 1996). In many developing countries, farmers also rely largely on local seed sources for their staple crops. Over 95 percent of rice cultivated in Nepal, 85 percent of durum wheat and 98 percent of barley in Morocco, and 50 percent of maize seeds in Mexico still come from local farmer sources (Mellas 2000, Upadhyay et al. 2001).

The programme for the development of strategies for in situ conservation and sustainable use of plant genetic resources in dryland areas of Africa is a collaborative one involving IFAD, IPGRI, FAO and national institutions in Mali and Zimbabwe, ranging from the lead plant genetic resources institution, breeders, and extension services, to universities and NGOs working directly with communities.

The overall objective of the action–research programme is to develop project instrumentalities designed to address genetic erosion caused by drought and desertification in the dryland ecologies of Africa. The aim is to mitigate the impact of temporary drought induced conditions in the desert margins through the development of comprehensive community–based strategies for monitoring, analysing, assessing and addressing the desertification phenomenon.

As a part of the research, the programme develops and applies socioeconomic analytical tools quantification of the cost of inaction and loss of biodiversity due to desertification. The project also studies incremental benefits of intervention to mitigate desertification. The research involves the identification of viable community–based mechanisms for promoting the sustainability of traditional production and seed supply systems. This is intended to provide farmers secure access to their traditionally preferred genetic material, which could otherwise be eliminated by droughts. This project is intended to have strong links to investment projects, and to provide outputs that could support investment and development projects.

The programme comprises four main components: (1)site identification, selection, mapping, genetic inventory, and analysis of these areas at different levels of risk of desertification; (2)development of community and participatory methodologies for monitoring and maintenance of genetic resources vulnerable to desertification; (3)action research in selected project sites for testing models of community–based gene banks and on–farm seed production systems; and (4)development of tools of cost benefit analysis for use in project design for capturing costs due to desertification and benefits from intervention to conserve and utilize endangered traditional crop genetic resources.

The project has made substantial progress toward the achievement of its objectives. Factors contributing toward this progress are described in the following sections.

2. THE ROLE OF GENETIC DIVERSITY IN COPING STRATEGIES OF FARMERS

Genetic diversity has a mixture of private and public attributes. Since farmers are not likely to be willing and able to maintain diversity solely for its public attributes, their involvement in in situ conservation on–farm relies largely on the strength of the private benefits (Smale et al. in press). The role of genetic diversity in farmers' coping strategies is therefore a major focus of this project. Participatory diagnostics carried out so far in eight villages in Mali and four villages in Zimbabwe indicate that farmers' reasons for using diversity can be grouped into three main natural resource management strategies.

One, is to fulfill different uses. Crop diversity played a role in maintaining a balanced diet, varietal diversity could be the result of different ways of preparing them, e.g., in Tsholotsho, Zimbabwe, some varieties of sorghum are good for making porridge, some are suitable for brewing beer, some have stalks that are good for chewing and others for fodder. Distinct varieties were sometimes maintained for market. Two, to optimize use of a diverse resource base. Farmers explained that differences in water regimes of soils within the farm, e.g., valley bottoms and uplands, and even within individual plots, were a major factor in determining where to grow each variety. And, three, to cope with unpredictability. For example, rainfall was consistently mentioned as the most important of these factors. All three of these strategies are very important in marginal areas, where limiting factors are more critical and more varied. Self–sufficiency is also more important in marginal areas where transaction costs are high.

Morphological characterization, both by farmers and researchers, is expected to shed more light on these strategies, by identifying specific traits of importance to farmers. On–station characterization has already been carried out on sorghum in both countries, but the data needs further analysis. Farmer participatory characterization of sorghum, cowpea, millet has been carried out in Zimbabwe and is currently being conducted in Mali.

Molecular characterization carried out in Zimbabwe (yet to be analyzed) and that being carried out in Mali should shed further light on a level of diversity that farmers may have little awareness of, but which may be important in providing resistance to diseases, diversity for yield levels, etc.

3. FACTORS INFLUENCING GENETIC RESOURCES AND THEIR EROSION

Least cost in situ conservation will occur at sites that are most highly ranked in terms of public benefits and where, because private benefits farmers obtain from growing genetically diverse varieties is greatest, public interventions to encourage them to do so will be least. There, private and social costs will both be relatively low (Smale et al. 2002).

Over the past forty years rainfall in Mali has decreased steadily. During participatory diagnostics, farmers have repeatedly and consistently identified drought as one of the most common causes of loss of varieties. They also identify a long list of specific varieties that have been lost, most of which are long duration varieties.

Surprisingly, a nationwide collecting mission and subsequent morphological characterization carried out by the project in Mali, comparing the present varieties with varieties from a collecting mission by Orstom, ICRISAT and IBPGR of twenty years ago, provides only marginal support for this theory (Traoré et al. 2000). While the varieties with the very longest duration to maturity did appear to have disappeared, this was not statistically significant. Also less than 30 percent of varieties found in the dry north twenty years ago are presumed lost compared to over 70 percent in the more humid south. The major cause of the high erosion in the south is the increased area under maize and other cash crops such as cotton, which do not grow in the northernmost zones.

Farmers practice a range of seed selection methods. In Zimbabwe, some farmers selected while the crop was still in the field, others select when the harvest is on the threshing and drying floor, while some select even after threshing.

The project has recorded traditional seed storage practices at all the project sites. There is considerable variation in the methods used, both in terms of containers used, whether these are sealed and in additives used to protect the seed from humidity, pests and diseases. Comparative tests of these methods are being carried out at several sites.

4. ASSESSMENT OF FARMERS IN MANAGING GENETIC RESOURCE

While farmers clearly have advanced management strategies involving varietal diversity of all their major crops, some weaknesses have been identified. The diversity that farmers are growing does not always appear to result from deliberate, well–informed choice of variety. The harsh and highly variable climate in the study areas means that a variety needs to be tested for several years before it can be considered well evaluated for stable yield. Farmers therefore tend to stick with material that they have had for a long time and limit their search for new varieties. Nevertheless, farmers have been eager to obtain new material at seed diversity fairs in both Mali and Zimbabwe.

Preliminary analysis of morphological characterization data for sorghum in Zimbabwe (e.g., Appa Rao et al. 1987) indicates that while a farmer variety name generally implies the presence of one to three clearly defined traits, many other traits may vary very considerably from one farmer to the next, even when the variety has the same name. A survey of farmers in Mali also indicated that most were not necessarily aware that material with the same name could be morphologically and genetically diverse. This seriously limits farmers' ability to exchange information, experience and seed, compounding the problems of access to seed. This may have major implications for the development activities that the project proposes. Further analysis indicated that the number of farmer variety names found in a region was also a poor indicator of the overall morphological and genetic diversity in sorghum in Zimbabwe. In Mali and Zimbabwe, additional socioeconomic and varietal data are being gathered to obtain a sufficiently large data set to draw conclusions about the social equity of varietal distribution. The statistical analyses were carried out in the statistical environment "R" (Venables et al. 2001).

5. IMPACTS OF CONTINUED DETERIORATION OF GENETIC DIVERSITY

The future food supply of all societies depends on the exploitation of genetic recombination and allelic diversity for crop improvement, and many of the world's farmers depend directly on the harvests of the genetic diversity they sow for food. The impact of continued genetic erosion or subutilization of the genetic diversity for the individual farmer is largely a function of the role of genetic diversity as described previously.

Two developments in the last couple of years demonstrate the importance of conservation in medium term strategies. In Mali, the price of cotton was so low in 2000 that many did not plant cotton in 2001. However, as this study has shown, over 70 percent of the sorghum variety names previously grown in this area have since disappeared and farmers' alternatives are therefore limited.

In Zimbabwe, the economic crisis has meant that smallholders are unable to buy the hybrid maize seed, fertilizer and pesticides to grow hybrid maize. The loss of earlier genetic diversity of crops such as pearl millet and sorghum means that many are now planting second and third generation hybrid maize with resulting declines in yield.

6. TESTING MODELS FOR COMMUNITY–BASED ACTIVITIES TO IMPROVE SEEDS

The project is testing a number of community–based activities that aim to strengthen grassroots organizations to mitigate the impact of temporary drought–induced conditions. The project has developed a methodology for participatory characterization of farmers' diversity to ensure that priorities and perceptions of traits of different socioeconomic groups of farmers is incorporated in seed multiplication and breeding efforts and in the evaluation of the impacts of new project instruments developed.

Seed diversity fairs facilitate seed exchange between farmers and communities and ensure access to a diversity of seed. They are also an opportunity for farmers to share knowledge about local crops and varieties. The concept was first introduced to Africa in Zimbabwe in the early 1990's and several seed fairs have been organized by the project. The project brought the concept to Mali, where three seed diversity fairs were held last year and three are being organized this year. IPGRI carried out an evaluation of their impact one and a half years after the first had been held. This evaluation set figures on the impact and found that the fairs did effectively increase diversity present in farmers' fields and proposed several improvements to the way they are organised.

Farmer field schools in Zimbabwe and farmer schools on Biodiversity in Mali are on–farm participatory diagnosis and experimentation processes jointly carried out by farmers, researchers and extension agents. The project has extended this methodology, which is well known in IPM, to plant genetic resources management.

In Mali, the project is promoting an improved seed storage method developed by the Institut d'economie rurale (IER) for cowpea and Bambara groundnut, addressing a major constraint in cowpea production and conservation. Other improved seed storage methods are being tested and compared to traditional methods.

7. ASSESSMENT OF POTENTIAL PROJECT IMPACTS TO ASSIST CAPACITY OF FARMERS

Over the last ten years, seed diversity fairs have been growing in popularity in rural areas of the developing world as a means for on–farm conservation of plant genetic resources. Seed diversity fairs are a community–based strategy to promote agrobiodiversity by creating an opportunity for farmers to meet and exchange planting material and associated knowledge. IPGRI carried out one of the first studies to look at the impacts of a seed diversity fair conducted in Mali. The main objective was to measure its effects on the agrobiodiversity of smallholders with a survey approach, one year after the fair. Ten farmers were interviewed during the survey.

The results indicated a clear increase in diversity in farmers' fields a year after the fair. Four of the ten farmers interviewed had tested the new material obtained at the fair and would continue growing it. The survey also identified a gender component: men only gave material to other men at the fair, while women farmers gave seed to both men and women, suggesting a need to emphasise the presence of women at the fairs. Success of seed diversity fairs seems to be dependent on the preparation and conceptualization of the event. Also, when the names do not denote a clear genotype, exchange of information and experiences is seriously hampered, and in the case where farmers are not aware of the possible confusion, can even create misunderstandings with substantial consequences.

8. SUMMARY

The overall objective of this programme is to develop interventions designed to address genetic erosion caused by drought and desertification in the dryland ecologies of Africa. The aim is to mitigate the impact of temporary drought–induced conditions through the development of comprehensive community–based strategies for monitoring, analysing, assessing and addressing the desertification phenomenon.

The project attempts to understand the complex role that genetic diversity plays in coping strategies of farmers. This research focuses on a description of the crop genetic diversity, its extent and distribution, the socioeconomic situation, environmental factors, and on participatory diagnosis. In addition, we look at the positive and negative dynamics influencing the genetic resources and genetic erosion, including traditional management practices for selection, conservation and multiplication of the genetic resources.

The effectiveness of traditional management practices and the difficulties encountered by farmers in their management of genetic resources has been studied to better understand the impact of continued deterioration or subutilization of the genetic diversity. Finally, the project is testing models of community–based activities, such as community gene banks, improved seed storage systems, seed diversity fairs, farmer field schools, and on–farm seed production. The expected results of the project described in this paper is a series of project instruments that will enable development projects to support farmers in optimizing the benefits of their plant genetic diversity in marginal areas of Africa.

9. ACKNOWLEDGEMENTS

IPGRI wishes to thank IFAD for the important support that it has given to the development of strategies for in situ conservation and utilization of plant genetic resources in dryland areas of Africa.

10. REFERENCES

Appa Rao, S., and Mushonga, J.N. 1987. A catalogue of passport and characterization data of sorghum, pearl millet and finger millet germplasm from Zimbabwe. IBPGR, Rome.

Brush, S., 1991. A farmer–based approach to conserving crop germplasm. Economic Botany 45:153–65.

Bellon, M.R., 1996. The dynamics of crop infraspecific diversity: A conceptual framework at the farmer level. Economic Botany 50: 26–39.

Gower, J.C. 1971. A general coefficient of similarity and some of its properties. Biometrics 27: 857–871.

Mellas, H., 2000. Morocco. Seed supply systems: Data collection and analysis. Pages 155–156 In: Jarvis, D., Sthapit, B., and Sears, L. (Eds.) Conserving agricultural biodiversity in situ: A scientific basis for sustainable agriculture. Rome, Italy.

Smale, M., Bellon, M.R., Jarvis, D., and Sthapit, B. In press. Economic concepts for designing policies to conserve crop genetic resources on farms. Genetic Resources and Crop Evolution.

Shannon, C.E., and Weaver, W. 1949. The mathematical theory of communications. Urbana, Ill.: Univ. Illinois Press. 117 pages.

Struyff, A., Hubert, M., and Rousseeuw, P.J. 1996. Clustering in an object–oriented environment. Journal of Statistical Software 1(4): http://www.jstatsoft.org/v01/i04/.

Traoré, S.B., Reyniers, F.N., Vaksmann, M., Koné, B., Sidibé, A., Yoroté, A., Yattara, K., Kouressy, M. 2000. Adaptation à la sécheresse des ecotypes locaux de sorghos du Mali. Sécheresse 11: 227–237.

Upadhyay, M.P., Rijal, D., Gauchan, D., Baniya, B.K., Bajracharya, J., Rana, R.B., Joshy, K.D., Khatiwada, S. P., Shakya, D.M., Subedi, A., Sapkota, R.P., and Sthapit, B. 2001. Paper presented for the workshop, "In Situ Conservation of Agrobiodiversity: Scientific and Institutional Experiences and Implications for National Policies." International Potato Center (CIP), La Molina Peru, August 14–17.

Venables, W.N., Smith, D.M., and the R. Development Core Team. 2001. An Introduction to R. Notes on R: a programming environment for data analysis and graphics. Version 1.3.0, 2001-06-22. http://www.cran.r-project.org/.

Zimmerer, K.S., and Douches, D.S., 1991. Geographical approaches to native crop research and conservation: The partitioning of allelic diversity in Andean potatoes. Economic Botany 45: 176–189.

Chapter 5

A PARTICIPATORY APPROACH FOR CONSERVATION OF FOREST GENETIC RESOURCES

Lambert G. Ouedraogo, Haoua Sary, Christiane S. Yameogo–Gamene, and Oblé Neya
Centre National de Semences Forestiéres, Ouagadougou, Burkina Faso; email cnsf@fasonet.bf

1. INTRODUCTION

In Burkina Faso as well as in the other sahelian countries people rely on trees and shrubs for their daily life (Guinko 1984, 1985; Bognounou 1987, 1988; FAO 1987; Kessler and Boni 1990). Trees are used for human and animal food, for medicinal purposes, for energy (firewood and charcoal), for construction, for raw material, for soil fertilization, etc. Despite the importance of the forest resources to the livelihood of rural people, Burkina Faso loses 32,000 ha of forest each year (FAO 1995).

In order to sustainably use forests, there is a need to develop a rational method of management. The sustainable management of the ecosystems and particularly the forest genetic resources is very important for the preservation of the biodiversity and to insure a long–term utilization of the forest resources in the arid and semi–arid zones. To accomplish this both in situ and ex situ conservation of resources will be required.

The National Forest Seed Center (CNSF) of Burkina Faso has approached the rural population to both learn from and to enhance the value of their knowledge about seed storage which is important for the management program of the natural forest stands and also for the forest seed stands. It is important to note that forest seed storage and sowing are not usual in the rural area whilst they are for crops.

2. OBJECTIVES

The objective of this case study is to enhance the traditional knowledge of rural people in the areas of seed collection and storage and to enhance their opportunities for participation in forest management in order to better conserve and sustainably use the forests in the arid and semi–arid regions of the country. The case study has six specific objectives: (1)selecting and mapping of forest areas for management, (2)selecting the most important tree species whose number should be increased, (3)obtaining information on the best handling of forest seeds, (4)determining efficient and reliable methods to maintain forest seed viability and direct sowing efficiency, (5)identifying simple traditional storage methods which protect seeds and control parasite attacks during forest seed storage, and (6)evaluation of direct sowing impact on natural forest regeneration.

3. MATERIALS AND METHODS

Study and management areas were located in the sudanian and sahelian zones of Burkina Faso which is part of the semi–arid zone of West Africa (Guinko 1984, Flandez et al. 1993). The study sites were chosen in forest exploited areas and in fallow areas.

Direct sowing trials were established and monitored during 2 years (from July 1996 to July 1998) on three sites: (1)Tamsin village (province of Bazèga) located about 90 km south of Ouagadougou (fallow field area), (2)Cassou village (province of Ziro) located about 160 km south of Ouagadougou (forest exploited area with lateritic soil); and (3)Benega village (province of Sanguié) located about 170 km southwest of Ouagadougou (forest exploited area with a rich soil).

In these provinces, the average rainfall is about 600 mm to 1200 mm per year. The species were chosen by the farmers in collaboration with the forest technicians in charge of forest management in the region. The criteria taken into account for the species selection were: (1)the representiveness and the use importance of the species, (2)the ability of the species to grow in the area, and (3)the likelihood of acquiring enough seeds of the species for sowing. A total of 13 tree species were involved into different research paths seed collection and sowing operations were supervised by the National Tree Seed Center researchers.

The species used for direct sowing in Tamsin were *Detarium microcarpum*, *Daniella oliveri*, and *Acacia dudgeoni*. The total number of sowing holes was 608. The species used for direct sowing in Cassou were *Detarium microcarpum*, *Vitellaria paradoxa*, *Anogeissus leiocarpus*, *Afzelia africana*, and *Parkia biglobosa*. The total number of sowing holes was 480. The species used for direct sowing in Benega were *Detarium microcarpum*, *Daniella oliveri*, and *Parkia biglobosa*. The total number of sowing holes was 540.

For seed procurement an inventory was carried out with the farmers for all tree species populations around the villages on the basis of the vegetation map of region. This inventory permitted to select the best seed stands according to their surface cover, the density of trees, and their maturity stage. The areas of the seed stands were determined by a Global Positioning System (GPS). The data were computed and mapped with GIS. In collaboration with the farmers, at least 25 trees per species and stand were elected for seed collection. Each tree had to have a minimum distance of 100m from the one nearest to it. Each selected tree was marked by the local people so that they would be able to permanently find it. Fruits were collected from these trees by the farmers, stocked in bags and transported into the village for handling according to the National Tree Seed methods. They would then be stored using the traditional methods of local populations.

The following four species were used for studies on seed treatment: *Daniellia oliveri*, *Detarium microcarpum*, *Parkia biglobosa* and *Tamarindus indica*. Three treatments were used: (1)the seeds were packed in plastic bags and then placed in a loft, (2)the seeds were placed in a granary and then placed in a house, (3)the seeds were packed in plastic bags and then placed in a cool room at laboratory of CNSF.

The first and the second treatments were carried out in the villages. In addition to the preliminary assessment of the moisture content and germination, other assessments were performed every two months to monitor the viability of the seeds.

To control the parasite attacks during storage, the seeds were mixed with materials or plants resistant to insects and then were placed in granary in a house. The study was carried out in 12 villages with 6 forest species: *Acacia nilotica*, *Acacia senegal*, *Faidherbia albida*, *Khaya senegalensis*, *Parkia biglobosa* and *Ziziphus mauritiana*. The treatments used depended on the uses of the local population.

In order to simulate natural germination conditions, non–pretreated seeds were sowed by farmers in holes dug with hoes as they would be during routine agricultural sowing conditions. The interval between holes was estimated to be 4m x 4m on a line and between the lines. These intervals are commonly used in the villages' plantations. Each hole was identified by a small colored stake of wood of 25 to 30cm length.

Approximately one month after establishment, the trials were evaluated to determine the germination rate of each specie using an inventory sheet. At the 3rd, 9th, 12th, 17th, 21st and the 24th month, the survival rates were calculated. Height and diameter of seedlings were measured only at the 15th month.

4. RESULTS AND DISCUSSION

4.1 Seed Procurement

In Tamsin, seeds were collected from 120 trees of *Detarium microcarpum* and 60 trees of *Acacia dudgeoni*; in Cassou seeds were collected from 242 trees of *Detarium microcarpum* and 25 trees of *Afzelia africana*; in Benega, seeds were collected from 200 trees of *Detarium microcarpum* and 70 trees of *Daniellia oliver*. The seeds were collected from 1–3 stands.

4.2 Germination Rates

The germination rates were variable according to the species and the trial locations. In Tamsin, germination rates after one month for *Detarium microcarpum* were 84 percent, for *Daniellia oliveri* 70 percent, and for *Acacia dudgeoni* 16 percent. In Cassou, the germination rate after one month for *Detarium microcarpum* was 50 percent, for *Vitellaria paradoxa* 0.0 percent, for *Anogeissus leiocarpus* 0.0 percent, for *Afzelia africana* 27 percent, and for *Parkia biglobosa* 24 percent. In Benega, the germination rate after one month for *Detarium microcarpum* was 88 percent, for *Daniella oliveri* 60 percent, and for *Parkia biglobosa* 71 percent.

For *Vitellaria paradoxa* and *Anogeissus leiocarpus* there was no seed germination. The relatively higher germination rates of Tamsin and Benega can be explained by the relatively poor environmental conditions of Cassou, which has left forests with lateritic soils. The very low germination rate obtained with *Anogeissus leocarpus* is likely due to the quality of its seeds, which are naturally infertile and has poor germination capacity (Von Maydell 1983, Kambou 1992). According to the local farmers, *Vitellaria paradoxa* seeds were dig up by birds and rodents, and it also has been demonstrated that this species produces recalcitrant seeds which prematurely lose their viability (Gaméné 1987). *Parkia biglobosa*, *Afzelia africana* and *Acacia dudgeoni* show a relatively high rate of germination when their seeds are not damaged.

4.3 Survival of Seedlings

The survival of seedlings varied after intervals of time. In Tamsin, the survival rate after one year for *Detarium microcarpum* was 41 percent, for *Daniellia oliveri* 21 percent, and for *Acacia dudgeoni* 36 percent. After two years the rates were 29 percent, 11 percent, and 2.5 percent, respectively. In Cassou, the survival rate after one year for *Detarium microcarpum* was 27 percent, for *Vitellaria paradoxa* 0.0 percent, for *Anogeissus leiocarpus* 0.0 percent, for *Afzelia africana* 11 percent, and for *Parkia biglobosa* 2 percent. All of these species suffered bush fire damage between 3 and 9 months of the experiment. After 24 months, the highest survival of seedlings was from *Detarium microcarpum* at 3.1 percent. In Benega, survival of seedlings after one year for *Detarium microcarpum* was 8.7 percent, for *Daniella oliveri* 3.7 percent, and for *Parkia biglobosa* 2 percent. It is important to note that all sites were damaged at approximately 17 months into the experiment by bush fires. Further, bush fires can affect the survival of seedlings. For example, after a bush fire the roots of some species generally continue to survive and even in unfavorable conditions they can produce a stem. In addition, during the dry season stems can dry and remerge during the next rainy season. In part, this accounts for some of the fluctuation of experimental data. Some species like *Detarium microcarpum*, *Daniella oliveri* and *Vitellaria paradoxa* were more resistant to bush fires and have a relatively good regeneration capacity. Seedlings' survival and growth were inhibited by fire frequency and intensity, the climatic conditions, insect attacks and by browsing animals.

4.4 Seedling Growth

At 15 months, the average height for all species ranged from 0.0–16.59cm and the average diameter ranged from 0.0–2.2mm. In general, all of the species in Tamsin grew relatively well, followed by those in Benega. The tallest seedlings were observed for *Acacia dudgeoni*, *Daniellia oliveri* at Tamsin and *Detarium microcarpum* (in the three sites). In Cassou, the low height growth of *Vitellaria paradoxa* (3.8cm) is a result of the soil conditions at Cassou and the necessity for this species to develop roots during the first years of growth. Overall, the lower height growths and diameters were obtained at Cassou, and the best regeneration occurred at Benega and Tamsin. The best growth was observed in the follow fields where fire intensity was less.

4.5 Parasite Control Methods During Storage

Several local and traditional methods utilizing either materials or plants were used to control parasites during storage. The materials used were ash, potash, sand, petrol, oil, salt and residue of shea butter. The plants used to control parasites were chili pepper (fruits), *Cymbopogon schoenanthu* (leaves and stem), *Azadirachta indica* (leaves and seeds), *Hyptis spicigera* (fruits and stem), *Cassia mimosoides* (stems and leaves), tobacco (leaves), *Boscia senegalensi* (leaves), *Pericopsis laxiflora* (leaves), *Khaya senegalensi* (bark), and *Luffa cylindrica* (fruits). The materials and parasite control plants can be used individually on the seeds or used in combination to control insect attacks.

4.6 Storage Methods for Seed Viability

Proper storage methods are essential for maintaining seed viability, and must include regulation of moisture content as well as protection against termites and seed eating species. For example, seeds of *Daniellia oliveri* could be stored for 10 months in a granary placed in a house and maintain good germination capacity which allowed them to be distributed for planting. However, storage for 6 months only in the loft reduced germination due to an increase of the moisture content (the seeds were humidified by water) and probably also due to termite attacks. Other species performed better when stored in plastic bags in a cool room, etc.

4.7 Controlling Insect Attacks During Seed Storage

The following species' seeds were protected from insects:

Faidherbia albida. The seeds of these species were stored in two different villages, Djigo and Zorkoum. With 87 percent as the initial germination percentage, the seeds in Djigo mixed with ash or tobacco gave 60 percent of germination after 6 months storage while at Zorkoum 73 percent of germination was obtained with the seeds mixed with ash (the initial germination for those seeds was 88 percent).

Acacia nilotica var. adansonii. The seeds were stored in 3 different villages. In Seytenga after 6 months of storage the percentage of germination was 77 percent for the seeds mixed with neem leaves and 80 percent when mixed with tobacco. The initial germination rates were 3 percent and 4 percent, respectively. In Pouytenga, after 6 months of storage the results obtained were more or less satisfactory; 53 percent and 56 percent, respectively, for seeds mixed with the ash and *Hyptis spicigea* and the ash and *Cymbopogon schoenonthus*. The initial germination of the seeds was 76 percent. In Manga, the best rate of germination (75 percent) after 6 months of storage was obtained with the seeds mixed with *Cymbopogon schoenonthus*. The initial germination was 84 percent.

Acacia senegal. The seeds were stored in Kirbou. Their initial germination was 90 percent. For the first 3 months the germination was 90 percent. For the first 3 months the germination for the different treatment was 82 percent for the seeds mixed with ash and 89 percent for the seeds mixed with *Cymbopogon schoemonthus*. But after 4 months storage the quality of the seeds had decreased and the germination percentage was under 30 percent.

Zizuphus mauritania. The seeds were stored at Falagountou and Dabokitila. Their initial germination were respectively 48 percent and 46 percent. After 3 months of storage 10 percent and 20 percent of germination was obtained.

Parkia biglobosa. The seeds were stored in 3 localities. The seeds of the first locality (Niogho) initially had a good germination rate (83 percent) but after 5 months of storage all of the treatment yielded germination rates under 20 percent. The seeds at Tiokoura village had an initial germination rate of 77 percent; after 6 months of storage the germination rates were 74 percent for the seeds mixed with neem leaves and 68 percent for the seeds mixed with ash. The seeds stored at Toussiana had a 95 percent initial germination rate. After 5 months of storage the best results were obtained with the seeds mixed with tobacco (88 percent) and those mixed with *Hyptis spicigea* (85 percent).

Khaya senegalensis. The seeds of this species were stored in Kopienga with an initial germination rate of 60 percent. The germination levels obtained after 3 months of storage were low (10 percent).

5. CONCLUSION

Some traditional methods used for crops seed storage are also suitable for forest seed management. Depending of the species, the seeds could be stored in granaries as well as in lofts. The list of materials and plants used for traditional storage analyzed here is not exhaustive but further investigation should continue if we want to utilize the knowledge of local people and build upon it for their benefit.

In this study, we have only used granaries as containers for the storage; other containers could give results different from those we have obtained. Of the seven inert materials used, ash and potash from the stems of cereals are those which significantly protect the quality of the seeds. Based on these results, we advise their utilization for the local storage of forest seeds in rural areas. They are simple to use and have a low cost. According to CEEMAT (1988), the ash is the most frequently used material for traditional storage of seeds. The ash plays an abrasive and asphyxiation role on the predators of the seeds but the amount of product to use should be studied further.

Ten plants which protect against insect predation have been studied. The best ones for use in traditional storage are *Azadirachta indica*, *Cymbopogon schoenantus*, *Hyptis spicigera* and tobacco. Bognounou (1994) as will as Bognounou (1994) also mentioned *Hyptis spicigera*, *Khaya senegalensis*, *Cymbopogon schoenantus*, *Azadirachta indica*, *Cassia nigricans* and *Vitellaria paradoxa* as plants used by rural people in Burkina Faso for crop seeds storage. Many insect–repulsive plants have been used for seed conservation but for most their modes of action against pests in not known.

To obtain the best results in direct sowing we recommend: (1)reduce the storage duration of recalcitrant seeds like *Vitellaria paradoxa* or improve their storage conditions in order to get the best germination rate, (2)sow 3 seeds by hole in order to obtain a viable plant, (3)sow early in order to develop the root system by the first rainfall season, and (4)efficiently protect against grazing and bush fires during the three first years.

The results of this study demonstrate that participatory approaches between people from local communities, scientists, and government representatives can work and achieve better understanding of problems concerning forest seed storage where there is a dearth of knowledge. As stated in the beginning of this paper, forest seed storage and sowing are not usual in the rural areas whilst they are for crops. Hence, the scientific community has had to rely on local people to share their knowledge of seed crop storage in order for the scientific community to obtain a better understanding of forest seed storage and, by so doing, pass newly acquired knowledge back to local people.

6. REFERENCES

Bognounou, O. 1987. Importance socio–economique des essences locales de la Haute Volta: Leur importance dans l'alimentation en pays mossi. Notes et Documents Voltaiques 11: 82–91

Bognounou, O. 1988. De quelques utilisations traditionnelles du karité = Butyrospermum paradoxum (Gaetner F) Hepper subespèces parkii, Sapotaceae: Arbres à usages multiples. Séminaire national sur la valorisation du karité pour le développement national. Bilan et perspectives. Ouagadougou, IRBET/ISN–IDR.

Bognounou, O. 1994. Quelques techniques et plantes utilisées dans la protection naturelle des cultures et des récoltes au Burkina Faso.

CEEMAT. 1988. Conservation des graines en régions chaudes. Ouagadougou. 545 pages.

FAO. 1995. Evaluation des ressources forestières 1990. Pays tropicaux.
Etude FAO forêts n° 112, Rome, Italy. 42 pages.

FAO. 1987. Étude sur la contribution du secteur forestier à l'économie du Burkina Faso. Rapport de synthèse du projet préparé pour le gouvernement du Burkina Faso. FAO/TCP/BKF/5761(I). 86 pages.

Flandez S., Ouédraogo M.K and Guigma P. 1993: Aménagement des forêts naturelles et semis directs. Pages 420–430 In: Somé, L.M et De Kam M. (Editors) Proceedings IUFRO Working Group on Seed Problems. Ouagadougou, Burkina Faso.

Gaméné, C. S. 1987. Contribution à la maîtrise des méthodes simples de prétraitement et de conservation des semences de quelques espèces récoltées au Burkina Faso. Mémoire de fin d'études, IDR, Univ. Ouagadougou. 94 pages.

Guinko, S. 1984. Végétation de la Haute Volta. Thèse de Doctorat d'Etat es Sc. Nat. Univ. Bordeaux III. France. 394 pages.

Guinko, S. 1985. Contribution a l'étude de la végétation et de la flore du Burkina Faso: Les reliques boisées ou bois sacrés. Bois et Forêts des Tropiques. Numéro 208 : 29–36.

Kambou, S. 1992. Contribution à l'étude de la biologie florale et de la régénération de Anogeissus leiocarpus au Burkina Faso. Mémoire de DEA. Univ. de Ouagadougou. 124 pages.

Kessler, J. J. and Boni, J. 1990. L'agroforesterie au Burkina Faso. Bilan et analyse. Tropical Resource Management Paper 1. Wageningen Agricultural University. The Netherlands. 102 pages.

Renes, G. J. B. 1991. Regeneration capacity and porductivity of natural forest in Burkina Faso. Ecology and Management 41: 291–308

Von Maydell, H.J. 1983. Arbres et arbustes du Sahel; Leurs caractéristiques et leurs utilisations. GTZ, Eschborn. 531 pages.

Chapter 6

AGRICULTURE AND BIODIVERSITY CONSERVATION IN AFRICA THROUGH INDIGENOUS KNOWLEDGE

M.B.K. Dakoh
Department of Environmental Science, University of Botswana, Gaborone, Botswana; email darkohmb@mopipi.ub.bw

1. INTRODUCTION

It is increasingly becoming apparent that strategies are needed to help policy makers in Africa reconcile the task of conserving biodiversity while at the same time increasing agricultural productivity. In this paper, I explore some of the complementary activities between agriculture and biodiversity. I also discuss some of the characteristics and advantages of indigenous management systems in the management of agriculture and biodiversity. In the past, indigenous knowledge or local farmer knowledge about production systems has been largely overlooked, though, farmers are seen as partners to conserve and manage biodiversity whether for nature reserves, or to improve crops and livestock yields. I maintain that local knowledge systems, traditions, institutions and environmental conditions are fundamental to biodiversity conservation and management. As a matter of policy, I advocate the promotion of agricultural programs and biodiversity conservation projects that incorporate indigenous knowledge in their design and implementation. A blend of modern science and indigenous knowledge will be required to face the challenges of increasing agricultural production and managing the environment on a sustainable basis in the decades ahead in Africa. It is also important to ensure that the relevant local communities are given control over and access to land and other resources as well as management responsibility for the natural areas or resources upon which their continued prosperity depends. Furthermore, development assistance to agriculture and biodiversity conservation and land management will be most beneficial when it attempts to enhance but not replace existing agricultural systems.

Habitat conservation has two parts: (1)safeguarding natural habitats for wild species and populations and managing habitats that have been modified for human use, such as farmland (Srivastava et al. 1996); and (2)understanding that economic growth and poverty alleviation hinge in large measure on managing biodiversity in habitats transformed by humans such as farmland, rangeland, and forests exploited for timber and non–timber forest products. How farmers and livestock owners manipulate natural resources is therefore critical to the question of conserving biodiversity.

All plants and animals that contribute directly or indirectly to raising crops and livestock are part of "agrodiversity." Agrodiversity includes crops and their varieties as well as livestock and the different breeds that have been developed over the ages to suit different cultural needs and to fit a wide array of environments. Some managed habitats such as agroforestry systems can be surprisingly rich in both species and varieties.

2. BIODIVERSITY, PEOPLE AND AGRICULTURE

Biodiversity has been heavily utilized in Africa to produce food and other products (Perrings 2000, Darkoh 2001), however, much of the continent's remaining biodiversity will be lost unless future needs can be met from areas already cultivated or grazed If biodiversity is to survive in the next half century in Africa, some dramatic changes will be needed in the ways that crops and livestock are raised. How agriculture is transformed and intensified in a sustainable manner will be a key to how many species and how much genetic variation will survive into the next century in Africa.

As a practice, agriculture is often perceived as a threat to biodiversity rather than as an integral part of it. This perception arises because raising livestock and crops inevitably alters vast expanses of habitat. Because of population growth and other factors farmers often move into forests, pastoralists are squeezed into ever shrinking spaces and sometimes overgraze the land, and high–input modern farming practices frequently pollute the water and soil with chemicals. All of these activities trigger widespread and potentially dangerous loss of biodiversity. Alternatively, some land use systems and agricultural practices enhance biodiversity within managed landscapes (Richards 1985, Brokensha 1986, Gorse and Steeds 1987, Altieri 1983, 1987; Warren 1991, 1993; Srivastava et al. 1996). For example, the judicious use of livestock waste as organic manure enhances the species diversity of microfauna (Bohac and Poharzhevsky 1987, Srivastava 1996).

Some African land use practices emphasize biodiversity conservation and sustainable utilization over high productivity in their resource management (WWF et al. 1993, Lalonde 1993). The practices include carefully managed crop species and cultivar's diversity, staggered planting and harvesting dates, mixed cropping, relay cropping, cultivars mixers within plots and planting of scattered plots in a variety of microenvironments. They also have numerous insect control techniques. Cultivators use techniques such as intercropping to stabilize yield fluctuations, facilitate pest, weed, and disease control, provide soil cover far longer than single culture crops, and optimize available soil moisture for crop production. For example, the Dogon of Mali and the Wakara of Ukara Island in lake Victoria, Tanzania, create elaborate manure and compost piles and carefully tend and modify the structure and quality of their soils. The Akamba of Kenya and several other communities in the drylands of eastern and western Africa make extensive use of intercropping and multicropping. Gorse and Steeds (1987) demonstrated how mixed agricultural, forestry, and pastoral traditional production systems of such Sudano–Sahelian local communities as the Hausa, the Mossi, the Serer, the Saakolle, the Fulbe, the Tuareg, the Maures, and the Tubu are tailored to sustainable resource utilization and biodiversity conservation. All these practices support the contention that the fates of agriculture and biodiversity are intertwined.

Agriculture also has a direct role in conserving wider biodiversity. Wild species are essential for agricultural improvement because they are sources of new economic plants and animals and provide important services such as pollination and pest control. Wildlife habitats provide environmental services to agriculture such as protecting watersheds that improve year–round water availability that could be useful for irrigation, among other things. Relatively undisturbed habitats also contain wild populations of domesticated plants, and these populations contain useful genes that are often absent in the domesticated gene pool. When agriculture causes excessive soil erosion, loss of forage cover, or pollution of streams, it threatens species and often diminishes biodiversity. If species and genetic variation are diminished by human activities options are closed for improving agriculture and many other activities that are essential for human survival and economic growth. It should be possible to regulate resource exploitation and secure at least some of the remaining habitats for endangered plants and animals only if agriculture is intensified in a sustainable manner. Doing so should reduce pressure on forests, grasslands, and other habitats important for wildlife.

3. AGRICULTURAL INTENSIFICATION: ENEMY OR FRIEND OF BIODIVERSITY?

Africa's population is currently over 800 million and it is expected to at least double before it stabilizes. But demand for food and other agricultural products is likely to triple within the next fifty years. Many people look to agricultural intensification to meet this demand, and many people also think of agricultural intensification in terms of more purchased inputs such as fertilizers, pesticides, herbicides, machinery, and their associated environmental impacts. As African societies become more urban and income levels rise, consumption patterns will shift up the food chain. As more livestock products are consumed, greater demands will be placed on African landscapes to produce feed for cattle, sheep and goats. This process is unfolding especially in the drylands of Eastern and Southern Africa, and is likely to exert further pressure on biodiversity unless a concerted effort is made to adopt more environmentally friendly and sound agricultural practices. How agriculture is transformed in the next century will be crucially important to the future prospects of biodiversity. Elsewhere, I have argued that agroecological intensification or sustainable agricultural intensification is the solution of the problem of increasing productivity and managing land on a sustainable basis in the decades ahead in Africa (Darkoh 2002).

Such a solution might consist of more rational and economical use of nutrients, space, and energy; greater recycling of nutrients; better use of biological resources to raise and maintain crop and livestock yields; greater use of indigenous knowledge to improve livelihoods and the environment; the development of environmental corridors in landscapes transformed by agriculture; and greater reliance on agrodiversity, agroforestry, and organic farming systems.

Although intensification alone will neither solve Africa's food problem nor guarantee the survival of wild biodiversity, it at least creates a possibility for both. If agricultural intensification is not pursued, then as Srivastava et al. (1996) have pointed out the only way production can increase is by clearing and destroying much of Africa's remaining wild heritage of plants and animals.

In advocating agroecological intensification or sustainable agricultural intensification, I am not talking about simply going back to "traditional" practices, whatever that term connotes in the context of evolving agricultural systems. A blend of modern science and indigenous knowledge will be required to face the challenges of increasing agricultural production and managing the environment in a sustainable manner in the decades ahead in

Africa (Brokensha 1986, Horwith et al. 1989, Darkoh 1996, Srivastava et al. 1996).

In the arid, semi–arid and forest lands of Africa indigenous knowledge systems, traditions, institutions and environmental conditions are not static but evolving. They are therefore fundamental to agricultural improvement and biodiversity conservation. The magnitude of the environmental problems which exist today in the drylands of Africa are beyond the ability of modern science to grasp and provide all the needed solutions. There are pertinent socioeconomic and cultural issues whose consideration is essential for the success of conservation strategies, plans and policies but whose nuances modern science is unable or ill–equipped to cope with (e.g., customs, values and traditions). The tasks of raising agricultural productivity, modifying or limiting activities destructive of biodiversity can be less cumbersome if efforts are made to respect, reinforce, and further develop indigenous practices or traditions that are already consistent with sustainable resource utilization and biodiversity conservation.

4. TAPPING INDIGENOUS KNOWLEDGE

Indigenous knowledge can be defined as the knowledge indigenous peoples or any other defined community have developed over time and handed down from generation to generation, often by word of mouth (Mathias–Mundy 1993, Berkes 1993, Lalonde 1993). Warren (1991, 1993) uses "1K" as a synonym for local knowledge which is unique to a given culture or society. To be indigenous, knowledge must be culturally active, stored, communicated and used by its community members (Mathias–Mundy 1993, Morin–Labatut and Aktar 1992). In contrast to the connotation of "traditional," 1K is dynamic and changes constantly through indigenous experimentation and stimuli from outside. Local people generate new knowledge every day by observing their environment and developing and testing new technologies through trial and error and deliberate experiments (Howes 1979, Brokensha and Warren 1980, Richards 1985, Atte 1989, Chambers et al. 1989, Eden 1990, Gamser et al. 1990, Altieri 1991, Haverkort, et al 1991, Darkoh 1992 and 1996). They also integrate and adapt new technologies into their knowledge systems, often intuitively and by use of oral observation (Brokensha 1986, Horwith et al. 1989, Pacey and Thrupp 1989, Cullis and Pacey 1992).

Indigenous knowledge as it relates to agriculture and biodiversity conservation comprises a wide range of accumulated local experience about natural resource use and management techniques, institutional and organizational arrangements as well as beliefs and values. Indigenous knowledge has been available in Africa to outsiders for some time, but until recently much of this information was largely ignored for a variety of reasons (WWF et al. 1993).

First, it was assumed that during the colonial era Africans had little knowledge to impart and that African land–use practices were primitive and destructive relative to European and Asian techniques. Widespread practices such as slash and burn agriculture, for example, were cited as destructive of forests and of soil fertility, and responsible for environmental degradation.

Second, over time a stereotyped image evolved of African societies as timeless, unreflective, tradition–bound, and as passive adaptors to their environment rather than as analysts and manipulators of it. Top–down development projects, which ignored the existence of local expertise, were a logical consequence.

Third, post colonial educated Africans were supportive of northern models of development. Only recently has there been substantial interest among African elites in indigenous knowledge and skills.

Recently, many techniques once labeled "primitive" have been vindicated by modern soil science, particularly those associated with shifting cultivation. Some examples of rational African land use practices have been cited already but there are more. One example includes the incomplete clearing of fields, leaving stumps and some large trees, and keeping root systems intact to help bind the soil when the rains come, thus reducing the potential for erosion. Another example of appropriate traditional practices, the ancient technique of shallow ploughing, has more recently become minimum tillage (Lal 1979). It is a superior technique in low quality rain intensive soils. After 30 years of testing alternatives, Belgian agronomists eventually embraced "slash and burn" as the optimum clearing and cultivation method. Burning proved in forest zones to be more effective in soil enrichment than burial of green compost. The species and varieties grown in the swiddens are in a state of continuous adaptation, and in many places the crops are enriched by the exchange of genes with wild or weedy relatives. Altieri and Merrick (1987) contend that "maintenance of traditional agroecosystems is the only sensible strategy to preserve in situ repositories of crop germplasm."

Indigenous systems of agriculture have been adapted to a wide variety of local conditions, produced a diverse and reliable food supply, reduced the incidence of disease and insect problems, used labor efficiently, intensified production with limited resources and earned maximum returns with low levels of technology (McNeely 1995). They make use of a wide range of species and land races that vary in their reaction to diseases and insect pests, as well as to different conditions of soil, rainfall, and sunlight. They provide sustainable yields by drawing on centuries of accumulated experience by farmers who have not depended on scientific information, external inputs, capital, credits or markets.

But with growing populations and food demands, steps need to be taken to enhance the productivity of lands under indigenous systems of agriculture. In arid, semi–arid and forest lands of Africa, modern agricultural development should take existing indigenous systems as starting points and use modern agricultural science to improve on their productivity. The essential element, as McNeely (1995) has suggested, is to design self–sustaining agrosystems that assure the maintenance of the local genetic diversity available to farmers, thereby enabling rural communities to maintain control over their production system.

As a matter of policy, since in the past local farmer knowledge about production systems has been largely ignored and it is only recently that farmers are beginning to be seen as partners in efforts to improve agriculture and manage biodiversity, it should be a high priority for agricultural research programs and development projects to include elements that seek to incorporate farmers in the design and implementation of research and development efforts. The benefits of such collaboration far outweigh the costs of taking time and trouble to improve the channels of communication between research establishment and the intended beneficiaries: farmers and livestock managers. Some specific contributions of indigenous knowledge to sustainable agriculture include (Srivastava et al. 1996): (1)crop varieties adopted to harsh growing conditions, (2)cropping patterns that minimize the build up of diseases, (3)breeds of livestock that tolerate poor feed or resist diseases and pests that afflict imported breeds, and (4)intimate knowledge of botanical resources in forests and other habitats that could provide leads for plant domestication or other uses.

Promoting local or indigenous knowledge for agricultural development should include renewed efforts to demarcate and protect the territories of indigenous groups. This does not imply that indigenous people should be cut off from the outside world and discouraged from changing their cultural systems. They should be allowed to do so at their own pace, however, and with their lands intact. In this manner, information on biodiversity resources can be shared between indigenous groups and the national society. Too often "development" arrives at the door of indigenous lands and destroys the rich and unique knowledge systems that have coevolved with the environment.

5. INDIGENOUS KNOWLEDGE AND BIODIVERSITY CONSERVATION

In applying indigenous knowledge to biodiversity conservation programs, planners need to understand indigenous knowledge and processes that conserve or advance biodiversity and use these as the basis for developing and implementing indigenously rooted biodiversity conservation programs. Indigenous knowledge should be used within a biodiversity program from its inception. This not only integrates local knowledge with program planning, but also increases the likelihood of the program being sustainable. Forest conservation or regeneration programs could reference and exploit indigenous traditions of tree care and cultivation instead of basing them exclusively on non–African models. The initiative for conservation programs however can come from the community itself or from outsiders. Community–initiated projects promise an enhanced survival chance in the long run, because of the fact that community members have taken the initiative and demonstrated their interests in the projects.

On the other hand, it might be difficult for a community to get a project going without outside help and some initial funds. If outsiders take the initiative, they should limit their role to facilitators or consultants rather than as actors. National and international agencies, personnel and NGOs could be key catalytic agents and facilitators in developing and implementing biodiversity conservation programs.

Indigenous knowledge can help advance biodiversity conservation efforts in a variety of ways (WWF et al. 1993): (1)when program planners understand the full range of fauna and flora that different indigenous groups use, the planners can better assist the transfer of knowledge among groups; (2)local knowledge can be tapped as a shortcut to compiling an inventory of biological resources, as has been done successfully in the conduct of soil and crop surveys in Southern Sudan; (3)local knowledge can serve as a basis for environmental monitoring and as an early warning system of ecological changes; (4)local observers can act as monitors of faunal and floral growth, distribution and change; and (5)indigenous ideology can serve as a resource for conservation metaphors and practices. Many African conservation practices are unintelligible without reference to ideas that are labeled as beliefs and enshrined in religion, folk media and other indigenous forms of communication and practices.

6. IN SEARCH OF AN AGENDA FOR ACTION

With rising populations, increasing demands for food and the risks of land degradation, desertification and other environmental problems posed by current practices of land "extensification," one option open to Africans in the years ahead is agroecological intensification or sustainable intensification of agriculture. In order to avoid the mistakes of the past where indigenous knowledge and systems of agriculture and biodiversity conservation were either ignored or downplayed, the success of any rural development project in Africa will depend on how agricultural research and development projects incorporate indigenous knowledge in their design and implementation efforts.

Introducing new technologies and trivializing or marginalizing local knowledge and practices that are ingrained in a people's culture and which have been proven to work successfully is an exercise in futility which achieves nothing but alienation of the very people the innovation is supposed to help. Nevertheless, indigenous knowledge alone will not suffice to ensure increased productivity or prevent biodiversity loss. Any strategy for sustainable agricultural intensification in Africa will have to incorporate a blend of modern scientific approaches and some traditional methods in order to face the challenges of increasing agricultural production and managing the environment on a sustainable basis in the decades ahead.

I believe that no conservation project will be successful in the arid and semi–arid lands of Africa unless the local people are empowered and involved. Local people need the rights to self–determination and to set their own development agenda. As Friedmann (1992) has discussed, empowerment emphasizes autonomy in the decision making of territorially organized communities, local self–reliance (but not autarchy), direct (participatory) democracy, and experiential social learning. It is important to eschew paternalism and ensure that the relevant local communities are given control of and access to land and its productive resources as well as management responsibility for the natural areas upon which their continued prosperity depends. Community participation is meaningless unless visualized in terms of what Slocum et al. (1995) call a process where the beneficiaries (local communities) influence the direction and execution of the development projects. This means involvement of people in their own development from decision making and implementation to evaluation and sharing of benefits and efforts in programs. Governments should consider returning at least some nationalized resource systems, such as forests and wildlife, to community based tenure systems which are often more cost effective. Putting resource management back in the hands of local communities means putting power into the hands of those making local decisions. It also helps governments divest themselves of responsibility for functions they cannot adequately fulfill (McNeely 1995). Local initiatives in biodiversity conservation should be encouraged and supported and outsiders should "listen to the people" and try to limit their role to facilitators or consultants rather than as actors. Development assistance to agriculture, biodiversity conservation and land management will be most beneficial when it attempts to enhance but not replace existing indigenous practices. The lessons derived from the repeated failures of the "transformation" or so–called "modern" approaches to conservation in the recent past suggest that resource management and conservation strategies based upon local communities and grass roots initiatives that are enhanced by modern science and technology are more likely to be acceptable and successful in the arid and semi–arid lands of Africa.

Planners also need to be aware that strict protection does not always lead to biodiversity (Mc Neely 1995). To conserve and better use biodiversity, it is essential to preserve and manage a diverse array of habitats, ranging from areas with little human disturbance to environments that are managed for a variety of products. This is because a species cannot thrive if its surrounding environment is destroyed or seriously impaired. A broad landscape planning approach to conservation is required to ensure biodiversity and environmental sustainability. Multiple use areas bringing benefits to local people that combine conservation with sustainable land use need to be considered as practical and possible alternatives to establishment of protected areas and buffer zones that are often in conflict with local people over rights of access to land and other natural resources. It is important that biodiversity conservation be extended beyond buffer zones and protected areas to include all elements of the landscape and all ecosystems (WWF et al. 1993, Mc Neely 1995).

At the same time, there are threats to sustainable agriculture in Africa. Traditional agriculture is currently under threat from the new global consumer culture, which is spreading through television, trade and other means. Management systems that were effective for centuries have been displaced by systems of exploitation that bring short-term profits for a few and long–term costs for many (McNeely 1995, Darkoh 2001). Modern farming technology is removing innovation from the farm and placing it instead in the laboratory. The uniform varieties produced at the research center, with their dependence on chemical fertilizers and pesticides, are displacing farm-bred varieties. Once these traditional varieties are gone, the knowledge of their cultivation and use is also lost.

Furthermore, neither is the "museum approach" to conservation sufficient. Fencing off ecosystems valued for their diversity as protected areas, keeping plants in botanic gardens, and storing germplasm in seed banks is hardly an adaptive long–term solution (McNeely 1995). It seems apparent that preserving genetic variety is pointless unless the farming system that produced it is also preserved, along with its climate and soil and the accumulated knowledge of its cultivation and use.

In other words, the ultimate goal of tapping IK should be its application in development to the benefit of indigenous peoples and local communities. IK can be used in development in several ways (Mathias–Mundy 1993, WWF et al. 1993).

First, it can facilitate development effort. Understanding IK enhances the cultural appropriateness of projects and increases their feasibility and sustainability. Considering IK during the planning stage can help to avoid costly mistakes and provide insightful clues on what should be and needs to be done. Second, it can be a valuable source of appropriate technologies. Local practices and technologies provide less costly alternatives to introduced high technology methods and can be adapted to the local environment and draw on local resources. They can also be blended with international technologies. Third, it can reduce costs. Local technologies and know–how are often cheaper than introduced ones. Fourth, it can facilitate local people's participation in development. IK is familiar to local people and easier to understand than introduced technologies. Local people are therefore more likely to participate in projects that build on and use IK because they will be able to retain control over processes that affect them. Fifth, it can aid sustainability. Using local resources and relying on local peoples' knowledge means that fewer inputs are required from outside. And indigenous methods tested and refined over centuries may be inherently less damaging to the environment than introduced approaches. Sixth, IK can be the basis for self–sufficiency and self–determination. Using IK to increase the use of local resources reduces the dependency on introduced technologies and imports and can lead to enhanced self–determination.

7. CONCLUSION

Tapping indigenous knowledge to enhance development cuts across many aspects of the dynamic interplay between biodiversity and sustainability of agricultural production. The expansion of the agricultural frontier, population growth, and an emphasis on maximizing agricultural output using potent agrochemicals are exerting pressure on habitats for wild plants and animals in Africa. In many cases the conversion of wildlife habitats for farmland and adoption of "modern" farming practices that rely heavily on purchased fertilizers, pesticides and herbicides have triggered a loss of biodiversity and extinction of species (Darkoh 2001).

An effective strategy to conserve biodiversity and further agricultural development in view of the rapidly rising population and food demands (against the background of increasing scarcity of land for intensification of farming) is agroecological intensification or sustainable agricultural intensification, especially in habitats transformed by humans such as farmlands. Various policies and practices can be devised to promote intensification of agriculture and biodiversity conservation in managed landscapes and to enhance better use of biodiversity in agriculture. Without sustainable agricultural intensification, little biodiversity of African countries will survive the next century. Further, without deliberate efforts in all agricultural research and development programs to include components that seek to incorporate indigenous knowledge in the design and implementation efforts, few agricultural development or biodiversity conservation projects will succeed in Africa. As already stated, a blend of modern science and indigenous knowledge will be required to face the challenges of increasing agricultural production and managing the environment on a sustainable basis in the years that lie ahead. To achieve this, there is a need, first and foremost, to understand the indigenous traditional farming and pastoral systems more fully and to try to improve them with the help of farmers, herders and modern technology.

8. REFERENCES

Abromovitz, J.N. 1994. Biodiversity and gender issues: Recognising common ground. Pages 75–91 In: Harcourt, W. (Editor) Feminist Perspectives on Sustainable Development, London: Zed Press.

Altieri, M.A. 1990 Why study traditional agriculture? Pages 551–564 In: Carrol, C.R. Vandemeer, J.H., and Rosset, P. (Editors) Agroecology, New York: MCGraw Hill.

Altieri, M.A. 1987 The significance of diversity in the maintenance of the sustainability of traditional agroecosystems, ILEIA Newsletter 3: 3–7.

Altieri, M.A. 1993 Agroecology: The Scientific Basis of Alternative Agriculture. Boulder, Colorado: West View Press.

Altieri, M.A., and.Merrick, L.C. 1987. In situ conservation of crop genetic resources through maintenance of traditional farming systems. Economic Botany 41: 86 – 96.

Atte, O.D. 1989 Indigenous knowledge as a key to local level development: Possibilities, constraints and planning issues in the context of Africa. Seminar on Reviving Local Self–Reliance: Challenges for Rural/ Regional Development in Eastern and Southern Africa, February 21–24, Arusha, Tanzania.

Berkes, F. 1993. Traditional ecological knowledge in perspective. Pages 121–132 In: Inglis, J.T. (Editor) Traditional Ecological Knowledge: Concepts and Cases. International Program on Traditional Ecological Knowledge, Ottawa, Ontario Canada: Canadian Museum of Nature.

Bohac, J., and Pokarzhevsky, K. 1987. Effect of manure and NPK on soil macrofauna in chernozem soil. Pages 1–2, 45–61 In: Szegi, J. (Editor) Soil Biology and Conservation of

Biosphere. Proceedings of the Ninth International Symposium, Budapest: Akademiai Kiado.

Brokensha, D.1986. Local management systems and sustainability. Paper prepared for the Annual Meeting of Society for Economic Anthropology, Riverside, California, April 3–4.

Brokensha, D., Warren D.M., and Werner, O. 1980. Indigenous Knowledge Systems and Development. Lanham, Maryland: University Press of America.

Chambers, R., 1983. Rural Development: Putting the Last First. Harlow: Longman.

Cullis, A., and Pacey, A. 1992. A Development Dialogue, Rainwater Harvesting in Turkana. London: Intermediate Technology Publications.

Darkoh, M.B.K. 1992. Planning arid lands development in Africa: Some reflections from the ringside. Pages 41–64 In Hjort–af–Ornas, A. (Editor) Security in African Drylands: Research , Development and Policy. Uppsala: EPOS, Uppsala University.

Darkoh, M.B.K. 1996. Towards an adaptive and community–based approach to the management of natural resources in the drylands of Sub–Saharan Africa. Pages 73–99 In Hjort–af Ornas, A. (Editor) Approaching Nature from Local Communities, EPOS, Linkoping University, Sweden.

Darkoh, M.B.K. 2001. Agriculture and biodiversity in the drylands of Africa. Pages 38–50 In: Wiseman, R., and Hopkins, L (Editors) Sowing the Seeds of Sustainability, Glands: IUCN.

Eden, M. J. 1990. Ecology and Land Management in Amazonia. London: Belhaven Press.

Friedmann, J. 1992. Empowerment, the Politics of Alternative Development. Cambridge Massachusetts & Oxford UK: Blackwell Publishers

Gorse, J. E., and Steeds, D.R. 1987. Desertification in the Sahelian and Sudanian Zones of West Africa. Washington D. C.: The Word Bank.

Havercourt, B.T. van der KampTand, Waters–Bayer, A. (Editors). 1991. Joining Farmers Experiments. Experiences in Participatory Technology Development. London: Intermediate Technology Publications.

Horwith, B.J., Windle, P.N., MacDonald, E. F., Parker, J.K., Ruby, A.M., and Elfring, C. 1989. The role of technology in enhancing low-resource agriculture in Africa. Agriculture and Human Values 4: .68–84.

Howes, M. 1979. The uses of indigenous technical knowledge in development, IDS Bulletin, 10: 12–23.

Gamser, M., Appleton, H., and Carter, N. (Editors). 1990. Tinker, Tiller, Technical Change. Technologies from the People. London: Intermediate Technology Publications.

Lal, R. 1979. The role of physical properties in maintaining productivity in soils in the tropics. In: Lal, R., and. Greenland, J.J. (Editors) Soil Physical Properties and Crop Production in the Tropics. Chichester, England: Wiley and Sons.

Lalonde, A. 1993. African indigenous knowledge and its relevance to sustainable development. In: Inglish, J.T. (Editor) Traditional Ecological Knowledge. Concepts and Cases. International Program on Traditional Ecological Knowledge, Ottawa, Ontario Canada: Canadian Museum of Nature.

Mathias–Mundy, E. 1993. Indigenous knowledge for rural development with an emphasis on sustainable living. Keynote paper presented at the 20th Waigani Seminar on Environment and Development at the University of Papua New Guinea, 22–27 August, 1993, Port Moresby. PNG.

McNeely, J.A. 1995. Biodiversity conservation and traditional agroecosystems. In Saunier, R.E., and Meganck, R.A. (Editors) Conservation of Biodiversity and New Regional Planning, Glands: IUCN.

Morin–Labatut, G. , and Aktar, S. 1992. Traditional environmental knowledge: A resource to manage and share, Development (Journal of SID) 4: 24–29.

Pacey, A., and Thrupp, L. (Editors). 1989. Farmer First: Farmer Innovation and Agricultural Research. London: Intermediate Technology Publications.
Perrings, C. (Editor) 2000. Biodiversity Conservation in Sub–Saharan Africa, Edward Elgar, Cheltenham, UK.
Richards, P. 1985. Indigenous Agricultural Revolution. London: Hutchinson & Co.
Slocum, R., Wichart, L., Rocheleau, D., and Thomas–Slater, B. (Editors) 1995. Power, Process and Participation: Tools for Change. London: Intermediate Technology Publications, London.
Srivastava, J.P., Smith, N.J., and Forno, D.A. (Editors) 1996. Biodiversity and Agricultural Intensification. Washington D. C.: The World Bank.
WWF, The Nature Conservancy, and USAID. 1993. African Biodiversity: Foundation for the Future. Beltsville, Maryland: Professional Printing Incorporated.
Warren, D.M. 1991. Using Indigenous Knowledge in Agricultural Development. Washington D. C: The World Bank.
Warren, D.M. 1993. Using indigenous knowledge for sustainable dryland management: A global perspective. Paper presented at the International Workshop on Listening to the People: Social Aspects of Dryland Management ,UNEP , Nairobi, December.

Chapter 7

PARTICIPATION OF LOCAL COMMUNITIES IN THE MANAGEMENT OF WETLANDS IN MAGADI AREA, KENYA

Nathan N. Gichuki and Jane M. Macharia
Department of Wetland Resources, National Museums of Kenya, P.O. Box 40658, Nairobi, Kenya; email nmk@museums.or.ke, cbd@museums.or.ke

1. INTRODUCTION

Wetlands are lands that are transitional between terrestrial and aquatic systems where the water is usually at or near the surface of the land or land that is covered with shallow water (Cowardin et al. 1979, Roggeri 1995). In the context of dry lands, wetlands are areas that are permanently, seasonally or occasionally water logged with fresh or saline water that supports characteristic animals and plants. In the dry lands of eastern Africa, wetlands cover about 3 percent of the total land area and include shallow lakes and margins of deep lakes, swamps and marshes found on upper flood plains of major rivers, coastal river flood plains and high mountain peat bogs and tarns (Omoding 1995). These ecosystems support valuable biodiversity, including large numbers of mammals, reptiles, fishes and birds as well as diverse plant communities (Denny 1993). They also provide valuable resources and environmental benefits, such as biomass cropping for livestock pasture, water supply, agriculture, fisheries and subsistence hunting of wildlife that sustain local economies and communities (Shumway 1999).

During the last three decades, rapid economic development in Kenya has led to deforestation of watersheds, increased water abstraction, impoundment of rivers and reclamation of wetlands in arid zones (Mungai 1992). Wetlands have been prime targets for development activities, primarily because of their fresh water, forage for livestock and fertile soil for agriculture. In addition, there has been a growing population of subsistence and commercial farmers as well as traders in the emerging urban centers. Lack of community participation in land use planning and in resolving resource ownership and land–tenure rights have led to alienation of local communities (Cernea 1983, Barrow 1996). Similarly, the open–access nature of wetland resources in dry areas of Kenya as well as limited awareness of the full range benefits and functions of wetland ecosystems have all contributed to poor community participation in wetland management and the unsustainable use of natural resources in general (Tiega 2001).

The purpose of this study was to determine wetland characteristics in Magadi (a semi–arid area in southern Kenya), assess their values, patterns of use and threats by local communities and determine the factors that influence community participation and success of local wetland management initiatives. The guiding principle was that well managed wetlands can help to reduce poverty and increase prosperity of rural communities in dry lands of Africa.

2. STUDY AREA

This study was conducted in Magadi area, which is located in southern Kenya and borders Arusha region in northern Tanzania southern Kenya (Figure 1). The area lies between latitudes 36° 05'E and 37° 55'E, and between longitudes 01° 10'S and 03° 01'S.

Eastern Rift Valley traverses Magadi area from north to south with Nguruman escarpment forming its western fringe. Land elevation ranges from 600m to 2300m above sea level. Mount Kilimanjaro, which is the highest mountain in Africa is about 100km away and has a dominating influence on climate. The climate of Magadi is essentially semi–arid. The annual rainfall is 765mm and varies with land elevation. Rainfall distribution is bimodal with long rains occurring in March–May and short rains in October–December. The highlands such as Loita Hills and Nguruman escarpment receive higher than average rainfall and hence serve as important water catchment areas.

The drainage pattern in Magadi is influenced by the Rift Valley and Nguruman escarpment. Seasonal streams combine to form large rivers that flow into major drainage basins. These basins are Lake Magadi, Lake Natron and River Ewaso Ngiro (Figure 1). The area has consistently high ambient temperature (26–33°C) and high potential evapo–transpiration, which ranges from 1700mm to 2500mm per year. Water resources in the area are therefore scarce. The lakes in Magadi are alkaline or salty and their water is not useful for water supply or fishery. Freshwater wetlands (springs and streams), however, are valuable sources of fresh water for people and livestock.

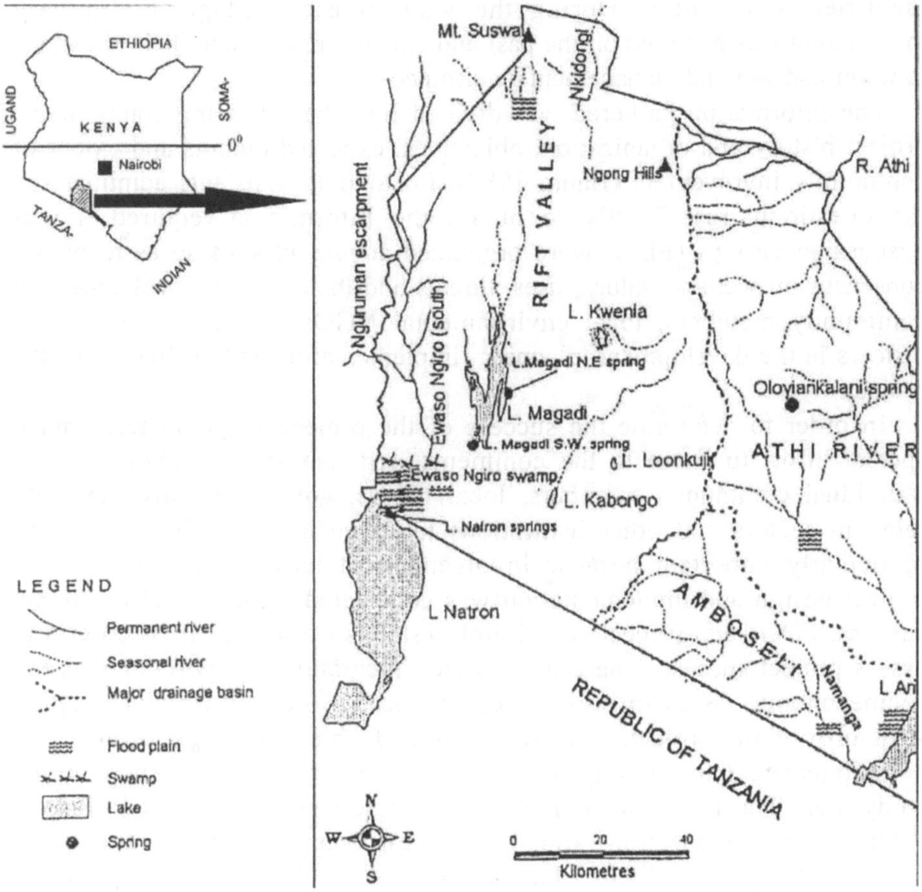

Figure 1. Magadi Area in Southern Kenya and Study Area

3. MATERIALS AND METHODS

This study was carried out in a number of villages (known as manyattas in Maasai language) in Magadi Division of Kajiado District during the period between October 1999 and November 2001. Fieldwork entailed gathering data on local population and economy, wetland characteristics, wetland values, uses and threats, current wetland conservation and management practices, community participation and success of local wetland management initiatives. Large scale maps (1:50,000) as well as aerial photographs were used to locate the wetlands, market centers and villages. Field visits were then made to each wetland where more villages were identified and visited. During the visit to each village, preliminary information was gathered on the past and existing project and their relevance to water and wetland management determined.

The information gathered was divided into the following components: project history and organization, objectives, expected outputs and extent of community involvement (Baum 1982). Local field assistants administered the questionnaires directly to obtain the information required. Focus Discussion Groups (FDG) were organized at project sites so as to obtain consensus on wetland values, uses, threats and the extent of involvement of community members, local environmental NGOs and local government officers in the development, planning, implementation and evaluation of the project.

In order to determine the success of the project, the field team made special effort to listen to the comments, criticism and suggestions from individual community members, local NGOs, government officers from relevant sectors and other institutions involved in the project. This was particularly important because involvement of local stakeholders in the project design and implementation was considered to be crucial to project success. FDGs and project staff jointly established a number of criteria to assess project success. The criteria were: integration of environmental and socioeconomic needs in project design, local ownership (implementation structure) of the project, achievement set of objectives, and linkage with complimentary initiatives by other stakeholders or interested groups in the study area. The ideas, methods and tools described by the UNDP–Project Design, Implementation and Evaluation Manual (1999) and FAO Community Forestry Field Manual (Davis–Case 1990) were particularly helpful.

4. RESULTS

4.1 Population and Socioeconomic Activities

The study area (Magadi Division) was inhabited by 35,000 people whose settlements were spread over an area of 2749 km^2 (Republic of Kenya 1999). This semi–arid area had a population density of 12.7 persons per km^{-2}, with most settlements being concentrated around freshwater sources, urban centers and along major roads. The main urban centers were Magadi,

Nguruman, Olkirimatian and Shompole. The field team located and visited 37 villages, which contained a total of 11,000 people. However, only 17 villages had water or wetland management projects either completed, in progress or abandoned. A total of 28 projects were selected for basic evaluation.

The majority of the inhabitants (90 percent) were the Maasai people, who practice nomadic pastoralism. The remaining 10 percent of the inhabitants consisted of immigrant communities such as the Kikuyu, Kamba, Luo, Luhya and Somali peoples. There were also small numbers of Chala (Wasanje) people from northern Tanzania who were engaged in farming activities. These were also Europeans who were working for religious and other non–governmental organizations. The immigrant communities occupied urban centers where they run small businesses or were formally employed. They also inhabited high potential land where they grew crops and reared livestock on a small scale, including bee keeping and fish farming.

The principal economic activities were livestock rearing, agriculture, mining of soda ash, tourism and trade. The Maasai people were mainly engaged in the livestock rearing and to some extent charcoal trade and tourism while immigrant communities were involved in agriculture, fish farming, mining of soda ash, tourism and trade. The introduction of modern farming practices appeared to reduce the capacity of the soil to remain productive. This situation was aggravated by the physical deterioration of the soil due to prevailing cultivation practices, especially on land leased out for intensive agriculture.

4.2 Wetlands and Their Characteristics

The wetlands in Magadi area occurred in a variety of landscape positions. The low gradient of the land (600–700m above see level) and low water permeability of the underlying rocks created favorable conditions for formation of wetlands. These conditions resulted in impended drainage.

Landscape positions included: (1)low lying areas subject to flooding, (2)gentle slopes in areas of ground water discharge or seepage, (3)isolated depressions surrounded by uplands, (4)flat areas without drainage outlets, and (5)flat or sloping areas adjacent to a body of water.

The hydrological conditions of wetlands in the study area varied from permanent inundation by shallow water (e.g., Lakes Magadi and Natron) through permanent soil saturation (e.g., Shompole swamp) to periodic inundation (e.g., Lake Kwenia) or periodic soil saturation (e.g., Ewaso Ngiro flood plain). The wetlands were small, often linear (along water courses) and separated by large areas of dry land. They occurred in natural depressions, on valley bottoms of rivers and streams, and at the edges of static water bodies, such as lakes and dams.

Five types of wetlands were identified in Magadi Division. These were 4 lakes (mostly alkaline), 4 marshes and swamps, 2 mudflats and floodplain grasslands, 34 springs and streams, and 27 human–made wetlands. The artificial wetlands were dams and pans, salt evaporation and water wastewater treat lagoons, fish ponds and irrigated fields. Some of these wetlands were permanent while others were seasonal. The diversity of wetlands in this semi–arid area was influenced by local topography, soils, hydrology as well as human and animal activities, especially the activities of warthogs and elephants.

The distribution of wetlands in the Magadi area was influenced by the drainage pattern, geological characteristics and rainfall, especially its variability in time and space. The majority of the wetlands occurred in the Rift Valley, an internal drainage basin with an average area of 7,260 km^2. Permanently and periodically inundated wetlands covered an area of 11,298 ha while seasonal wetlands covered an area of 10,641 ha. Temporary wetlands, which formed in small depressions and road sides occurred in all parts of the division during heavy rains.

The main wetlands occurring in the Rift Valley basin were the hypersaline Lakes Magadi and Natron, and the associated mudflats, salt evaporation pans and hot springs. There were also the moderately alkaline Lake Kwenia and the alkaline Lakes Kabongo and Loonkujit. Marshes, swamps floodplain grasslands and some human–made wetlands occurred in Ewaso Ngiro River basin. The river has a catchment area of 8,536 km^2 and flows for 80 km along the western edge of the Rift Valley in Magadi Division. The river is 40 km wide and receives additional water from rivers Pegassi, Sampu, Entasopia and Oloibortoto, which originate on Nguruman escarpment. River Ewaso Ngiro is characterized by high seasonal floods which provide water and supplies of nutrients to Nguruman swamp, Ewaso Ngiro swamp and Shompole swamp. The river forms an extensive floodplain, covering a total area of 17,845 km^2.

4.3 Wetland Values and Uses

Wetland resources as well as their attributes and functions contributed to the social and economic development of the local community in Magadi Division. Seventy–two percent of 412 persons interviewed during this study in different villages indicated that freshwater wetlands were important for their own livelihood. They also recognized that alkaline water wetlands were important to the community for generating income through mining of soda ash, conservation of wildlife and tourism. Specific socioeconomic values were accorded to different wetlands on the basis of available resources to individuals and to the community as a whole.

The most highly ranked resources (as a group) were pasture for livestock and for mining and salt licks; this was followed (as a group) by agriculture and tourism; and then (as a group) by fisheries and aquaculture, food products, medicinal products, and water supply. The most highly ranked wetland functions were ground water recharge, followed by flood control and regulation, and sediment retention. Values such as ground water discharge, soil erosion control, and water treatment were ranked lower.

Wetlands were considered valuable for subsistence and to some extent small–scale commercial agriculture. The availability of freshwater during the dry season permitted irrigation of food and horticultural crops, especially in Nguruman area. Moist and fertile soil in reclaimed marshes and on flood plains, where nutrients were replenished by seasonal floods allowed flood–recession agriculture. Wetland–based agriculture, primarily for production of vegetables and fruits took place on the lower slopes of Nguruman escarpment while flood recession agriculture for production of traditional areas (millet and sorghum) took place in the Ewaso Ngiro floodplain.

Swamp fishery was practiced in all permanently or seasonally inundated wetlands by both the indigenous Maasai people and the immigrant communities in Magadi Division. Two local youth groups carried out artesenal fishing in Shompole and Pegassi swamps and all along River Ewaso Ngiro. Large dams also offered opportunities for subsistence and sport fishing. In addition, four primary schools and 12 land owners had established fish ponds in the areas that were endowed with abundant freshwater. The main types of fish obtained from freshwater wetlands, including the river, marshes, dams and fish ponds were Tilapia, Common carp and Lungfishes. The fishes caught were consumed locally while the surplus was smoked or sun dried and sold in the local markets. It was estimated that $6000 USD worth of dried fish were generated from Magadi Division per year. Apart from the food value, fish had an ecological value by controlling vectors of human and livestock diseases, ranging from mosquitoes, leeches to snails.

The success and development of the livestock industry in the Magadi area was to a great extent dependent on the existence of freshwater wetlands. The herbaceous and woody vegetation found in marshes, swamps and flood plain grasslands remained productive all year round and hence constituted an important grazing resource for the pastoralist Maasai community. The major wetlands that were used for livestock grazing were Lake Kwenia, Lake Kabongo and all the marshes and swamps found within the Ewaso Ngiro River flood plain. These wetlands covered an area of approximately 18,000 ha and supported 180,000 heads of cattle per year. Livestock owners led their animals to wetlands where nutritious forage and water were available during dry season when upland pasture reserves were exhausted. The wetlands provided pasture security for the livestock farmers during difficult periods of the year. Flooding of lowland areas (wetlands), however, forced livestock farmers to move to upland areas where health conditions were better for people and livestock. Wetlands were, however, not only valued for pasture but also for soda ash (Magadi salt) mainly found in Lakes Magadi and Kabongo.

Communities that lived in Magadi area obtained a wide range of food and medicinal products from wetlands. Ponds and fresh leaves of several species of Acacia shrubs were used to prepare special traditional food, which was also known to have curative effects. Tubers of sedges and water lilies were collected by women and consumed by the family during periods of drought. Fruits of various trees and shrub species, such as *Rhus natalensis*, *Ficus natalensis* and *Ficus sycomorus* were used for a variety of purposes, including food or making alcoholic drinks. Seeds of wild cereals, such as wild millet and sorghum were collected from Ewaso Ngiro and Nguruman flood plain, threshed and stored or cooked for consumption. These indigenous foods, which were obtained from wetlands, sustained the community during periods of severe drought and famine. A wide range of medicinal plants, which were used for treatment of human and livestock diseases were also obtained from wetlands. In fact, wetlands served as natural stores of medicine that sustained the health of people and livestock.

In Magadi area, water is a critical resource that is needed for social and economic development of the local communities. Surface water was scarce and mainly occurred in permanent swamps, rivers and streams as well as natural springs and human–made dams. Although wetlands covered 8 percent of the surface area of the division, nearly 50 percent of them contained water that was too alkaline or salty, for human and livestock use. In addition, half the wetland area in Magadi Division had water seasonally, periodically or temporarily. About 50 percent of the population in the study area depended on River Ewaso Ngiro and its tributaries for water supply to meet their domestic, livestock and farm needs. Permanent wetlands also played an important role in recharging with water underground aquifers. Most of the water resources in the study area was underground and was accessed through shallow wells and boreholes. About 35 percent of the local population depended on underground water to meet their domestic and livestock needs. Underground water was expensive to exploit and hence some families (15 percent) depended on roof and rock catchments to harvest rain water. Piped water from Nguruman escarpment supplied several villages, institutions and towns in Olkirimatian plain and Magadi town, where large quantities of water were needed by the soda ash processing factory and the associated urban settlements.

Mining of minerals was an important economic activity. Mining of soda ash was mined out in Lake Magadi, which had large quantities of solid trona of sodium bicarbonate, sodium chloride and other chemical compounds. The lease holders, Magadi Soda Company, produced a wide range of marketable products, such as table and livestock salt, glass, industrial acids and food preservatives. These products were mainly for export. The indigenous community also collected raw salt and soda for livestock consumption and for barter trade with neighboring communities in northern Tanzania. The soda ash (Magadi) was used by local people for curing tobacco and for treatment of some human and livestock ailments. Sand was mined in dry river beds while building stones and marble plates were mined in a few areas with suitable rock formations.

Magadi area is rich in wildlife, particularly large mammals and birds. The floods plains and woodland areas supported elephant, buffalo, lion, zebra, giraffe and wildebeest. The wetlands, such as Shompole and Nguruman swamp as well as Lakes Magadi and Natron supported diverse communities of waterbirds. In fact, Lake Magadi supported up to 300,00 flamingoes and 65,000 small shorebirds, some of which were migrants from Europe and western Asia to Africa. The flood plain grasslands along the Ewaso Ngiro River supported several tens of ostriches and other game birds. The northern shores of Lake Natron had large numbers of flamingoes with their young and this lake is in fact the principal breeding area of flamingoes in Eastern Africa. Small numbers of tourists visited Lake Magadi every day and occasionally moved to Nguruman for spot fishing and to Shompole for flamingo viewing in the Lake Natron. The wetlands in the study area offered ample potential for conservation of biodiversity and development of ecotourism.

The wetlands in Magadi Division provided a wide range of environmental benefits to local people. The local people recognized that wetlands retained sediment and filtered water, thereby making it suitable for domestic and livestock use. Marginal wetlands protected river banks and lakes shores against soil erosion by water current. The wetlands that occurred along Ewaso Ngiro River were useful in controlling floods, which would otherwise destroy downstream settlements in Shompole area. Excess water collecting in wetlands during the rains or heavy river floods was stored and slowly released to springs or underground aquifers. This function of wetlands as well as that of treating waste water were not well understood and hence appreciated by the local community.

4.4 Threats to Wetlands and Water Resources

Development activities that impact on wetlands and water resources were apparently increasing in Magadi Division. The structural and functional integrity of aquatic ecosystems in the division was gradually being threatened by land–based human and livestock activities. The rapidly growing population and rapid expansion of villages near major permanent wetlands and other water sources, especially in Shompole and Nguruman areas, had serious implications on the ecological integrity of River Ewso Ngiro Basin ecosystem.

The major threats to wetlands in Magadi Division were siltation, overgrazing, water abstraction for irrigation and pollution from emerging market centers and villages (Table 4). Cultivation on steep slopes had led to destruction of natural vegetation cover and consequently triggering the processes of soil erosion in upland areas and soil deposition in lowland areas and in wetlands. While replenishment of soil nutrients was necessary for regeneration of wetland vegetation, suspended silt reduced water quality. Heavy grazing and browsing of natural vegetation and ground trampling by livestock was apparent in upland areas where the animals were concentrated during the wet season.

Major threats to wetlands in the Magadi Division included siltation (Nguruman and Ewaso Ngiro flood plain), water withdrawal (River Pagasse, Oloibortoto and Entasopia), overgrazing and trampling (Shompole and Nguruman swamps), cultivation and burning (Upper Ewaso Ngiro and Nguruman catchments), pollution (River Entasopiaand Oloibortoto), and mining for salt and sand (Lake Magadi and River Pegassi).

4.5 Conservation and Management of Wetlands

Land tenure and ownership strongly influenced wetlands management. Wetlands occurred on state, private, communal or trust land. State lands that contained some wetlands were forest reserves, national reserves, national parks and land that was reserved for public utilities, such as schools, livestock markets and rangeland research. Wetlands in those areas were managed by central government through the departments of Forestry, water and agriculture and through national agencies, such as Kenya Wildlife Service (KWS) and Kenya Agricultural Research Institute (KARI).

The wetlands that occurred on private lands were marshes, water storage dams, springs and fish ponds. Individual land owners were responsible for managing wetlands on their land, including access and utilization of water, fish and plant products. Some farmers who owned permanent dams or fish ponds leased them out for fishing by artesenal fishermen. The later paid a fee based on the total weight of fish captured from the wetland. In addition, individual land owners charged a fee per head of livestock grazing or drinking water from a wetland on their land.

Nearly 70 percent of the land in Magadi Division, however, was not adjudicated and was therefore owned by the community and used communally. Two group ranches, Okirimatian and Shompole, owned most of the land in the division. Title needs had been issued by government in the names of the two group ranches. Families and individuals that comprised a group ranch used the land and the associated resources communally. The Okirimatian Group Ranch which owned most of the land in Nguruman, had apportioned to individual members small parcels of agricultural land. While a few families settled and cultivated crops on their portion of the land, most families leased their land to small–scale commercial farmers who grew fruits and vegetables through irrigation for the export market.

Access and use of wetland resources in trust lands or communally owned land was regulated by customary laws. These laws regulated the use of water and pasture, and forbid burning of trees, deliberate burning of the marsh and drainage of the wetlands for any purpose whatsoever. This study established that some permanent springs, swamps and marshes were regarded as sacred sites by the indigenous Maasai community. Access and use of such sites was strictly regulated and was only permitted during community ceremonies and rites.

The Maasai community employed several traditional strategies that resulted in conservation of wetlands and sustainable utilization of the rangeland. This study identified five strategies: (1)movement between upland and lowland areas, (2)avoiding manipulation of wetlands, (3)promoting wetland zonation, (4)diversifying livestock to optimize forage utilization, and (5)controlling livestock numbers to avoid overgrazing and land degradation. Movement of people and their livestock from the lowland to upland areas allowed wetlands to recover or regenerate after heavy grazing during dry season. Customary rules forbid wetland drainage and indiscriminate harvesting of plant resources. Rangeland utilization rules promoted zonation of wetlands to avoid overgrazing and trampling of the land by large numbers of livestock. Family livestock was diversified (cattle, donkeys, goats and sheep) to optimize forage utilization at different positions or heights. The community adopted livestock management system that allowed individuals to loan or share livestock among relatives and friends. This practice ensured security of livestock during difficult periods and minimized overgrazing and land degradation in a particular area.

4.6 Community Participation in Wetland Management Initiatives and Lessons Learned

During this study, a total of 412 respondents, comprising 200 adult men, 140 adult women and 72 young persons were interviewed in 17 different villages. Data were gathered on community attributes that were likely to influence individual participation in wetland or water management projects and hence affect project success. The community attributes that were investigated were gender, age, level of education, family size and family income relative to the annual income of field–based government officers working in the Magadi area.

Gender had significant influence on community participation in different stages of project cycle. More men participated in project development and planning but less in implementation than women and youth. Participation in project monitoring and evaluation was generally poor but more so by women than men and young persons. The stronger involvement of women and youth in project implementation was a positive social change towards women and youth breaking cultural barriers.

Community participation in wetland and water management initiatives varied with the age of individuals. More old persons (>51 years) and young persons (<21 years) participated in project development (proposal writing) than middle–aged persons, perhaps because the latter were busy in other productive activities. The project planning or strategy formulation stage attracted the participation of more middle–aged persons than the old and the youth. Project implementation attracted the participation of a greater percentage of middle–aged persons and the youth than the older persons. Project monitoring and evaluation attracted more middle–aged persons and the youth than the older persons.

The level of education of individuals also influenced their participation in water and wetland management projects. The results of this study did not reveal any significant differences among mature persons with secondary, primary and pre–primary levels of education in their participation in project development and planning stages. However, more persons with primary and pre–primary levels of education participated in project implementation than those persons with secondary level of education. In contrast, more persons with secondary education participated in project monitoring and education than those persons with pre–primary and primary levels of education.

Family size also affected individual participation in project activities, especially that of women. More persons with families of 7–9 children participated in project development than those with family sizes of 1–3 children. This pattern was also shown in project planning with more persons with family size of 7–9 children participating than those with smaller family sizes. Project implementation also attracted the participation of more individuals with family sizes of 4–6 children and 7–9 children than those persons with smaller family sizes. Project monitoring and evaluation attracted low participation by individuals of all family size categories and there was no significant difference in their participation.

Family income also had an influence on individual participation in water and wetland management initiatives in Magadi area. More persons with low and moderate income participated in project development, planning and implementation than those with relatively higher income. This was probably because persons with relatively high incomes were capable of developing their own water resources by constructing water storage dams, tanks and preserving wetlands that were important sources of water and forage for their livestock. Some of the rich residents had private boreholes and wind or solar powered pumps. There was no difference between the number of persons with relatively high, moderate and low income that participated in project monitoring and evaluation. Lower participation of all community categories in project monitoring and evaluation implied that this activity was accorded low priority.

During this study, 28 community–based projects were selected: 12 in Nguruman, 7 in Okirimatian, 4 in Magadi and 5 in Shompole. Six projects involved in different aspects of water management, including construction of rock catchment for rainwater harvesting, fencing of freshwater springs, waste water drainage and apportionment of water to farmers and pastoralists were selected. These projects were operational for a period of approximately three years and were largely donor supported.

Three projects, which involved seeding with grass denuded areas and introduction of suitable browse plants, especially for goats, were selected. These projects were donor supported and were operational for a duration of five years each. There were five fisheries management projects that provided material support to community groups involved in artesenal fishing, especially in Shompole area; and that provided technical and financial support to fish farmers in Nguruman and Okirimatian areas. The projects were of a short duration (3–4 years) and were mainly supported by Kenya Government and external donors.

Six agricultural projects, which involved cultivation of subsistence and horticultural crops were selected for evaluation. The projects were selected because they involved use of irrigation water from springs, streams and dams, particularly in Nguruman area and in River Pegassi basin. The projects were executed by the community with support of Kenya Government and external donors. The projects were expected to operate for a period of 5–10 years, during which food production in their areas of operation was expected to triple.

Afforestation projects involved establishment of tree nurseries, enrichment of degraded woodland and control of soil erosion in denuded sites. Eight projects, which were implemented by the local community with technical support from Kenya Government's Forest Department and financial support from external donors, were selected for evaluation. All the projects were expected to operate for a period of five years.

The success of 28 community–based projects dealing with water management, pasture improvement, fisheries development, agriculture and forestry were evaluated according to set criteria of water management, pasture implementation, fisheries management, and afforestation. On average, 57 percent of the projects evaluated met all the criteria while 43 percent did not. There was fairly strong community involvement and ownership of the projects. Environmental concerns and livelihood needs were also well integrated in the project design but less so in project implementation. About half of all the projects selected achieved their set objectives and delivered some outputs. However, links between the projects and the other parallel or complimentary initiatives were weak. Overall, the most successful projects were in agriculture, particularly in horticulture and pasture improvement.

5. CONCLUSION

Wetlands in arid areas of Africa are poorly known, perhaps because of their variable and ill–defined nature (Hines 1996). They also constitute only a small percentage of the landscape and are often not taken into account in planning by rangeland managers. However, the resources and benefits that wetlands provide to communities living in semi–arid areas sustain local economies and livelihoods. In Magadi, wetlands covered about 8 percent of the land area of some 1749 km^2 (Gichuki et al. 2001). The major wetlands occurred in River Ewaso Ngiro basin and in the Rift Valley, where alkaline lakes were located. The diversity of wetland habitats in the area was not unique and in fact the structure, functioning and biota were similar to those of wetlands in other semi–arid regions of East Africa (Howard 1995, Omoding 1995).

The values of wetland resources and their system attributes (benefits) to local communities affords them greater importance than their size would suggest. Most of the wetlands in Magadi area, for example, were highly valued by the local community for at least some of the resources and attributes. The most important resources in Magadi area were freshwater, pasture for livestock, moist fertile soil for agriculture and soda ash. The Maasai people were totally dependent on permanent wetland systems, especially the Ewaso Ngiro and Amboseli swamps as dry season sources of water and pasture for their livestock. (Gichuki et al. 2001). Both permanent and periodically inundated freshwater wetlands had a major cultural influence because they determined grazing patterns of livestock, cropping cycles and food availability, not only within the immediate surroundings of the wetland but also further away.

The importance of wetlands in determining the distribution of human population was clearly apparent in Magadi Division and in other semi–arid areas of southern Kenya. About 65 percent of the population in study area lived in River Ewaso Ngiro basin where freshwater and fertile land for agriculture and pasture for livestock were available all year around. The proximity of human settlements to wetlands, however, created serious environmental problems, especially through overgrazing, wetland drainage, extraction of large amounts of water for domestic and farm use, and pollution of water resources. These threats were clearly apparent in Nguruman and Magadi towns where settlements were permanent. Further, it was noted that most urban centers in the study area had no proper sanitation and waste water was often directly disposed off into wetlands or open pits. Water supply and sanitation are priority areas for community development not only in Magadi area but also in other semi–arid areas in Africa (Shumway 1999).

Community participation in different aspects of water and wetland management initiatives in dry lands should aim at ensuring sustainability, reducing public expenditure burden and improving efficiency, equity and standards of project performance or service delivery (World Bank 2001). In Magadi area, community participation in different stages of the project cycle were influenced by basic socioeconomic factors, such as family size, level of education and family income. Apparently, there was greater participation of the youth and women in project implementation than in other stages of the project cycle. Inadequate community participation in project development and planning seriously impacted on project progress and success. The vision of the community was lacking in projects that failed to achieve their objectives and those that could not be sustained after the donors exited

Integration of environmental concerns and livelihood needs is crucial for project sustainability (World Bank 1998). However, community groups interviewed in Magadi area had limited understanding of environmental issues and hence did not appreciate the need for inclusion of wetlands conservation in the project strategy. For the community, visible outputs were priority while establishing a sustainable system of resource use was of secondary importance. This misconception of project goal could have contributed to reduced participation in project implementation.

In conclusion, this study has shown that arid zone wetlands are diverse and important for maintaining environmental stability. They protect the landscape against natural degradation and conserve water and valuable biodiversity. They also provide essential resources that sustain local economies and livelihoods of communities in dry areas of Africa. The full range of wetland benefits, however, is not well understood and appreciated. Approaches to management have been focused on a few resources (e.g., water, land for agriculture or minerals). This narrow approach has led to unsustainable exploitation of wetland resources, thereby threatening ecological integrity of these ecosystems. Arid zone wetlands, however, have immense potential for supporting sustainable development and fighting poverty (Hirji and Ibrekk 2001). Local communities should be empowered to participate in shaping their own destiny, through management of environment and resources according to their need and visions of the future. External donor support has had positive impacts on peoples' livelihood but also serious negative impacts on the environment in many dry areas of Kenya, primary because of narrow focus, conflict with community visions and lack of sustainability of local programs. Meaningful community support should therefore be long enough to produce sustainable outputs, integrate environmental concerns and livelihood issues and establish mechanisms for long–term management of environment and natural resources.

6. REFERENCES

Barrow, E.G.C. 1996. The drylands of Africa. Local participation in tree management. Initiatives Publishers, Nairobi. 268 pages.

Baum, W.C. 1982. The project cycle. World Bank, Washington, D.C.

Cernea, M.M. 1983. Culture and organization: The social sustainability of induced development. Sustainable Development 1:18–29.

Cowardin, C.M., Carter, V., Golet, F.C., and LaRoe, E.T. 1979. Classification of wetlands and deepwater habitats of the United States. FW/OBS–79/31. United States Fish and Wildlife Service, Biological Services Program, Washington, D.C.

Davis–Case, D. 1990. The community's toolbox. The idea, methods and tools for participatory assessment, monitoring and evaluation in community forestry. FAO Community Forestry Manual 2, Rome: FAO. 146 pages.

Denny, P. 1993. Eastern Africa. Pages 32–46 In: Whigham, P.F. (Editor) Wetlands of the World. Volume 1: Inventory, Ecology and Management: Dordrecht, Boston and London, Kluwer Academic Publishers.

Gichuki, N.N., Oyieke, H.A., and Ndiritu, G.G. 2001. Assessment and monitoring of wetlands for conservation and development in dry lands: A case study of Kajiado District, Kenya. Pages 97–111 In: Finlayson, C.M., Davidson, N.C. and Stevenson, N.J. (Editors) Wetland Inventory, Assessment and Monitoring: Practical Techniques and Identification of Major Issues. Supervising Scientist Report No. 161, Department of Environment and Heritage, Darwin, Australia.

Hines, C.J.H. 1996. Wetlands in arid lands: Values, threats and conservation. Pages 307–313 In: Beilfuss, R.D., Tarboton, T.R.W, and Gichuki, N.N. (Editors). Proceedings of African Crane and Wetland Workshop. International Crane Foundation, Wisconsin, USA.

Hirji, R., and Ibrekk, H.O. (Editors) 2001. Linkages among water, environment and dimensions of poverty. Environment Strategy Paper No. 2, World Bank, Washington, D.C.

Howard, G.W. 1995. Biodiversity issues in African wetlands. Page 85 In: Bennun, L.A., Aman, R.A., and Crafter, S.A. (Editors) Conservation of Biodiversity in Africa: Local Initiatives and Institutional Roles. Proceedings of International Conference on Conservation of Biodiversity in Africa, National Museums of Kenya, Nairobi.

Mungai, P.M. 1992. Wetlands and dry land management in Kenya. Pages 167–171 In: Crafter, S.A., Njuguna, S.G., and Howard, G.W. (Editors) Wetlands of Kenya. Proceedings of National Wetlands Conference, IUCN Communications Division, Rue Mauverney 28, Gland, Switzerland.

Omoding, J. 1995. Habitat classification scheme for East Africa. In: Howard, G.W. (editor). Mechanisms for Conserving Wetland Biodiversity in East Africa. IUCN Eastern Africa Regional Office, Nairobi.

Republic of Kenya, 1999. National Population Census Report: Kajiado District. Government Printer, Nairobi.

Roggeri, H. 1995. Tropical freshwater wetlands: Guide to current knowledge and sustainable management. Kluwer Academic Publishers, Dordrecht, Boston and London. 349 pages.

Shumway, C. 1999. Forgotten Waters: Freshwater and Marine Ecosystems in Africa. Strategies for Biodiversity Conservation and Sustainable Development. Boston University, Alexandria.

Tiega, A. 2001. Priorities for wetland biodiversity conservation in Africa. Pages 112–120, In: Finlayson, C.M., Davidson, N.C. and Stevenson, N.J. (Editors). Wetlands, Inventory, Assessment and Monitoring: Practical Techniques and Identification of Major Issues. Supervising Scientist Report No. 161, Department of Environment and Heritage, Darwin, Australia.

United Nations Development Programme. 1999. Programming Manual: Project design, implementation and evaluation. N.Y. 188 pages.

World Bank. 1998. Integrating Freshwater Biodiversity Conservation with Development: Some Emerging Lessons. Environment Department Paper No. 61, Washington, D.C.

World Bank. 2001. Natural Resources, Poverty and Sustainability. Pages 153–162 In: Making Sustainable Commitments, A World Bank Policy Paper, Washington, D.C.

Chapter 8

THE VALUE OF LOCAL AND INDIGENOUS KNOWLEDGE FOR THE DEVELOPMENT OF INFORMATION SYSTEMS FOR CONSERVATION MANAGEMENT

[1]K. Kellner and [2]O.J.H. Bosch
[1]School of Environmental Sciences and Development, Potchefstroom University for CHE, Potchefstroom, South Africa; email erestore@intekom.co.za; [2]School of Natural and Rural Systems Management, University of Queensland (Gatton), Gatton, Queensland, Australia

1. INTRODUCTION

Human and natural interactions in the earth's vast drylands have global–scale influences. Dryland ecosystems play a major role in global biophysical processes by reflecting and absorbing solar radiation, maintaining a balance of atmospheric constituents, and sustaining biomass and biodiversity. A continuously increasing human population and demands for food and water, has lead to increasing pressures on the functioning of natural ecosystems and subsequently the maintenance of genetic, species and ecosystem diversity. The development of technologies and practices for the conservation and promotion of sustainable systems to enhance biodiversity requires full consideration of the social, cultural, ecological, political and economic values which different communities place on species and ecosystems (WOCAT 2000). These call for better communication, linkages and coordination between stakeholders at all levels of decision making. However, effective decision making relies on better informed individuals and organizations, which makes the availability of knowledge to achieve different goals, a prerequisite for sustainable management of any system.

Wise decision making will be dependent on the quality and availability of relevant knowledge. Knowledge is power and sharing appropriate knowledge lie in the root of developing systems that address problems in sustainable conservation management (Hambly and Angura 1996). Local and indigenous knowledge can serve as a valuable basis for interpreting information and data, and for solving problems identified by scientists, policy makers and resource managers. This knowledge is, however not always well documented or easily accessible to others in the same area with the same problems and challenges. Large investments are made in a top–down approach to conserve natural resource systems, often not taking the local knowledge and experiences into account. One of the challenges in finding solutions to these pressing problems is, to find methodologies and tools that provide the linkages between traditional knowledge of land–use management systems and modern scientific methods and technology. The development of inventories and information systems where indigenous and scientific knowledge are incorporated into single expert knowledge systems, will not only help in the awareness, education, training and capacity building of agriculturalists and conservationists, but also support land–users and managers in their future decision making.

2. INDIGENOUS KNOWLEDGE AND ITS IMPORTANCE

There are several terms, which are synonymous with indigenous knowledge, such as local knowledge, indigenous skills, traditional knowledge or cultural knowledge. In practical terms, indigenous knowledge can be defined as: The ideas, experiences, practices and information that either have been generated locally or elsewhere, but have been transformed by local people and incorporated into their way of life that is unique to their given culture or society. Indigenous knowledge and beliefs have mainly been transmitted through oral tradition from generation to generation, or is obtained as a result of experience and trial and error or a combination of the two. This long time experience and experiment means that indigenous knowledge cannot be replaced quickly by other knowledge systems. It also encompasses the skills, experiences and insights of people, applied to maintain or improve their livelihoods (UNCCD 1999).

Indigenous knowledge is often in contrast with the international knowledge system generated by universities, research institutions and private firms. It however, forms the basis for local level decision making in agriculture, health care, food preparation, education, natural resource management, and a host of other activities in rural communities (Warren and Rajasekaran 1993). Indigenous knowledge is the information base for a society, which facilitates communication and decision making. Indigenous information systems are dynamic, and are continually influenced by internal creativity and experimentation as well as by contact with external systems. (Flavier et al. 1995).

In recent years, there is a greater attention to indigenous knowledge particularly in agriculture and natural resource management in attempts to create more appropriate and environmentally friendly technologies empowering people, such as land–users and farmers, to have greater control over their destinies and creating technologies that will be more justifiable in their specific socioeconomic situation. Indigenous knowledge techniques are of benefit for sustainable agriculture and conservation, because they are based on profound understanding of the local ecology, social structure, economy and culture of the community, which belong to it. This knowledge is often a part of the lives of the rural poor. Their livelihood depends almost entirely on specific skills and knowledge essential for their survival and therefore provides problem solving strategies. Moreover, indigenous knowledge is relatively cheap, locally available, and less destructive to the local environment and in keeping with the local community. It has however, been marginalized and even under attack for being backward, static and a hindrance to modernization.

Reasons why indigenous and traditional techniques are not applied can be summarized by: (1)a lack of awareness that environmental damage, soil degradation, desertification and biodiversity loss results from the loss of traditional knowledge; (2)a lack of information from technicians involved in national planning on the validity and benefits of indigenous knowledge; (3)lack of information on the role to be assigned to traditional techniques and their operation; (4)a lack of awareness of the innovative use of indigenous knowledge; (5)prejudices and dissemination of criticism of indigenous knowledge; and (6)a lack in communication and exchange of successful experiences. Most of the problems mentioned above result from a lack of information and communication (UNCCD 2000).

Indigenous knowledge therefore, is a part of the global knowledge. It has a value and relevance in itself and must therefore be recognized, appreciated and properly documented so that it can be preserved, transferred, or adopted elsewhere. Developing appropriate information systems in the form of Decision Support Systems (DSS), including databases and expert systems, as instruments and tools for the capturing and dissemination, as well as application of indigenous knowledge in development and conservation management practices will reduce the risk of valuable indigenous knowledge becoming extinct. Information systems will enhance the facilitation and sharing of knowledge, experiences, practices and innovations among all stake holders, and create partnerships in sustainable management processes (WOCAT 2000).

3. SOURCES OF INDIGENOUS KNOWLEDGE

Land–users and pastoral communities have years of knowledge and developed risk aversion strategies to survive in harsh environments and fragile ecology rangeland areas (Squires et al. 1992). These differ by area according to environmental conditions and people's activities, and take their socio–economic and cultural demands into consideration. Local populations know their land and are good observers. Their technologies have been used for centuries to guide environmental and livelihood planning and action, often long before scientific knowledge attempted to understand the processes of environmental change and development (Kiyunyu and Swantz 1996). It is certain that only through participatory processes, this knowledge can be identified and explored. According to Meyerhoff (1991), it is more effective to work with target groups rather than individuals, without ignoring the respect of the elderly. Effective tools and processes are therefore, required to capture the different forms of information that exist. Equally important is the need for processes to "make sense" of all this information, and transform it into knowledge that will be useful and useable for management. Such a "making sense" process requires the creation of co–learning environments, during which different stakeholders can share their own understandings with each other, and through discussion help to understand the context in which different pieces of information are provided. Management practices and problems, such as degradation, can be formulated differently, depending on the social system and values of the people in that community. Analysis and interpretations can therefore not be standardized into general theories. Transformation and contextualisation of new knowledge is an important step before local and indigenous knowledge and experiences can be documented. It is also important to carefully examine and evaluate the usefulness of these experiences and knowledge, before they are incorporated into any information system.

4. VALUE OF LOCAL AND INDGENOUS KNOWLEDGE

A profound loss of indigenous peoples and rural groups and their knowledge has been accompanied by the neglect and marginalization of their practices and beliefs, often figured as inferior forms of knowledge to be replaced by universalized knowledge derived from the western scientific tradition. Attempts to apply western tradition universally without regard for traditional knowledge systems has in many cases led to failures in sustainable resource use and the erosion of biological diversity (Meyerhoff

1991). The importance of the contribution that local and indigenous knowledge can make to the knowledge base for conservation of biodiversity and rangeland management in general, can therefore not be over–emphasized. Appreciation of indigenous knowledge plays an important role in the process of linking science and community action (Seely 1998). Unfortunately, the value of such knowledge systems is recognized only recently, mainly due to international and donor project funding requirements. It is often said: "What rural people know is regarded by some as primitive and unscientific," and managers and policy makers have to rely on the information from formal research and its transformation by extension in order to develop the useful knowledge required by them. The question arises, what information is useful in future planning and formal conservation management strategies. There are number of key characteristics that might be considered. The information must be reliable and predictive in the outcome of the strategy applied. The strategies should not only be known to community members but also be adopted and used by them. Scientific and modern technological inputs can be important to optimize the effectiveness and usefulness of certain the indigenous strategies. There are however some limitations. Bringing the researcher–extension–farmer (land–user) together at all times is practically difficult considering the existing bureaucracies and spatial, as well as academic distances, among the personnel of these groups. Hence, academically well trained and "research minded" extension personnel to validate farmer experiences and experiments are often needed. This will include understanding the rationale behind the farmer's experimentation, recording the mode of conducting certain experiments and identifying and contextualizing the farmer's evaluation criteria. Every collaborative initiative between farmer, researcher and/or extensionist is however an interactive learning experience whereby the capacity of each representative is strengthened. It creates an environment of respect between all parties and leads to increased participation and empowerment.

5. COMPATIBILITY OF INDIGENOUS KNOWLEDGE AND SCIENCE

In order to plan actions to abate deterioration of the livelihood of populations in arid and semi–arid systems that are degrading, it is necessary to study the changes in biophysical attributes in time and space and the consequent socio–economic reactions including the intervention measures taken by populations and scientists concerned, in a collaborative way. It is often said that indigenous and traditional knowledge is not "technologically competitive" and therefore "technologically inefficient and less productive" than modern technologies (UNCCD 2000). This critique is, however, not justified since there is no reason to treat traditional techniques as less competitive, inefficient or unproductive. Traditional technologies consider a series of contextual factors often omitted by modern techniques, and results differ. The procedures are not often apparent and need more work, however, this does not represent a negative feature. Traditional techniques are selected and accepted through the process of environmental, historical and social comparison. Their efficiency is appraised according to their validity over the long term, their contextual benefits and their full sustainability. Indigenous knowledge includes the accumulation of experiences through informal experiments and an intimate understanding of the environment in a given culture. Indigenous knowledge is not proposed because it contains less technology compared to the conventional ones, but because it has better results, technologically speaking, compared to the determined environmental and social contexts. The technologies may be very simple, but are refined and still appropriate and most of all, ecologically compatible and locally manageable.

It is necessary that the knowledge of conservation practices and biophysical conditions gained on either a traditional and/or scientific level should be a unique object of the current and future research activities. We need to include in the research, indigenous knowledge by understanding, explaining and evaluating it in scientific terms. It will only be adapted and/or reinforced for application in management strategies. Scientific experimentation is regarded by many as the most appropriate way to assess and evaluate best practices in conservation management. Science is often regarded as objective and free of social values, such as emotions, feelings, perceptions and intuition. Science is seen as dealing with problems in a more universal and generic way, whereas local knowledge can only achieve understanding in the particular context in which it was developed. The challenge is to combine these different forms of knowledge in such a way that it serves as a mechanism for integrating science and management. For this, "on–site", "on–farm" or "demonstration" research could play an important role, especially if the local and indigenous knowledge on social and economic considerations could be used to maximize the usefulness of the results from such scientific experiments (Kellner and Shabangu 2000). A prerequisite is that land–users perceive the same significant problems and consider the application of a certain technology worthwhile (Van Rooyen 1998). On–site research also gives the land–user a feeling of ownership and it has a high demonstration value. This type of community–based research involves all levels of understanding, promotes communication and bridges the gap between science, transfer of technology and community participation. Community–based research requires long–term commitment by all stakeholders. Difficulties to sustain long–term funding, especially in arid and semiarid environments, often lead to discontinuity in the research process and demoralization of the land managers. It is therefore, essential to develop processes and mechanisms whereby knowledge building processes such as through community–based research, adaptive management, monitoring, feedback and community dialogue ("making sense") can be institutionalized or embedded in the community, before funding is ceased. In this way knowledge building becomes an ongoing process, and the information system in which the knowledge is captured, evolves, as more information becomes available through either research or management. Community–based demonstration projects have to be sufficient in temporal and spatial scales to enhance the acceptance and application of the results obtained. The information or technology to be demonstrated, or educational goal sought, must be relevant to people living in the region for which the research is intended (Folliott et al. 1998). Community–based projects that are implemented must not utilize an ad hoc list of interventions and inputs, as they can be expensive, disappointing and difficult to carry out. The research methodology must be characterized by the application of an interactive bottom–up approach integrating the information necessary to support decisions in planning of conservation management, based on indigenous, environmental, as well as scientific knowledge and priorities. Involving research–minded community individuals, during the identification of research–based projects and technologies is highly recommended. Projects must be planned, implemented and assessed by those who intend making the project work.

6. SPECIAL TREATMENT OF LOCAL AND INDIGENOUS KNOWLEDGE

The importance of involving communities and indigenous knowledge in decision making processes has been used and documented widely. The problem however, is not only the identification and recognition of local knowledge, but how to document and capture this knowledge in a format that can be used by the land managers in future decision making processes. Reporting and information systems must be more responsive to and supportive of the local knowledge used in decision making. In capturing local and indigenous knowledge about ecological understanding and best practices it is important to pay special attention to the different time scales at which experiential knowledge was gained (Meyerhoff 1991). It is further essential to understand the different perceptions, mental models of different knowledge providers in the community with regard to issues such as biodiversity, degradation and sustainability. The age of the land–users; level of literacy and education; interest in the land and occupation are all, important factors that will determine the context within which interpretations are made. This necessitates multidisciplinary teams with environmental, economic and social scientists that can help to document and contextualize information in a more holistic way.

Efforts to improve the acceptability of local knowledge to decision makers must focus on establishing credibility (verification), the translation and calibration between local and scientific knowledge, and the simplification and demystification of complex, culturally specific information (Hambly and Angura 1996). Verification and validation of information can be based on literature, statistical analysis and scientific experimentation, as well as sampling theory and logical analysis of data collected by different methods, such as inventory data (WOCAT 2000). Solutions can be proven with positive results in the field, while education activities should involve the entire community (Meyerhoff 1991).

7. MAXIMIZING COMMUNITY INVOLVEMENT

When gathering information from communities, especially from poverty stricken, rural areas, it is important to inform them about their role in this participatory venture. Their realistic benefits from the information provided should be explained without creating unnecessary expectations. Their cultural habits should be respected and each contribution should be acknowledged. Development of information sharing protocols through participation is a prerequisite for ensuring a positive attitude towards knowledge sharing. Ongoing involvement is also facilitated when feedback mechanisms are in place to demonstrate the benefits of being involved.

Working relationships between all stakeholders, including governmental agencies, must be established. People involved in community–based projects tend to work on their own perceptions, which are often gained from the literature, rather than from a developed understanding of local ways of living and managing the environment, and of the problems the local people faced and why certain technologies had to be developed to overcome these problems (Squires 1992). Through working relationships, mutually advantageous links and communication (e.g., meetings, workshops, seminars and personal visits), an improved understanding of traditional problem solving practices and technologies, can be achieved (Meyerhoff 1991).

8. MAXIMIZING USEFULNESS

One of the most important aims in collating local and indigenous knowledge of best practices is to capture and transform it to a user–friendly and easily accessible form. This will also enhance the adoption of the knowledge/management options by the end–users. Effective information systems that serve as decision–making tools for conservation management strategies are often developed for first world situations with a high technological infrastructure. People in these countries are also not so aware of the natural resource as a day to day survival mechanism, whereas people in the developing world, are much more dependent on the land as a resource. Maintaining and improving the condition of the land in these regions is thus very important and will also lead to an improvement of the livelihoods, contribute to poverty alleviation and long–term sustainable development.

Identifying the correct type of information system and database that will serve the needs of a specific target group and help to identify the most appropriate technology for a certain conservation strategy, is therefore of great importance. Any information is data that has been refined and organized by processing and purposeful intelligence. The latter, purposeful intelligence is crucial to the definition because people provide the purpose and the intelligence produces true information (Whitten and Bentley 1998).

Information is a resource created from the data to serve the management and decision–making needs of the user. Information systems are used to transform the data into useful information. Information systems arrange data and processes for the purposes of supporting and improving, as well as fulfilling the problem solving and decision–making information needs of managers in the application of conservation management strategies. Information systems exist with or without the computer. The method, by which the analyzed data from research and transformed knowledge is transferred and disseminated, will depend on the availability of technology and the level of understanding. This can vary from simplified modes such as using stories, art, ceremony, ritual, community associations, social gatherings, farmer–farmer or farmer–scientist knowledge sharing exercises. It also includes exchange visits between all stake holders, on–site demonstration sites, workshops and seminars, photographic records, audiovisual and printed media, right up to sophisticated information systems with user–friendly front ends.

Information technology has however, significantly expanded to power and potential of most information systems. Information technology therefore, describes the combination of computer technology (hardware and software) with knowledge sharing technology. Various information systems and technologies exist, each serving the needs of different users.

One of these systems is the DSS. A DSS is an information system that provides its users with decision–orientated information whenever a decision making situation arises. It must be realized, that a DSS does not make decisions or solve problems, people do. Decision Support Systems are only concerned with providing useful information to support the decision making process. In particular, DSS are usually designed to support unstructured decisions, that is, those decision making situations that cannot be predicted in advance. DSS is based on the reality that the information needed to support decisions, have already been captured by information systems, such as participation processes in capturing local and indigenous knowledge. The DSS only provides the decision maker with tools to access that information and analyze it to make a decision, such as identification of problems and possible solutions. Since the process of recording and evaluating rangeland management, biodiversity or restoration best practices, often involves large data sets, the use of computerized databases is essential. The DSS must therefore include a "data warehouse" of as much possible information and solutions for decision–making, as possible. Data is a resource that must be controlled and managed so that it can be used in an integrated way in several information systems. Information systems must be built around a resource that will give the database developers (e.g., computer programmes), as well as the end–users, flexible access to the data. DSS can therefore also include an expert system whereby the knowledge and expertise of the problem solver or decision maker that is captured is made available. The expert system can also be used to simulate the "thinking" of that expert for those that have less expertise. Expert systems are implemented with artificial intelligence technology that captures and simulate the reasoning of the experts. Case–based reasoning (CBR) methodologies are an example of artificial intelligence technology that is used in the construction of a DSS for biodiversity and conservation management (Baretta 1991, Slade 1991, Bosch et al.1997). CBR is a methodology that has been developed to diagnose problems and find the best solutions to these problems by searching through a database of previous cases with similar problems. The underlying concept of CBR is "learning from experience" and therefore consists of a pool of knowledge, local and indigenous, as well as scientific of applied technologies. The database incorporated into a user–friendly the computer software package is based on questionnaires and qualitative or quantitative assessments (Kellner and Van der Merwe 2000, Kellner et al. 2002).

A questionnaire is completed together with the land–user, manager, scientist or technician that has implemented a certain technology. This background information is important to get some historical data about the type and extent of the problem, environmental conditions (such as soil and long term climatic data), as well as the technology and approach that was used. A quantitative or qualitative assessment and field studies of the vegetation and habitat condition is made by the scientists who develop the database in order to evaluate the effect of the specific technology that was applied over a certain period of time. Practical experience of the technology and the exchange of experiences by personal contact stimulate and enhance the excitement and collaboration between the two parties. Photos are taken of outcome of the technology applied, range condition and the type of implement and methods used during the implementation of the technology. These are included in the computerized database for consultation so that the user can assess the result of a certain technology visually. The scientists developing the database, to rank the effectiveness of the technology, carry out statistical analyses. Comparisons are made between technologies applied to address the same problem under the same conditions. It is very important that the quality of the data, as well as the description of the technology is of a very high standard (WOCAT 2000).

The consultative section of the knowledge system includes expert system methodologies that are developed in a way that the user will be guided to the most appropriate and applicable solution and technology that can be used to address a specific problem. A complete reference and information of the technology applied, as well as the person with address who has implemented the technology will be available to the user to create the possibility for personal contact. This will also enhance networking and sharing of experiences between land–users and between technicians and scientists on a more personal note. In this way, land–users do not have to rely on their own experiences, but can share in the pool of technologies and case studies that have been applied by many other land–users, technicians, agriculturalists, conservationists and scientists (Bosch et al. 1999).

The creation of a DSS and expert system as an information system can however, only be a reality if the people that provide the information and the users for which the DSS is developed, are successfully involved in all the phases of the system development. It is essential that the end–users (e.g., extension officers, consultants, policy makers) have a direct hand in their design to ensure ownership. The development of computerized DSS and databases requires specially skilled people who have the ability to understand and include environmental aspects into a form that can easily be accessed by the untrained end–user. Choosing the most applicable database and expert system for the particular type of knowledge and data is important. The specifications, limitations and type of outputs must be well understood before optimization and adoption of conservation management strategies is assured.

9. CONCLUSION

The overwhelming majority of the population in most developing countries are small–scale farmers, each working less than a few hectares of land. These farmers represent many languages and ethnic groups and in most cases, the knowledge systems of these farmers have never been recorded systematically in a written or any other form. Hence, this knowledge is not easily accessible to agricultural researchers, extension workers, development practitioners and land–users (Warren and Rajasekaran 1993). International and national development agencies are increasingly recognizing the value of indigenous knowledge and participatory approaches to decision making for sustainable approaches to development and biodiversity management. Participatory rapid appraisal (or related models) conducted through thematic workshops would provide appropriate forums and platforms for collaboration, exchange of information and documentation of local and indigenous knowledge and the scientific community. The involvement of non–governmental and community based organizations in the different phases of these processes to initiate community related projects is indispensable. NGOs and CBOs often bridge the gap between the cultural and socioeconomic differences that exist between local communities and researchers (UNCCD 1999). The development of information systems for conservation management that are mainly based on local scale indigenous knowledge can also be used for political and socioeconomic decisions at different administrative and policy levels. Development activities that work with and through indigenous knowledge and organizational structures have several important advantages over projects that operate outside them. They create favorable conditions to maintain, produce and reproduce knowledge and integrate it with modern technology and innovations. Indigenous knowledge provides the basis for grass–roots decision making, much of which take place at the land–user and community level through indigenous organizations and associations where problems are identified and solutions are determined. Solution seeking behavior is based on indigenous creativity, leading to experimentation and innovations as well as the appraisal and implementation of knowledge and appropriate technologies. Information systems based on already existing knowledge which has been evaluated and verified, will help in the decision making process. Policy actions in the years to come should therefore give priorities for recording, documenting and incorporating indigenous knowledge, technologies and skills, into computerized information systems.

Information systems developed for different aspects of natural resource conservation, such as biodiversity, desertification and climate change could also be incorporated in the implementation of national action programmes

(NAPs) of the respective UN Conventions that define these environmental aspects. These information systems will facilitate the compilation of research and information on indigenous and local knowledge into databases and in the developing of adequate methods and procedures for their compilation, storage and dissemination. Information systems can also help in setting up monitoring mechanisms of the implementation of conservation management technologies and will ensure that action–orientated research, based on indigenous knowledge, is carried out.

Appropriate ways of linking DSSs, databases and expert systems based on indigenous knowledge, especially for conservation of biodiversity and sustainable resource management, on local (national), regional and global networks, are essential. Indigenous knowledge information systems should also be compared to avoid duplication. This can be facilitated by the global network of indigenous knowledge resource centers (UNCCD 1999). Information systems can also strengthen the capacity of regional and national agricultural and conservation extension organizations. The aim should be to generate sustainable, dynamic and updated technological options, rather than technical packages of practices.

10. REFERENCES

Barletta, R. 1991. An introduction to case–based reasoning. Artificial Intelligence Expert 6: 42–49.

Bosch, O.J.H., Gibson, R.S., Kellner, K. and Allen, W.J. 1997. Using case–based reasoning methodology to maximise the use of knowledge to solve specific rangeland problems. Journal of Arid Environments 35: 349–557.

Convention to Combat Desertification (UNCCD). 1999. WWW.UNCCD.INT. Science and Technology, ICCD/COP (3)/CST 2–5.

Convention to Combat Desertification (UNCCD). 2000. WWW.UNCCD.INT. Science and Technology, ICCD/COP (4)/CST 2–5

Folliott, P.F., Fisher, J.T., Sachs, M., deBoer, D. W., Dawson, O. and Fulbright, T.E. 1998. Role of demonstration projects in combating desertification. Journal of Arid Environments 39: 155–163.

Hambly, H. and Angura, T.O. (Eds.). 1996. Report of the Workshop on Grassroots Indicators for Sustainable and Equitable Development. Grassroots Indicators for Desertification: Experience and Perspectives from Eastern and Southern Africa. International Development Research Centre, Ottawa. 168 pages.

Kellner, K. and Shabangu, A. 2000. Community participation in rangeland restoration. Pages 46–60 In: Kellner, K. (Ed.) Desert Margins Program. Proceedings of a National workshop: Appropriate Restoration Technologies in South Africa, 1998. Potchefstroom University for CHE, Potchefstroom, South Africa. 116 pages.

Kellner, K. and Van der Merwe. 2000. Data base and expert system for rangeland restoration: A community participatory approach. Pages 156–165 In: Ringrose, S. and Chanda, R. (Eds.) Towards Sustainable Management in the Kalahari Region: Some Essential

Background and Critical Issues. Directorate of Research and Development, University of Botswana, Gaborone, Botswana. 304 pages.

Kellner, K., Bosch, O.J.H., Barac, A. and Van der Merwe, J.P.A. 2002. EcoRestore–A Decision Support System for Rangeland Restoration. Proceedings of the Third World Network of Scientific Organization (TWNSO) Conference. Centre for Environmental Studies and Research, Sultan Qaboos University, Sultanate of Oman. 56 pages.

Kiyunyu, L and Swantz, M.L. 1996. Research methodologies for identifying and validating grassroots indicators. Pages 60–82 In: Hambly, H. and Angura, T.O. (Eds.) Grassroots Indicators for Desertification: Experience and Perspectives from Eastern and Southern Africa. International Development Research Centre, Ottawa. 168 pages. Meyerhoff, E. 1991. Taking Stock: Changing Livelihoods in an Agropastoral Community. African Centre for Technology Studies. Biomass Users Network, Harare, Zimbabwe. 58 pages.

Seely, M.K. 1998. Can Science and community action be connected to combat desertification? Journal of Arid Environments 39: 267–278.

Slade, S. 1991. Case–based reasoning: A research paradigm. Artificial Intelligence Magazine 12: 42–54.

Squires, V.R., Mann, T.L. and Andrew, M.H. 1992. Problems in implementing improved range management on common lands in Africa: An Australian perspective. Journal of the Grassland Society of Southern Africa 9: 1–7.

Van Rooyen, A.F.. 1998. Combating desertification in the southern Kalahari: Connecting science with community action in South Africa. Journal of Arid Environments 39: 285–297.

Warren, D.M. and Rajasekaran, B. 1993. Putting local knowledge to good use. International Agricultural Development 13: 8–10.

Whitten, J.L. and Bentley, L.D. 1998. System Analysis and design methods. Fourth Edition. Irwin McGraw–Hill, New York. 724 pages

WOCAT. 2000. World Overview of Conservation Approaches and Technologies. Centre for Development and Environment (CDE). Hallerstrasse 12, CH–3012, Berne, Switzerland.

Chapter 9

BEST PRACTICES IN THE WORLD'S OLDEST DESERT

M.K Seely and J.R Henschel
Desert Research Foundation of Namibia, Gobabeb Training and Research Centre, P.O. Box 20232, Windhoek, Namibia; email drfn@drfn.org.na

1. INTRODUCTION

Promoting best practices for conservation and sustainable use of biodiversity of global significance in the semiarid areas of Namibia is being undertaken by a loose public, private, nongovernmental (NGO) partnership. This is being led by the public sector that manages a number of established national parks and more recently, in partnership with the NGO community, is promoting community based natural resource management through conservancies on commercial and communal farming lands. The private sector is primarily involved in tourism that provides the foreign exchange income motivating the public sector, with its many alternative social responsibilities, to retain their interest in biodiversity. In the arid areas of Namibia, a similar situation prevails with differences of emphasis but not overall pattern.

Namibia is situated on the western coast of southern Africa with landscape elements ranging from hyper arid along the southern Atlantic coast, through arid and semiarid areas extending to dry subhumid in the northeast. From southwest to northeast the rainfall varies from winter to summer dominance. Moreover, the country is situated on the western escarpment of southern Africa. The result of these geographic factors is a varied landscape with a mosaic of high biodiversity caused by different factors with differing patterns throughout the country (Barnard 1998). Of particular note is the winter rainfall southern Namib desert where a combination of climatic and geographic factors serve to create a biodiversity "hotspot" based on the northward extension of the Nama Karoo floral assemblage (Myers et al. 2000). An adjacent area of high biodiversity is the Central Namib Desert that supports a high diversity of invertebrates, reptiles and lichens, sustained by the supplementary water source represented by frequent fogs (Seely 1992).

The public sector concerned with biodiversity, the Ministry of Environment and Tourism (MET), provides strong leadership in the maintenance and development of protected areas. Ranging from one national park to a range of game parks, reserves and recreational areas, protected areas cover over 11 percent of the land surface of Namibia (Barnard 1998). On the other hand, these protected areas are predominantly located in the more arid areas with no agricultural potential, and not all the biomes of Namibia are represented. The MET is involved in protection, research and tourism activities in these protected areas. Most recently, Namibia has signed an agreement with South Africa to form a cross border peace park joining protected arid areas on both sides of the Orange River.

The private sector became involved in biodiversity protection in the 1970's when new legislation allowed commercial farmers, with freehold tenure, ownership over game resident on their farms. In accordance with stipulated regulations, neighboring farmers could join together to form conservancies and thereby gain the rights to harvest their game. This gave rise to a variety of approaches ranging from game viewing and bird watching to trophy hunting of protected species such as elephant (de Vos 1979).

The nongovernmental sector, working with rural communities living on communally managed land, became a major force for biodiversity protection after independence in Namibia in 1990 (Barnard et al. 1998). With changed legislation in the mid 1990's, communities, supported by NGOs, were allowed to form conservancies that gave them rights over and benefits from the wildlife and tourism activities on conservancy land (Turner 1996). This conservancy movement is gaining momentum and providing the mechanism for de facto control by communities over other resources such as grazing and water. In the absence of a communal land act, and continued inequitable distribution and access to land in Namibia, the conservancy movement appears to be providing communities with a mechanism to protect their biodiversity while contributing to the prevention of further land degradation and proving alternative income generating potential to alleviate poverty (Swanson et al. 1996).

For forty years, the NGO Desert Research Foundation of Namibia (DRFN) (formerly the Desert Ecological Research Unit), in partnership with the MET, has coordinated and undertaken research in the Namib Desert based at Gobabeb. Several years ago, these two institutions embarked on a joint venture and the Gobabeb Training and Research Centre was formed.

Since so much of the protected areas of Namibia falls within the Namib Desert, the results of this research have served as the basis for understanding the biodiversity of arid Namibia (Seely et al. 2000). In turn, this is a source of information for policy and decision makers as well as managers of these areas (Seely et al. 2002). It is also a source of interpretive information for the growing tourism industry for which the desert is the second most important destination after the Etosha National Park.

This case study focuses on the Desert Research Foundation of Namibia (DRFN) in its base at the Gobabeb Training and Research Centre and its research and dissemination of information in support of biodiversity conservation in arid, western Namibia. These institutions, DRFN and Gobabeb Centre, participate as members of the Biodiversity Task Force established by government and play a particular role in long–term research on biodiversity in Namibia. Under the aegis of the task force's Environmental Observatories Network of Namibia (EONN), an Environmental Observatory has been established at Gobabeb that has become a member of the International Long Term Ecological Research Network (ILTER) (Henschel et al. 2000). The remainder of this paper focuses on lessons learned in science, public policy and management, use and management of biodiversity by local populations, sustainable use of natural resources, capacity building and partnerships.

2. SCIENTIFIC LESSONS LEARNED

Originally research focused on finding explanations for the high species richness of tenebrionid beetles in the Namib (Koch 1962), compared with any other desert in the world, when the rest of Namibia and Africa was focused on the charismatic megafauna. This approach continually evolved covering a wide range of basic and applied research (Seely et al. 2000). For example, the importance of fog as the primary water source for the especially rich and diverse invertebrate fauna and lichen flora was recognized (Seely and Hamilton 1976, Seely et al. 1998). This, supported by a wide variety of environmental research has led to current studies on fog harvesting and its application by people (Henschel et al. 1998). The behavior and role of ephemeral rivers or 'linear oases' that cross the desert while supporting a riparian forest with its associated fauna, farming opportunities and recharging underground aquifers upon which coastal development depends, was another major contribution (Jacobson et al. 1995). The formation and dynamics of sand dunes and their role in the biotic system (Lancaster 1989, Seely 1990), changes in the microenvironment caused by a variety of rock weathering agents (Goudie 1972), and the impact of off–road vehicles on the topography and biodiversity of the desert surface (Seely and Hamilton 1978, Daneel 1992) were other important studies. Studies on vegetation have provided a variety of basic and applied results focusing on ephemeral and perennial grasses near the base of the food chain (Seely 1978, Yeaton 1988, Burke 1997, Hachfeld 2000), the endemic family of Welwitschiaceae (Bornman 1978, Henschel and Seely 2000) and the variety of species upon which the local people depend, for cash and other necessities (van den Eynden 1992, Henschel et al. 2002). These studies in all have yielded over 1400 publications (Henschel and Henschel 2001; see also www.drfn.org/pubs).

3. PUBLIC POLICY AND MANAGEMENT

Lessons learned surrounding biodiversity in arid areas and public policy issues are not unique. Much of the Namib Desert is set aside as "parks," not to protect the biodiversity, but for political reasons, for mining and because it was viewed as useless, particularly for agriculture. Change in public policy has been driven by changing global perceptions of landscape and the value of biodiversity which fuels international tourism. The tourist value of the Namib Desert in particular has been enhanced by research, and dissemination of the results of research, on the unique biodiversity of this hyper arid, coastal area (Seely 1992, Lovegrove 1993). Even today, as vast protected areas that served as a buffer for coastal diamond mining are opened up to alternative use, their status as a biodiversity hotspot influences public policy and plans for future use (Pallett 1995).

At the national level, Namibia has a very active policy setting arm in the Directorate of Environmental Affairs in the MET. Preparing first a "Green Plan" with a chapter on biodiversity, this directorate has taken the lead in having the government ratify the Convention on Biodiversity, as well as the Convention to Combat Desertification, Ramsar and CITES, and integrating them into National Development Plans. The public sector has an increased awareness of biodiversity although a draft Environmental Management Act languishes.

Management lessons for this arid area are indirectly research driven. The Namib Desert was originally considered by most as an open, rugged space to test vehicles and people against the environment. As information on the biodiversity and relative fragility of the landscape became available, and the number of researchers attracted to the area was noticed, management directives increased. This was partly in support of enhancing a tourism attraction and partly to protect the biodiversity. Currently, as in many parts of the world, tourism activities are attracted by the relatively unspoiled open spaces while increasing tourism is itself impacting on the landscape. Although a variety of management tools have been applied to the development of the Namib Desert, for example, requiring environment impact assessments for infrastructure development (Tarr 1996), economic considerations, not biodiversity, remain paramount.

4. USE AND MANAGEMENT BY LOCAL POPULATIONS

The definition of local populations in the Namib Desert is not entirely clear. The only people in Namibia living within a park, the Topnaars, live along the ephemeral Kuiseb River in the Namib Desert (Botelle and Kowalsi 1995). Elsewhere in Namibia in communal farming lands, community based natural resource management is being developed, but the modalities for developing this approach within a park have yet to be established (Dausab and Jones 1997). Moreover, the Topnaars have very close ties with the port town of Walvis Bay where most of their population lives. Researchers working at Gobabeb on the lower Kuiseb River live close to and amongst the local Topnaar people and both parties have learned many lessons from each other, directly and indirectly. Similarly, urban coastal residents have shared many experiences with the researchers.

Currently, several research projects are being undertaken directly with the Topnaars. These focus on developing their management and markets for indigenous fruit products (Henschel et al. 2002), community based tourism (Kooitjie S. and Dausab R. personal communication) and harvesting fog (Henschel et al. 1998). Growing from the efforts of the Topnaars and researchers, several developmental NGOs are now involved, with particular reference to marketing. The DRFN, collaborating with the Department of Water Affairs, is working with all communities living within or using water from the ephemeral Kuiseb River basin to pilot Basin Management Committees, a key component promoted in the draft Water Act (Amoomo 2000, van Langenhove G. and Botes A. personal communication). The research base established in this basin and the long–term interactions among the wider community enhance this development. From a similar approach, concerns about desertification are being addressed by forming partnerships with rural communities throughout Namibia and establishing sound foundations for sustainable natural resource management (Seely and Jacobson 1994, Seely 1998, Seely and Wöhl 2001).

5. SUSTAINABLE USE OF RESOURCES FOR TOURISM

Most of the unique biodiversity of the Namib desert lends itself to support tourism in the area. Tourism is also based on the landscape that is inextricably linked to the biodiversity. This begs the question of how this diverse resource can be exploited for tourism without destroying it. Management of tourism in Namibia is very diffuse and can be guided by public policy, rules and regulations, but is implemented mainly by the private sector. With a current focus on low volume tourism the balance with environmental protection could be maintained. As the numbers of tourists continue to increase, even though not extending to mass tourism except on a very local and temporal basis near coastal towns, the landscape and its associated biodiversity are in jeopardy. At the same time as basic research yielding new, interesting information concerning the biodiversity of the Namib is decreasing, dissemination of existing information to government and the tourism industry is increasing. This shift in balance between research information, and policy and tourism is of particular concern for the biodiversity of the Namib because future management to be effective must be based on current research.

6. CAPACITY BUILDING AND PARTNERSHIPS

Interest in and concern for biodiversity in this arid area has been influenced by a changing relationship between the public sector, the private sector, and the NGOs because of their constantly shifting interests, capabilities, and resources. Although the public and NGO sectors shared the Gobabeb research facility with the Namib–Naukluft Park since 1970, the beginnings of a partnership only arose when the Joint Venture Agreement was signed and the Gobabeb Centre was formed enhancing partnerships and capacities of all involved. The Topnaar community, the Southern African Development Community Environment and Land Management Sector (SADC–ELMS) and a Namibian tertiary education institution are represented on the board of trustees. Similarly, the public and NGO sectors were guiding tourism development; the private sector tourism industry, including the Topnaar, was based on their support. However, rather than a true partnership to date, each sector has proceeded relatively independently. Interest in the biodiversity and landscape has varied among sectors, but has remained the central point for direct or indirect interaction.

Similarly, capacity building has been addressed differently by the three main sectors concerned with biodiversity in the Namib Desert. Since independence in 1990, capacity building in the public sector concerned with biodiversity has, at the local level, focused on management and law enforcement. At the national level, the MET is responsible for the UN conventions including the Convention on Biodiversity. The country program has consolidated available information in the form of publications (e.g., Barnard 1998), has established a multidisciplinary task force and in general has contributed to establishing partnerships amongst concerned public and NGO institutions (Barnard and Shikongo 2002). Capacity building within relevant institutions is beginning to emerge from these efforts. In the private sector, the requirement of formal qualifications for tourism operators is a driving force to increase capacity building. In the NGO sector, specific efforts for capacity building with respect to biodiversity have been a long tradition and this focus continues. A variety of hands–on training opportunities with DRFN and a variety of partners takes place at Gobabeb and throughout Namibia, many established on an annual basis (Seely et al. 2000a, 2000b). These range from one week courses in arid zone ecology for university and polytechnic students to ten week courses in environmental problem solving for sustainable development. In a young developing country like Namibia, this latter approach has resulted in many graduates assuming positions in the public and private sector with enhanced capacity and understanding of the relevance of biodiversity.

7. CONCLUSION

The future of biodiversity management in Namibia lies in continuing to strengthen the association between conservation, research, tourism, management, training and dissemination of awareness and information. It involves and reaches a wide variety of people ranging from rural communities to top level politicians. An interactive partnership between public, private and NGO enterprises that serves the interests of them all is the driving force that keeps these diverse aspects going. From initial lessons learned and partnerships formed in the Namib Desert, a best practice was formulated and is being applied in other regions of Namibia. This demonstrates the substantial value that can be gained from science being relayed, related to, applied to, and influenced by the environmental problems of the day.

8. REFERENCES

Amoomo, H., Elago, P., Gaseb, N., Hoveka, V., Khairabes, M., Mbangula, E., Maharukua, V., Mukuya, S., Ndjeula, G., Noongo, E., Shinedima, R., and Zaaruka, B. 2000. Determining the water reserve for the Kuiseb River. In: Hamilton, W.J, Klintenberg, P., Montgomery, S., and Seely, M (Eds.) DRFN Occasional Paper No 11. Windhoek, Namibia. 61 pages.

Barnard, P. (Ed.) 1998. Biological Diversity in Namibia–A Country Study. Namibian National Biodiversity Task Force, Windhoek. 325 pages.

Barnard, P., Brown, C. J., Jarvis, A. M., Robertson, A., and van Rooyen, L. 1998. Extending the Namibian protected area network to safeguard hotspots of endemism and diversity. Biodiversity and Conservation 7: 531–547.

Barnard, P., Shikongo, S., and Zeidler, J. 2002. Biodiversity and Development in Namibia: Namibia's Ten Year Strategic Plan of Action for Sustainable Development Through Biodiversity Conservation (2001–2010). Directorate of Environmental Affairs, Ministry of Environment and Tourism,Windhoek. 138 pages.

Bornman, C.H. 1978. Welwitschia. Struik Publishers, Cape Town. 71 pages.

Botelle, A., and Kowalski, K. 1995. Changing Resource Use in Namibia's Lower Kuiseb River valley: Perceptions from the Topnaar Community. Institute of Southern African Studies at the University of Lesotho and the Social Sciences Division at the University of Namibia, Desert Research Foundation of Namibia, Windhoek. 145 pages.

Burke, A. 1997. Is the grass always greener at certain sites? A contribution to patch dynamics in the central Namib desert. Tropical Ecology 38: 125–127.

Daneel, J.L. 1992. The Impact of Off–Road Vehicle Traffic on the Gravel Plains of the Central Namib Desert. Namibia. Master of Science thesis, University of Natal, Pietermaritzburg. 117 pages.

Dausab, R., and Jones, R. 1997. Report on the Topnaar–Richtersveld Exchange. Desert Research Foundation of Namibia, Gobabeb. 17 pages.

de Vos, A. 1979. The need for an action programme on wildlife conservation in arid and semi–arid regions of Africa. Journal of Arid Environments 2: 369–372.

Goudie, A.S. 1972. Climate, weathering, crust formation, dunes, and fluvial features of the central Namib desert, near Gobabeb, South West Africa. Madoqua, Series II 1: 15–31.

Hachfeld, B. 2000. Rain, fog and species richness in the Central Namib Desert in the exceptional rainy season of 1999/2000. Dinteria 26: 113–146.

Henschel, J.R., and Henschel, I.A. 2001. DRFN Bibliography 2001, 1418 Publications. Desert Research Foundation of Namibia, Windhoek. 37 pages.

Henschel, J.R, Dausab R., Moser, P., and Pallett, J. (Eds). In press. NARA Natural Resource Management of the Aonin !nara Fruit as the Basis for Development of the !Khuiseb Topnaar. Desert Research Foundation of Namibia and Namibian Scientific Society, Windhoek.

Henschel, J.R., Mtuleni, V., Gruntkowski, N., Seely, M.K., and Shanyengana, S.E. 1998. Namfog: Namibian Application of Fog–Collecting Systems, Phase I: Evaluation of Fog–Water Harvesting. Desert Research Foundation of Namibia, Occasional Paper No. 8. 81 pages.

Henschel, J.R., and Seely, M.K. 2000. Long–term growth patterns of *Welwitschia mirabilis*, a long–lived plant of the Namib Desert (including a bibliography). Plant Ecology 150: 7–26.

Henschel, J.R, Seely, M.K., and Zeidler, J. 2000. Long–term ecological research at Gobabeb: Gaining and applying knowledge about a highly variable environment. Journal of the Namibia Scientific Society 48: 89–115.

Jacobson, P.J., Jacobson, K.M., and Seely, M.K. 1995. Ephemeral Rivers and Their Catchments: Sustaining People and Development in Western Namibia. Desert Research Foundation of Namibia, Windhoek. 167 pages.

Koch, C. 1962. The Tenebrionidae of southern Africa. 31: Comprehensive notes on the tenebrionid fauna of the Namib Desert. Annals of the Transvaal Museum 24: 61–106.

Lancaster, N. 1989. The Namib Sand Sea. A.A.Balkema, Rotterdam. 179 pages.

Lovegrove, B. 1993. The Living Deserts of Southern Africa. Fernwood Press, Cape Town. 224 pages.

Myers, N., Mittermeier, R.A, Mittermeier, C.G, da Fonseca, G.A.B., and Kent J. 2000. Biodiversity hotspots for conservation priorities. Nature 403: 853–858.

Pallett, J. (Ed.) 1995. The Sperrgebiet Namibia's Least Known Wilderness. Desert Research Foundation of Namibia (DRFN) and NAMDEB Diamond Corporation (Pty) Ltd., Windhoek. 84 pages.

Seely, M.K. 1978. Grassland productivity: The desert end of the curve. South African Journal of Science 74: 295–297.

Seely, M.K. (Ed.) 1990. Namib Ecology: 25 years of Namib Research. Transvaal Museum Monograph 7, Transvaal Museum, Pretoria. 223 pages.

Seely, M. 1992. The Namib: Natural History of an Ancient Desert. Shell Namibia Ltd. Windhoek, Namibia. 108 pages.

Seely, M.K. 1998. Can science and community action connect to combat desertification? Journal of Arid Environments 39: 267–277.

Seely, M.K., and Hamilton, W.J. 1976. Fog catchment sand trenches constructed by Tenebrionid beetles, Lepidochora, from the Namib Desert. Science 193: 484–486.

Seely, M.K. and Hamilton, W.J. 1978. Durability of vehicle tracks on three Namib Desert substrates. South African Journal of Wildlife Research 8: 107–111.

Seely, M.K., Henschel, J.R., and Robertson, M. 1998. The ecology of fog in Namib sand dunes. Pages 183–186 In: Schemenauer, R.S. and Bridgman, H. (Eds.) Proceedings: 1st International Conference on Fog and Fog Collection. Vancouver, Canada.

Seely, M.K., Henschel, J.R., Zeidler, J., and Shanyengana, E.S.C. 2000. Namib research: Its development at Gobabeb and implications for Namibia. Journal of the Namibia Scientific Society 48: 62–88.

Seely, M.K., and Jacobson, K.M. 1994. Desertification and Namibia: A perspective. Journal of African Zoology 108: 21–36.

Seely, M., and Wöhl, H. 2001. Research competence in combating desertification. Entwicklung and Ländlicher Raum 35: 15–19.

Seely, M.K, Zeidler, J., Henschel, J.R, and Barnard, P. In press. Creative problem solving in support of biodiversity conservation. Journal of Arid Environments.

Swanson, T., Barnes, J., and de Jager, J.V. 1996. Conflicts in wildlife conservation–the role of property rights in their resolution: The southern African conservancies. Pages 191–331 In: Swanson, T.M., Fernandez Ugalde, C., and Luxmoore, R.A (Eds.) A Survey of Wildlife Management Regimes for Sustainable Utilization. World Conservation Monitoring Centre, Cambridge.

Tarr, P.W. 1996. Namibia's environmental assessment policy. Namibia Environment 1: 135–138.

Turner, S.D. 1996. Conservancies in Namibia: A Model for Successful Common Property Resource Management. Multidisciplinary Research Centre, University of Namibia, Windhoek, Namibia. 42 pages.

Van den Eynden, V., Vernemmen, P., and Van Damme, P. 1992. The Ethnobotany of the Topnaar. Universiteit Gent, Gent. 145 pages.

Yeaton, R.I. 1988. Structure and function of the Namib dune grasslands: Characteristics of the environmental gradients and species distributions. Journal of Ecology 76: 744–758.

Chapter 10

A COMMENTARY ON STRATEGIES AND INCENTIVES TO IMPROVE BIODIVERSITY IN ARID AND SEMI–ARID ZONES

Hassan M. Hassan
Lead Environmental Consultant for the World Bank Washington D.C., USA; email Hhassan@worldbank.org

1. INTRODUCTION

Biodiversity in arid and semi–arid zones is the total variability of living organisms and the genes they contain from all ecosystems and the ecological complexes of which they are part; this includes diversity within species, between species, and ecosystems. Untold numbers of arid and semi–arid zone species have become extinct in the past 100 years, and every arid zone on the earth can be cited in the list of habitats containing rare and endangered species.

Only limited research on biological diversity, species composition, characteristics and endemism has been conducted in the world's arid and semi–arid zones. Much of what has been done has been in developed countries such as Australia, the central Asian countries of the former Soviet Union, and the United States. Developing countries such as Kenya and Namibia have also made significant contributions to understanding arid zone ecology. However, biodiversity in the arid zone remains a largely unknown domain. Identification of endangered species tends to be concentrated on popular game animals (e.g., white rhino, cheetah, elephant), rather than the hundreds of thousands of small animals, plants, birds, fish, and so on. Also overlooked are the microorganisms which constitute the fundamental building–blocks in arid and semi–arid production systems. Knowledge of arid and semi–arid zone ecosystems and their genetic characteristics is an important tool in efforts to conserve biodiversity and combat drought and desertification.

Sustaining biodiversity in the arid and semi–arid zone has been mainly fought as a defensive battle: a fight against desertification. Although people living in that zone have been fighting a battle for survival, it is important to think about the solution as offensive rather than defensive: proactive rather than reactive because it essential to look for win–win solutions and develop approaches, strategies and incentives for positive actions rather than survival solutions. The solutions should aim at improving the peoples' livelihood in a sustainable manner while conserving biodiversity.

The concept of an expanding desert has been found not to be based on strong foundation, and the idea of fighting an ever spreading desert was found to be unrealistic. Arid and semi–arid zones have a high degree of resilience and that through proper policies, incentives, institutional and investment measures, arid and semi–arid zones populations could be well adapted to reverse the situation.

2. LOSS OF BIODIVERSITY

Loss of species diversity is of global concern because biodiversity plays a critical role in maintaining the integrity of arid and semi–arid areas. The disappearance of plants and animals from the arid and semi–arid zones limit development options and increases the vulnerability of the zones to drought and desertification. Loss of biodiversity in arid and semi–arid zone manifests itself in a number of ways:

1. Deforestation: Forests have a major role in sustaining biodiversity and alleviating poverty particularly in arid and semi–arid zones. Ninety percent of the world's 1.2 billion extremely poor people live in the arid and semi–arid zones and depend mainly on natural resources and forests for their survival. Deforestation in these zones is a major cause for biodiversity loss. Deforestation has a direct impact on soil erosion and loss of soil fertility; loss of forests' canopies lead to less interception of rain water and loss of humus and under story reduces the capacity of the soil for water absorption.
2. Soil Erosion: Due to the fragile nature of the arid and semi–arid zones and the exposure to extreme weather conditions which is becoming a characteristic of those zones, soil erosion has become a major leading to loss of biodiversity. Frequent droughts followed by flash floods combined with inappropriate land use practices lead to soil wash–out by water and wind.

3. Loss of soil fertility: Soils in the arid and semi–arid zones experience loss of fertility due to salinization and loss of organic contents from wash–out of top soil in areas of irrigated agriculture. Use of heavy machinery in mechanized agriculture in the zones of fragile soils causes compaction and denudation and loss of biodiversity.
4. Loss of vegetative cover: Loss of vegetative cover is both a cause and a consequence of desertification and loss of biodiversity. Prolonged droughts and inappropriate land use practices, particularly overgrazing in marginal lands, leads to loss of both genetic and species biodiversity.
5. Habitat conversion: In many arid and semi–arid lands, particularly in the Sahellian zone in Africa, people tend to convert habitats in fragile and marginal areas to agriculture and urban development in response to short–term crisis. Those conversions may have been appropriate if not for the frequent droughts experienced lately in the Sahel.
6. Water degradation: Soil erosion leads to water pollution and degradation and destruction of aquatic biodiversity which in turn leads to soil erosion and thus perpetuates repeating vicious cycles.
7. Sand dunes: In many areas of Northern Africa and the Sahel wind erosion creates huge sand dunes that move across the landscape as bodies covering agriculture and urban infrastructures destroying vegetation and biodiversity.

3. SYMPTOMS OF BIODIVERSITY LOSS

Many times people confuse the underlying causes of biodiversity loss with the symptoms that manifest themselves as consequences of the causes. It is true those biophysical processes and natural phenomena are major factors in desertification and biodiversity loss, but these are not the underlying causes. For example overgrazing pastoral lands is usually mistaken for the underlying cause of desertification. I submit that overgrazing is only a symptom, and that really the underlying causes are inappropriate policies and unfavorable markets that lead people to overstocking and thus overgraze these fragile lands. Over–cultivation is caused by inappropriate land use practices, which are usually driven by inappropriate policies and unfavorable markets. Land conflicts are usually caused by unclear tenure system. Use of inappropriate technologies is really caused by failures in markets and institutions. Deforestation and loss of vegetative cover are results rather than causes. Excessive and frequent drought is a natural phenomenon but most of the time becomes part of the vicious cycle of desertification.

4. UNDERLYING CAUSES FOR BIODIVERSITY LOSS

Effective actions to improve the livelihood of arid and semi–arid zone populations, while conserving biodiversity, require government commitment, political will, proper strategies and incentives and capacity. These are usually lacking in many arid and semi–arid zones. Environmental concerns are often the domain of environmental agencies that are usually not involved in setting national macroeconomic and sector policies. Participation of arid and semi–arid zone populations is crucial to improve their livelihood and conserve biodiversity in the area. If the policies and interventions of governments and donors in arid and semi–arid zone are to succeed, they have to be based on the knowledge, aspirations, desires, priorities, and decisions of the people living there. These communities have strong interest in the preservation of natural resources that provide their and their children survival.

The geographical isolation of many of the arid and semi–arid zones from major population centers and markets has often blocked the access of many of them to services, technologies, and markets; and therefore undermined their potentials for development. Transport costs of inputs and outputs are usually high in those geographically isolated zones and profitability becomes questionable and variable.

Arid and semi–arid resource use and biodiversity conservation depends first and foremost on the millions of communities–based household and individual actions. It is at the local level that resource use decisions are taken. It is at that level that awareness, strategies, institutional reforms and incentives are to be made or taken in order to secure sustainable development and conservation of biodiversity. However, national and international policies are still decisive in shaping the strategies and incentive measures. Thus actions to support improved and sustainable natural resource management and biodiversity conservation at national and international levels require: (1)concern for the welfare of arid and semi–arid populations, for the resource degradation and for natural phenomena such as frequent droughts; (2)a political will to address these problems; (3)developing, implementing and enforcing appropriate policies, institutional and legal reforms; and (4)active participation and ownership by the local population.

In summary, the underlying causes of biodiversity loss in the arid and semiarid zones are:

1. Poverty: Poor people, especially those living in the arid and semi–arid zones, depend heavily and directly on genetic, species and ecosystem biodiversity for their livelihood. Poor rural populations derive a

significant share of their income from the exploitation of natural resources. However, even though the poor are dependent on natural resources, they generally use less of these resources than the more affluent. The poor women in developing countries tend to be disproportionately affected by the degradation of natural resources and the loss of biodiversity because it is they who tend to be primarily involved in collection of fuel wood, fodder and water. The linkages between poverty and natural resource degradation, particularly biodiversity loss involves difficult tradeoffs between short–term and long–term benefits.

2. Policy failures: Macroeconomic conditions and general economic policies play an important role in shaping the possibilities for natural resource and environmental management including biodiversity conservation. This is particularly true in arid and semi–arid zones where activities are directly influenced by macroeconomic policies such as exchange rates, trade policies, and subsidies.
3. Market failures: More often than not, market failures lead to unsustainable patterns of resource use and consumption. More work is needed in market reforms to make biodiversity products more competitive through such measures as the repeal of restrictive licensing rules and the creation of certain incentives. The poor populations of the arid and semi–arid zones are often unable to participate in markets due to their inability to meet the quality and supply requirements. Market reforms need to include ways and means to improve the local population production skills, technologies and capacities to allow them to compete in markets. Market reforms also need to address these externalities through developing incentive measures such as regulations and market–based instruments (MBIs) to respond to these externalities.
4. Institution failures: Appropriate institutional and social settings provide a necessary prerequisite for achieving resource conservation and sustainable development. Inappropriate institutional settings will inevitably have the opposite effect. Key institutional considerations include the patterns of laws, policies, tenure, and organizational arrangements that ultimately shape both societal and household decision making and resource utilization. These include international and national policy environments which determine the incentive structures for sustainable development.
5. Unclear and insecure land tenure systems: The absence or uncertainty of tenure systems can be a major deterrent to long–term sustainable resource use patterns. Moreover, inequitable distribution of land and transformation of communal land into de facto open access resources

through nationalization and ambiguities in usufruct rights of land, water, and trees can accentuate poverty and environmental degradation.

6. Short–term versus long–term plans: The combination of market demands and the need to balance various local and international political interests frequently leave governments with little room to maneuver; accordingly, national economic policy frameworks may not provide an enabling environment for sustainable development in those fragile zones. For example, macroeconomic policies may preempt resource flows to low–priority areas and fail to provide incentives to households and communities to undertake long–term investment.
7. Lack of participation and ownership by the local population, particularly women: Without participation and ownership by the proper stakeholders, development often is less effectual. In sub–Saharan Africa and many other parts of the developing world, women are increasingly becoming responsible for decisions involving natural resource management. Biodiversity conservation and natural resource management programs that fail to integrate women into the decision making and management process have not achieved their desired objectives.

5. REMOVAL OF PERVERSE INCENTIVES

According to the Biodiversity Convention (CBD) "An incentive is a specific inducement designed and implemented to influence government bodies, business, non–governmental organizations and local people to conserve biological diversity or use its components in a sustainable manner. Incentives usually take the form of new policies, laws or economic or social programs."

To avoid and or mitigate biodiversity loss, it is important to remove all perverse incentives that encourage unsustainable resource exploitation. Perverse incentives include below cost pricing of agriculture products, and unsustainable use of forestry, water, energy, fisheries, mining and transport. They also include marketing systems that encourage a narrower range of species. The removal of perverse incentives could be a very cost–effective approach to biodiversity conservation and sustainable development. Most of the time such reforms are opposed by special interest groups who benefit from the existing situation.

Perverse incentives also include inappropriate subsidies in developing and developed countries, inappropriate tax relief, inappropriate market restrictions, free distribution of seeds which are not linked to direct relief and safety nets. Lack of economic transparency could have an adverse effect on the market.

6. IMPORTANT ECONOMIC STRATEGIES AND INCENTIVES

Of the most important economic incentives are those that seek to address the underlying causes of biodiversity loss and natural resource degradation rather than the symptoms. They include: (1)well–defined and secure tenure systems and property rights which grantee long–term ownership and thus help resource owners develop long–term and sustainable plans, (2)better economic returns among given alternatives (other things being equal), (3)full participation and clear ownership by the local populations from the decision making to the implementation, (4)alternative income generation activities aimed at reducing pressure on biodiversity and natural resource exploitation, (5)diversification and development of innovations to traditional natural resource exploitation, (6)open access to markets and development of new markets, and (7)developing a general enabling environment.

7. REGULATIONS AND MARKET–BASED INSTRUMENTS (MBI)

Focused and targeted national and local regulations and directives allow authorities to directly impose and enforce activities that are beneficial to biodiversity and natural resource conservation and sustainable development. Regulations include standards, quotas and trade restrictions that are legally binding. Some examples of such regulations are emission standards and game hunting quotas. Economic and trade regulation and market–based instruments (MBIs) may include trade regulations, economically designed taxes and well–planned fees and charges.

Other more common MBIs include: (1)tradable permits for environmental goods and services such as fishing quotas which encourage economic efficiency in resource exploitation, especially when permits distributed to local communities may be traded for cash and so generate income while forcing the most economic use of the resource exploitation; (2)tradable offset taxes and credits, such as recent offset credits developed for carbon trade. (This is related to the Clean Development Mechanism (CDM) which was negotiated under the Kyoto Protocol of the Climate Change Convention.); (3)tradable development rights such as those referring to compensation of resource owners for not exploiting certain natural resource and which include debt for nature swaps; (4)upstream or down stream issues may be solved through some types of negotiated regulations and MBIs that grant the right of the down stream populations while safeguarding the right of the up stream populations; and (5)development and implementation of Environment Impact Assessment (EIA) for every development activity to guarantee conservation and sustainable utilization of natural resources and biodiversity.

8. SUPPORTING REFORMS

In addition to the policies, incentives and MBIs, other supporting reforms are also important. These include: (1)institutional restructuring and realignments; (2)capacity building within the responsible agencies; (3)development and dissemination of information technologies; (4)general education, training and awareness raising; (5)coordination among agencies and elimination of conflicts and avoiding overlaps; (6)internalization and integration of biodiversity issue into national and sectors' planning strategies; (7)development of appropriate research and extension services; (8)participation and ownership of all stakeholders; and (9)use of indigenous knowledge.

Policies, strategies and incentives may generate local and national funding opportunities. Some other forms of direct funding to conserve biodiversity include: (1)regional and sub regional resources, (2)national endowment funds, (3)support from development partners, (4)tradable swaps, and (5)Global Environment Facility (GEF).

9. BIBLIOGRAPHY

Ahmed, Y. J. and Kassas, M. 1987. Desertification: Financial Support for the Biosphere. UNEP, Nairobi.

GTZ. 1998. Land Tenure in Development Cooperation: Guiding Principles. GTZ, Germany.

Hassan, H.M., and Dregne, H.E. 1997. Natural Habitats and Ecosystems Management in Drylands: An Overview. World Bank, Washington, D.C.

OECD. 1999. Handbook of Incentive Measures for Biodiversity: Design and Implementation. OECD, Paris.

Oygard, R., Vedeld, T., and Aunes, J. 1999. Good Practices in Drylands Management, World Bank, Washington, D.C.

Russel, C.S., and Powell, P. T. 1996. Choosing Environmental Policy Tools: Theoretical Cautions and Practical Considerations. Inter–American Development Bank, Washington, D.C.

Steinfeld, H., De Haan, C., and Blackburn, H. 1998. Livestock–Environment Interaction: Issues and Options. World Bank, Washington, D.C.

WWF–MPO. 1999. Socioeconomic Root Causes of Biodiversity Loss: An Analytical Approach Paper for Case Studies.

Young, A. 1998. Land Resources–Now and for the Future. Cambridge University Press, UK.

Chapter 11

CONSERVATION AND SUSTAINABLE USE OF BIODIVERSITY IN THE TRANS–ALTAI GOBI DESERT OF MONGOLIA

Ch. Dugarjav[1] and B. Tsetseg[2]
[1]Institute of Botany and [2]Institute of Biology, Mongolia Academy of Sciences, Ulaanbaatar–51, Mongolia; email mbiol@mobinet.mn

1. INTRODUCTION

From ancient times Mongolians have had a custom to "worship" nature. The first law on the protection of environment was issued in the 12th century when Khaan Tooril of the Khereid aimag took over the protection of the Bogdkhan Mountain, which surrounds Ulaanbaatar, the present capital of Mongolia in the south. At present, there are 48 protected areas in Mongolia, covering 20.5 million hectares, 13.1 percent of the country's territory. Among others they include twelve Strictly Protected Areas (SPA) (Oyungerel 2001). Two of these 12 SPAs are situated in the arid zone of Mongolia. Of these two, the Great Gobi SPA established in 1975 and announced in 1991 by UNESCO's Man and the Biosphere Program as a Biosphere Reserve is of special interest. It is the largest protected area in Asia and the fifth largest one in the world. There are 410 species of plants and 170 species of animals and birds in the area (Gunin et al. 1998). Included in the SPA is the unique Trans–Altai Gobi desert.

The Trans–Altai Gobi desert is a combination of unique Central Asian deserts among which the extremely arid deserts have no analogues anywhere in Asia. In addition to its unique landscape and climate, the Trans–Altai Gobi supports a unique biodiversity as well. For example, the Snow Leopard, Gobi Bear, Asiatic Wild Ass, Wild Mountain Sheep, Goitered Gazelle and the world's only population of the Wild Bactrian Camel occur here. Moreover, several international projects have studied the ecology of this region. Conservation and population recovery projects for the Trans–Altai Gobi's rare and endangered animal species were implemented here. Consequently, this region serves as an excellent case study of improved methodology for conservation and management, and increasing public awareness and involvement of local population in conservation and management programs. It also exemplifies the efforts of international cooperation for the conservation of biodiversity of global significance in arid zones.

2. ECOLOGICAL CONDITIONS

2.1 Topography

Mongolia is situated in Asia between the Russian Federation and the People's Republic of China and covers an area of 1564 thousand square kilometers. It is the seventh largest country in Asia. The arid zone occupies one quarter of Mongolia's territory. Taken together with the semi–arid zone it covers one–third of the territory, known as the Mongolian part of the Gobi Desert. The arid zone lies to the south of the Mongolian Altai Mountains and the Gobi Altai mountain ranges and stretches to the Mongolian border with the People's Republic of China. This zone includes three desert regions: the Dzungarian Gobi, the Trans–Altai Gobi and the Alashani Gobi.

The Trans–Altai Gobi spreads over an area of 9.4 million hectares and lies between the Alashani Gobi and Dzungarian Gobi. It is the most arid region in Central Asia. Based on types of soil and vegetation and climatic conditions, it is divided into three sub–zones: the steppe–desert (the northern part), the true desert (the middle part) and the extremely arid desert (the southern part) (Vostokova et al. 1995). The relief of the Trans–Altai Gobi is a combination of mountain chains, ranges, massifs and valleys. The absolute altitudes vary from 3957 m (the range of Ikhe–Bogd in the Gobi Altai Mountain) to 700 m (depression to the north of the Gobi's Tian–Shan). Landscapes of the Trans–Altai Gobi reflect the major zonal and high altitude belt types of the Gobi Desert.

The Trans–Altai Gobi is unique with its richness in various ecosystems. According to Gunin (1990), there are 43 different ecosystems. Among them, the largest are the extreme arid deserts, which occupy 56.7 percent of the territory. The gravel–stony extreme arid desert (gammada) of Central Asia is widespread and accounts for 17.9 percent of the extreme arid deserts. The second in size are the true desert ecosystems (40.5 percent). The high mountain steppes occur only in the mountain regions and their area does not exceed 1 percent of the region's territory. Besides these, there are ecosystems with additional moisture supplied by subsoil waters and springs. They constitute 3.6 percent of the territory.

2.2 Climate

The climate in the region of the Trans–Altai Gobi is sharply continental. It is characterized with large amplitudes of diurnal and seasonal temperatures. For example, low winter and high summer air temperatures fluctuate from –34 to +40 ^{0}C, while soil temperatures range from –33 to +70 ^{0}C. Low precipitation (20–110 mm per annum, index of dryness –20), their uneven seasonal distribution (more than 80 percent of rainfall in summer) and the prevalence of western and northwestern winds are distinctively characteristics of this region (Gunin and Zolotokrylin 1986). Favorable and unfavorable climatic conditions for the existence of biota alternate cyclically with periods ranging from 5–7 and 10–14 years respectively. During unfavorable years, the level of precipitation does not reach its annual average. There have been some years when almost no rainfall was observed.

2.3 Soil

Soil cover of the Trans–Altai Gobi is weakly developed. There are yellow–brown soils in the north and gray–brown soils in the south. In the extreme arid soils, a light–yellow, dry and porous crust occurs under the compactly laid gravel–stony "cuirass". Within oases hydromorphic solonchaks with salt crusts, meadow–swamp, deeply–humic dark–colored meadow, meadow–takyr–like and meadow–desert sandy soils are common.

Based on moisture types, soils can be divided into three groups: the automorphic soils with typical atmospheric moistening, soils with additional surface moistening and soils with subsoil moistening. Content of moisture in the automorphic soils depends mainly on the quantity and quality of the rainfall. The lowest and annually unstable moisture content characterizes the extreme arid soils. Observations made during the period 1977–1997 recorded the minimum moisture at the depth of one meter as 190–225 m^3/ha and the maximum as 400–450 m^3/ha.

2.4 Water Resources

One of the most important factors for existence of life in deserts is the availability of water. The Trans–Altai Gobi belongs to the inland basin of Central Asia. There are no permanent rivers and lakes. However, due to rainfall temporary streams arise along the dry riverbeds called "saira." These areas are not very rich in water springs. Around 120 water springs are recorded, but most are temporary. Low mineral content (0.7–2.5 g l^{-1}) is a characteristic of most springs. In the absence of natural surface waters, ground waters play a crucial role in the Trans–Altai Gobi. Within this region three types of ground waters are found: subsoil, artesian, and fissure waters.

Livestock receive water mainly from ground waters through simple shallow mine wells. These wells are not easily accessible to wild animals. The only permanent water sources for them are the artesian and fissure waters originating as springs. These are located in intermountain hollows with deep water bearing horizons and in numerous zones of tectonic disturbances. There are also several oases in this region. Oases and springs with high discharge (3000–5000 l per hour or more) such as Ekhiin Gol and Shara–Khulsnii–Bulag are of great importance for supporting biodiversity. These also serve as refuges for the conservation of biodiversity.

Thus, the conditions for formation and functioning of ecosystems in the Trans–Altai Gobi desert are very extreme, giving rise to life adapted to harsh ecological conditions. Crucial factors for survival in the Trans–Altai Gobi are water and temperature.

Considering the importance of understanding arid ecosystem functioning in Mongolia, the Experimental Station of the Mongolian–Russian Biological Expedition was established in the oasis of Ekhiin Gol in 1977. Studies on desert climate, soil, flora and fauna have been carried out at this station along profiles stretching from the Mountain Shine–Jinst through the oasis Ekhiin Gol to the Mountain Tsagaan Bogd; this profile includes all three subzones of the Trans–Altai Gobi desert.

2.5 Economic Importance

The Trans–Altai Gobi region supplies 17.48 percent of the country's cashmere wool, 4.86 percent of camel wool and 18.73 percent of sheep wool (Mongolian Statistical Yearbook 1999). There is some expectation that deserts such as these with rich solar energy and potentially fertile soils may play an even more important role in Mongolia's economy.

3. BIODIVERSITY

Although the number of species, including the endemic species of animals and plants in Mongolia is lower than those in many other countries, the species assemblages and almost pristine ecosystems are rarely found elsewhere. Among these, the biodiversity of the Gobi Desert is exceptional because of its unique assemblage of biota endemic to arid Central Asian mountains and deserts. As mentioned earlier, populations of rare animals like the Wild Camel *Camelus ferus bactrianus*, Gobi Bear *Ursus arctors gobiensis*, and Asiatic Wild Ass *Equus hemionus* inhabit the Trans–Altai Gobi. Moreover, the less disturbed and mostly intact natural arid and semi–arid zones of Mongolia are of global and regional importance as model ecosystems. In terms of conservation, this region is a reservoir of many endangered species in Central Asia. It can thus serve as a donor of endangered and locally extinct species for reintroduction into neighboring ecosystems in need of rehabilitation (Shiirevdamba et al. 1998).

Sinitsin (1956) summarized the early data on the biological diversity of plants and animals in the arid zone of Mongolia. Since 1970, researchers of the Joint Mongolian–Russian Biological Expedition organized that year by the Mongolian and Russian Academies of Sciences have comprehensively explored Mongolia's nature and ecosystems including the arid zone, resulting in a rich collection of data. Short reviews of biodiversity given here are mostly based on these data.

3.1 Flora and Vegetation

The flora of the Trans–Altai Gobi desert is not too rich and varies between 304–410 species of plants depending on annual precipitation (Grubov 1955, Ulziikhutag 1989). The representatives of five multispecies families *Alteraceae, Poaceae, Chenopodiaceae, Fabaceae* and *Brassicaceae* constitute about a half (169 species) of the flora. Families *Asteraceae* and *Poaceae* are typically holarctic. The family of *Asteraceae* is chiefly presented by species of the xerophytic steppe and desert floras. However, the family of *Poaceae* comprises not only representatives of the arid land flora, but also includes representatives of the boreal humid flora. By number of its species, the family of *Chenopodiaceae* is the third among plant families dominating in the region (Grubov 1955). It prevails not only in species number, but many of its species contribute to biocenoses succession. The families of *Fabaceae* and *Brassicaceae* mainly comprise species and genera typical of steppes and deserts. Further, there are a number of families which consist of only 1–3 species (28 out of 45 families). Thus, the taxonomic heterogeneity of the flora indicates that in the historical past the flora of the region had been rich and diverse. The present flora comprises the survivors that adapted to the progressive aridity of the territory.

In the Trans–Altai Gobi, the desert flora accounts only for 45 percent of the total number of plant species. Despite this flora being poor in species composition, it is not young. For instance, species of old, distinctive and systematically isolated genera such as *Ephedra, Zygophyllum* and *Nitraria* grow in the Trans–Altai Gobi desert. Highly endemic genera and species of Central Asia occur throughout the Trans–Altai Gobi. Among the desert species in the Trans–Altai Gobi, two are endemic and six are very rare. These species do not occur elsewhere in the Gobi Desert.

The steppe flora growing in the Trans–Altai Gobi is impoverished characteristic of the Mongolian mountain steppe. The meadow flora constitutes 19 percent of the flora of the Trans–Altai Gobi and is represented by common meadow flora of river valleys. Thus, these plants indicate the presence of water there in the past. Being ecologically identical with meadow flora and also mesophytic, the tugai flora of oases' has different origins and phytogeography. Its basic component consists of species typical of river valleys in the eastern part of the ancient desert Mediterranean land. There are 42 species of weed plants in the Trans–Altai Gobi. The abundance of these species is a result of agricultural practices in oases. Weeds also occur widely in sites where wild animals gather.

Among the flora of this region, there are many useful plants: medicinal herbs and food, technical and decorative plants. For example, approximately 40 species of the Trans–Altai Gobi flora are used in Mongolian traditional medicine. Over 80 percent of the species are fodder plants for different wild animals and birds. Furthermore, the flora of this region is a depository of genes for xerophytic species adapted to the extreme harsh conditions of deserts.

There are all basic types of arid plants in the Trans Altai Gobi: trees, bushes, semi–bushes, perennial and annual grass. Vegetation communities of the desert are dominant in the Trans–Altai Gobi. Communities of meadow type vegetation can be found in wet locations near springs, in river valleys and oases. The special type of vegetation, "tugai" is patchy in the Trans–Altai Gobi. Arboreal and bushy plants of river valleys belong to this type. Thin groves of *Populus diversifolia* growing along "saira" and brushwoods formed by species of the genus *Tamarix* and multi–component brushwoods formed by species of the genera *Lycium* and *Nitraria* also belong to the tugai type. Tugai in combination with the meadow and desert vegetation forms the unique ecosystems of oases in the Trans–Altai Gobi.

The productivity and structure of vegetation communities in the Trans–Altai Gobi depend greatly on the amount of precipitation. In dry years the plant productivity decreases significantly because of the growth failure of annual and biennial plants that account for 40 percent the total number of species. Depending on a given year, the number varies between 8–15 species/m^2. The vegetation cover, in general, is not large and changes from 2.06 to 4.49 percent. The above ground phytomass varies between 133–295 kg/ha (Kazantseva 1988).

3.2 Fauna

The fauna of the Trans–Altai Gobi desert is unique and is of global importance. There are 16 species of reptiles (Munkhbayar and Borkin 1990), 159 species of birds (Sumiya 1995), 22 species of small mammals (Rogovin et al. 1986) and 11 species of large mammals (Sokolov and Jirnov 1986). In the region of oases, 886 species of insects, over 70 species of other soil invertebrates and eight species of armored ticks have been described (Ulypkhan 1981, Yanushev and Druk 1983, Bayartogtokh 1995).

In contrast to many other regions of Central Asia, the Trans–Altai Gobi is still inhabited by large mammals such as the Asiatic Wild Ass (*Equus hemionus* Pall.), Goitered or Black–tailed Gazelle (*Gazella subgutturosa* Guld.), Gobi Bear (*Ursus arctos gobiensis* L.), Wild Bactrian Camel (*Camelus ferus bactrianus* Prz.), Dhole or Red or Asiatic Wild Dog (*Cuon alpinus* Pall.) and Snow Leopard (*Uncia uncia* Schr.). At present, the Wild Horse is considered to be extinct and others are rare. They are all recorded in the Red Books of IUCN and Mongolia. To protect these threatened species, the Great Gobi Strictly Protected Area was established in 1975. Large wild animals of desert ecosystems are of a great research interest and therefore have been the subject of numerous publications (Bannikov 1954, Bold and Dulamtseren 1981, Sokolov and Jirnov 1986, Sokolov et al. 1996).

3.3 Microorganisms

In low–productive ecosystems like deserts where there is sparse vegetation and almost no litter, microorganisms and the lower plants (lichens and algae) play an important role in production processes. Unfortunately, knowledge on the diversity of microorganisms and lower plants in Trans–Altai Gobi desert is very limited. Some data are available on cyanobacteria, actinomycetes and algae.

Cyanobacteria with their ability to grow in extreme environments and fix atmospheric nitrogen are frequently encountered in desert zones (Schlegel 1993). Studies in the extreme arid saxaul desert formed on gray–brown gypsum–bearing soils and takyr–like soils of the Trans–Altai Gobi revealed that cyanobacteria were dominant. Fifteen species of cyanobacteria and 13 species of algae were recorded in the saxaul desert. In takyr–like soils, 14 species of cyanobacteria and 16 species of algae were recorded. *Chlorococcales* present in saxaul desert were absent in takyr–like soils and the diatom flora was also relatively rich in the latter (Novichkova–Ivanova 1980, 1988). Annual net primary productivity of hypolithic cyanobacteria in the extreme arid region of the Trans–Altai Gobi was estimated to vary between 43.96 and 706.27 mg $CO_2 \cdot m^{-2} \cdot year^{-1}$ (Allen et al. 1996).

In water samples taken from springs located in the Ekhiin Gol oasis, 42 species and subspecies of diatoms and green and euglenoid algae comprising 22 genera of 12 families were recorded. Tsetsegmaa (1986) registered 12 species and subspecies of cyanobacteria belonging to eight genera placed in five families.

Actinomycetes are one of the most widely distributed microorganisms in desert soils of Mongolia (Tsetseg 2002). Their spores (especially those of streptomycetes) resist drying thereby helping survival in extreme environments. Comparison of actinomycete populations the Dzungarian Gobi, the Trans–Altai Gobi and the Alashani Gobi revealed that the greatest numbers are in the soils of Dzungarian Gobi. The lowest numbers were observed in the extreme arid gray–brown soils of the Trans–Altai Gobi, which are extremely poor in humus (0.09 percent) and moisture (approximately 2 percent). Actinomycete populations decrease as the humus and moisture contents in soils decrease. They account for 30.2 to 33.5 percent of the total number of microorganisms in mountainous chestnut soils, 13.3 to 33.0 percent in desert–steppe brown soils, 6.2 to 20.0 percent in desert gray–brown to brown soils, and 4.7 to 10.9 percent in extreme arid gray–brown soils (Tsetseg et al. 1994, Tsetseg 2002). Unlike in the soils of forest–steppe and steppe zones, pigmented strains of actinomycetes are common in arid soils. In general species diversity in arid soils is higher than that of forest–steppe soils.

4. CONSERVATION AND SUSTAINABLE ACTIVITIES

4.1 Ecosystem Disturbance and Loss of Biodiversity

Mongolia could keep significantly large areas of intact natural environments because of a small population comprising approximately 3 million people with high concentrations in several large cities. Animal husbandry based on traditional mode of nomadic cattle breeding and traditional respect of Mongolians to nature play an important role in conservation. A great deal of these traditional values has been lost during the period of collectivization of herds.

Mongolia's arid zone is in the almost intact areas of the country. Nevertheless, some negative changes have been observed. The changes are: habitat alteration and loss; over–exploitation, nonnative species and disease introduction, pollution or toxic contamination of the soil, water and atmosphere, and climate change (Watson et al. 1995, Stedman–Edwards 1998). In Mongolia, over–exploitation mostly means over–grazing because of a large animal husbandry sector based on natural pastures. Approximately 30 million heads of livestock are grazing on 128 million hectares of pastures. In arid regions, the competition for water resources also is the most important factor influencing biodiversity loss. Since 1990, economic difficulties faced by Mongolia during its transition to a free–market economy and socioeconomic factors such as poverty and profit–seeking exploitation of nature have led to environmental degradation and biodiversity loss. Analyses on causes of ecosystem degradation and loss of biodiversity are presented in reports and methodological recommendations issued by the Mongolian–Russian Biological Expedition (Sokolov et al. 1983, Sokolov and Shagdarsuren 1989, Sokolov et al. 1990a, Sokolov et al. 1990b, Gunin et al. 1996) Data also can be found in Mongolia's first report on biodiversity (Shiirevdamba et al. 1998) and in the Mongolia's project "Conservation of the Great Gobi Desert and it's keystone species" (2000). However, some important information is presented here.

In the Trans–Altai Gobi region, especially in its more populated northern part, habitat alterations due to over–exploitation and climate change have played a crucial role in biodiversity loss and ecosystem degradation. According to the UNDP–GEF project (2000), the Great Gobi SPA is under increasing threats from: (1)habitat degradation along its northern border and throughout its buffer–zone, and (2)loss of ecosystem integrity due to mismanagement of limited water resources. There are also direct and indirect losses of important biodiversity as in cases of the Wild Bactrian Camel, Gobi Bear and Argali Sheep. However, the main causes for species decline are not clear at present.

The transition to a free market economy undertaken by Mongolia since 1990 has caused a significant level of poverty that has forced people to search for livelihood by exploiting natural resources. As a result, habitat degradation expressed as over grazing and over collection of trees and shrubs for fuel wood has dramatically increased in the SPA.

Factors causing the deterioration of ecosystems and loss of biodiversity are divided into two major groups: natural and anthropogenic. The most important natural factor is the intensified process of desertification. It is expressed as further aridization, soil degradation and salinization. The most important anthropogenic factors are over grazing, over collection of trees and shrubs, roads, cattle breeding and agriculture.

Over–grazing is responsible for the negative impact of human activity in the northern part of the Trans–Altai Gobi. As a result, only 9 percent of this territory is not disturbed. In the true desert and in the extreme arid desert, 45 and 91 percent of territories respectively are undisturbed. In the extreme desert, low productivity of pastures, the protection offered by Great Gobi SPA and its location near the frontier are all reasons for the large area of undisturbed land. Over collection of trees and shrubs is less important here, but the impact is important in areas with saxaul and gallery–like forests, and shrubs. The total area occupied by saxaul ecosystems is about 47 thousand square kilometers. At present 43 percent of it has deteriorated. Over collection of trees and shrubs also causes soil erosion.

In the oases of deserts, irrigated agriculture with application of mineral fertilizers is practiced. It has caused significant negative changes with reference to the formation of hydromorphic ecosystems with weeds and salinization of soils. For example, most part of soils in the oasis, Ekhiin Gol, where irrigated agriculture was developed is suffering from salinization. The use of all water springs for irrigated agriculture has resulted in the loss of globally important biodiversity. Roads and cattle breeding activities also have local impacts but are considered as less important in large areas.

4.2 Conservation and Sustainable Use Activities

In order to implement the recommendations of the CBD, Mongolia has developed the following activities: (1)institutional strengthening by renewal of old and development of new legislation; (2)organization of new protected areas, strengthening of old ones and establishment of protected areas network; (3)reintroduction of extinct species of animals and plants; (4)reforestation; (5)educating the public and making people aware of the values of biodiversity, its conservation and sustainable use; (6)active involvement of the scientific community and broad application of research results; (7)involvement of NGOs and private sector in the nature protection activities; (8)development of international cooperation and procurement of international funds, and (9)local benefit sharing programs.

All of the above activities are taking place in the Trans–Altai Gobi desert. However, four activities are the most important that played a crucial role in improvement of biodiversity conservation and sustainable use. These are: (1)renewal of legislation, (2)establishment and strengthening of the Great Gobi SPA, (3)development of international cooperation and procurement of international funds, and (4)active involvement of scientists and broad application of research results. These activities in close relation with each other and influencing each other would help further development of conservation activities in the area. Results achieved by these activities and lessons learned in the field of public policy and management, scientific research and assistance to local populations for the utilization and sustainable management of biodiversity resources are summarized below.

4.3 Public Policy and Management

Up to 1990, Mongolia had pursued a policy of intensive utilization of its natural resources for the acceleration of economic development. Nature protection was not a priority of the Government. Legislation concerning the use of natural resources was applied to a few cases only. Mongolia had no a clear long–term policy for conservation of biodiversity, the environment, restoration of disturbed environments and the sustainable use of natural resources (Shiirevdamba et al. 1998). The major management style in nature protection was the establishment of strictly protected areas and reserves and restriction of all human activities in them.

The Earth summit held in 1992 in Rio de Janeiro played a crucial role in raising the awareness amongst Mongolia's decisionmakers, the importance and necessity of biodiversity conservation and sustainable use of natural resources. In 1992, the Ministry for Nature Protection was transformed into the Ministry of Nature and the Environment. In 1993 Mongolia signed and ratified the CBD. In subsequent years, Mongolia has also signed and ratified the following globally important documents in the field of biodiversity conservation: the "Framework Convention on Climate Change" (1994), the "Convention on International Trade in Endangered Species of World Fauna and Flora" (1996), the "Vienna Convention for the Protection of the Ozone Layer" (1996), the "Montreal Protocol on the Substances that Deplete the Ozone Layer" (1996), the "UN Convention on Combating Drought and Desertification (1996)", the "Basel Convention on the Control of Transboundary Movement of Hazardous Wastes and their Disposal" (1997), the "Convention on Wetlands of International Importance Especially as Waterfowl Habitat" (1998), and the "Convention on Migratory Animals (1999)."

At the same time, the Ministry also started formulating the policy for Mongolia's nature protection and biodiversity conservation and its sustainable use within the country. The new economic system in the country and new approaches to the nature conservation underlined in the CBD had demanded not just a new policy, but also a new style of management and legislation to support that policy. Due to this, the development of new legislations and the updating old regulations were also started. Basic guidelines of the new policy are reflected in such directives as the "Mongolia's National Security Policy Orientation, 1995", the "National Development Strategy, 1996" and the "Ecological Policy Orientation of the Mongolian State, 1997". The essence of these documents are expressed in detail in the corpus of environmental laws, the National Action Plan for the Conservation of Biodiversity and the Program on Combating Desertification and the National Action Plan for Special Protected Areas.

The most important laws were the following: the Mongolian Law on Nature Protection, the Law on Special Protected Areas, the Law on Hunting, the Law on Plant Protection and the Mongolian Law on Forests, which were developed and approved during 1994–1998. These have provided the basis for all activities related to conservation, sustainable use and restoration of biological diversity and environments supporting them. For example, as a result of legislation, the harvesting of 133 endangered plant species has been prohibited. Saxaul forests, 60 percent of which has disappeared during the last 50 years, have started recovering naturally in some regions of the Gobi Desert due to the legal prohibition preventing their use. Furthermore, in Mongolia there are more than 20 laws and 180 regulations and resolutions that clearly reflect the duties and power of each administrative unit on environmental and biodiversity protection and conservation.

To develop international collaboration with neighboring countries in the field of protection and sustainable use of biological resources, Mongolia has signed agreements with the Russian Federation and People's Republic of China. Another important step in conservation policy and management was the strengthening of old and the establishment of new protected areas. Mongolian classification of protected areas was updated based on international classification of IUCN. Presently, in Mongolia there are four categories of protected areas: Natural and Historical Monuments, Nature Reserves, National Conservation Parks, and Strictly Protected Areas. Status of some protected areas was changed according to the new classification. The network of protected areas was established and a Special Protected Areas Bureau was set up to govern it. As a result, the number of protected areas has increased fourfold since 1990 covering 13 percent of the country's territory. However, there is a future plan to put about 30 percent of the territory under protection

Establishing protected areas networks has allowed the protection of 40 percent of habitats supporting more than 400 endangered or threatened plant species. Along with this, the strengthening of the Great Gobi SPA brings under protection the largest part of the fragile Gobi ecosystems that experience a continental climate and are the habitats for rare and endangered species. At present it covers fully the range of such endangered and extinct species as the Wild Camel, Gobi Bear, Wild Horse, Saiga Antelope and about 70 percent of the range of the Snow Leopard, Wild Ass and Wild Mountain Sheep (Shiirevdamba et al. 1998, Oyungerel 2001).

Policy aimed at the development of international cooperation and involvement of international funds has substantially helped enhancing all activities for nature protection in Mongolia. It was used to improve public awareness, conserve and restore endangered and rare animals and plants, strengthen the protected areas and assist local communities in earning their livelihood. For example, in the last decade several research and management projects supported by UNDP–GEF, the Wildlife Conservation Society, David Shepherd Conservation Foundation, Denver Zoological Foundation, Nature Conservation International and the Wilds have been implemented in the Great Gobi SPA. They have contributed significantly to the strengthening of the SPA's facilities.

The UNDP–GEF Mongolian Biodiversity Project (MBP pilot phase) built on the work of the UNEP project "Assistance to the Mongolian People's Republic in the Establishment of the Great Gobi National Park" implemented from 1979–1982 assisted the further development of the Great Gobi SPA. The main achievements of the project for the SPA were the development of preliminary model management plans for the Wild Camel and Gobi Bear, and captive breeding program for the Wild Bactrian Camel. At present, a new UNDP–GEF project titled "Conservation of the Great Gobi Desert and its Keystone Species" is going to be implemented. This project is based on new concepts and approaches developed in the CBD and its follow up such as the active involvement of local communities and benefit sharing, increasing the public awareness on biodiversity and nature protection issues. It will provide support for barrier removal in the sensitive buffer zone areas, more community participation, better awareness and information dissemination and increased technical capacity to implement modern ideas of integrated conservation and management in and around the protected area. Collaboration with Mazhongshan PNR situated in the Chinese People's Republic is also envisaged. Thus, the Great Gobi SPA established in 1975 as a refuge for rare animals of Central Asian deserts will gradually become a model SPA for research work, teaching, and ecotourism. It will also serve as a depository of rare animals and plants of intact and fragile arid ecosystems. During the conservation activities, it was apparent that the personnel of the Ministry and protected areas had no modern management skills and experience. In 2000, a project proposal aimed at the training of personnel in environmental management was discussed with the World Bank.

As a result of international cooperation and active participation of foreigners such as Ch. Oswald Foundation from Germany and Boumann Foundation from the Netherlands and the Dutch Government, the Wild Horse was reintroduced. In 1992, twenty wild horses were taken from Europe to Khustain Nuruu and Takhiin Tal in the Great Gobi SPA. In 1997, the horse population increased to 100 individuals.

4.4 Scientific Research and Lessons Learned

In order to conserve and use biodiversity in a sustainable manner and involve local populations into these activities, it is essential to have a basic understanding on ecosystems' functioning, their components, and the social and economic pressures that affect them. All of these can be achieved through scientific research and dissemination of knowledge generated.

The scientific community of Mongolia has played a crucial role in the evaluation of biological resources, ecology, inventory and monitoring, and environmental impact assessments and in conservation activities. It has prepared scientific grounds for all environmental laws and national programs for nature conservation, the Red Book of Mongolia (1997), reports of different local and international projects. These researchers have also participated in teaching programs on ecology and public awareness on conservation of biodiversity. Close cooperation between local and foreign researchers on one side, and between them and the Government of Mongolia, on the other has promoted the development of effective and fruitful measures for conservation and sustainable use of biodiversity and biological resources in Mongolia.

Essential role in this regard belongs to the Mongolian–Russian Complex Biological Expedition (MRCBE) that has explored Mongolia's biodiversity and environments during the last 30 years. Major activities have been aimed at: (1)the inventory of flora and fauna of main natural zones of Mongolia, (2)the development of measures for conservation and sustainable use of biodiversity; and (3)the monitoring of environments and ecological assessments. Based on the research results and recommendations of the MRCBE, the Great Gobi SPA was established in 1975. In subsequent years as a result of scientific research, the following protected areas were established in the desert region of Mongolia: Small Gobi SPA (1996), Natural Park Gobi Gurvansaikhan (1993) and Natural Reserves Sharga Mankhan (1993), Alag Khairkhan (1996), Zagiin Us (1996), Ikh Nart (1996), Ergeliin Zoo and Burkhan Buudai (1996).

From 1978–1982, Mongolian and Russian scientists carried out the ecological exploration of the Trans–Altai Gobi desert within the UNEP project "Assistance to the Mongolian People's Republic in the Establishment of the Great Gobi National Park." Data collected provided a valuable ecological baseline for evaluation of ecological status of the area and informed the development of a strategy for protection of endangered and rare species. Furthermore, the Expedition's research data supported the implementation of the UNDP–GEF Mongolian Biodiversity Project as well. Active participation of scientists is also envisaged in a new UNDP–GEF project titled "Conservation of the Great Gobi Desert and Its Keystone Species."

As a result of the Expedition's research, 20 different maps, 70 monographs and more than 4000 papers have been published in Russia, Mongolia and other ex–socialist countries. Their titles can be found in Dorofeyuk and Gunin (2000). The Expedition's publications have provided important references for the development of scientific justification for nature conservation and its sustainability. However, most scientific publications were made either in Russian or Mongolian and therefore, are not readily available to western scientists.

Despite the contribution of scientific research to biodiversity conservation and sustainable development, research institutions in Mongolia have not received any financial support to upgrade and strengthen their laboratory and field study facilities since 1990. Largely scientists have been using their own initiative and enthusiasm to conduct research. Unavailability of sophisticated facilities and methods has caused significant difficulties in the application of modern techniques to study biodiversity.

4.5 Assistance to Local Populations

Assistance to local populations in utilizing and managing biodiversity and using resources sustainably is a new approach to conservation of biodiversity in Mongolia. As mentioned earlier all human activities were restricted to protected areas. Shooting and collecting endangered species of animals and plants respectively were prohibited. There was no community participation in conservation activities. However, during the last decade some progress has been made towards the involvement of local populations in conservation activities and raising public awareness on sustainable use of bioresources.

In 1997, the "Mongolian Law on Buffer Zones of Protected Areas" was adopted. According to this law, a buffer zone of 1.8 million ha surrounding the Great Gobi SPA has been established. About 500 families are involved in grazing and small–scale agriculture in this zone. Poverty and profitable use of natural resources have become the main causes for the deterioration of ecosystems. To decrease these negative impacts several local benefit sharing programs were set up. For instance, the UNDP–GEF's Mongolian Biodiversity Project and the German Technical Assistance and Cooperation Agency's project titled "Environmental Protection and Buffer Zone Development" established the Small Projects Funds in order to increase the support for local communities within and around protected areas by improving their living conditions. At present, four small projects supported by Mongolian Biodiversity Project have been successfully implemented within and around the Great Gobi SPA: These are "Shower Facilities" in social service, "Kindergarten" in education service, "Solar Energy" in environment protection and, "Hand Made Carpet Factory" as a supplementary income project. Along with these, several workshops and teaching programs were conducted to increase public awareness of local community. Scientists have helped in teaching conservation and sustainable use of biodiversity to SPA's staff. They have provided recommendations on sustainable use of pastures, scientific information on globally important species of animals and plants, changing of ecosystems and number of species under threat from human activities and desertification. Scientists are also preparing recommendations on restoration of degraded ecosystems, preparation of fodder for wild animals to be used in winter and prediction of further changes in ecosystems.

4.6 Capacity Building and Partnerships

Mongolia has been confronted by a scarcity of information on the status of it biodiversity resources. This has a direct impact on capacity building and partnerships between institutions both locally and internationally. However, there is an expectation that a new UNDP–GEF project titled "Conservation of the Great Gobi Desert and Its Keystone Species" will help to identify problems of common concern and develop multi– and bilateral cooperation to strengthen research facilities, especially with the assistance of international funds. In Mongolia, there is no hope that sufficient internal funds will be available for biodiversity conservation in the near future. However, there is a wealth of essential research experience on natural environments and biodiversity of arid and semi–arid zones. This valuable potential in Mongolia could be harnessed to establish an International Training Center for environmental studies in arid and semi–arid zones under the auspice of the GEF and UNEP. The Center can provide research facilities, training and develop methodology for the biodiversity assessment of arid and semi–arid ecosystems and their sustainable use. This Center should employ international experts and eminent foreign scientists to enhance capacity building and worldwide dissemination of knowledge.

5. CONCLUSION

Mongolia, an Asian developing country with vulnerable ecological systems and unique biological diversity, has endeavored to develop integrated measures for the conservation and sustainable use of biodiversity in a unique region of the world, the arid Trans–Altai Gobi. A comprehensive account of the experience gained, especially in the last decade is given here. It is apparent that this achievement would not have been possible without international assistance.

The experience gained has demonstrated that the socioeconomic roots of biodiversity decline and deterioration of ecosystems are most difficult to overcome. Major difficulties encountered were the lack of modern research facilities and the shortage of researchers. Since 1990, Mongolian researchers had suffered from the shortage of government funding to conduct field and laboratory studies. Mongolia will not able to solve these problems by itself unless there is international cooperation and financial assistance.

In summary, Mongolia has learned the following lessons: (1)understanding what sustainable development means to the country and realizing that the conservation and use of natural resources in a sustainable manner is the basis for the existence of humankind; (2)the need for new policies, legislation and management style; (3)people without livelihoods are not interested in the protection of biodiversity; (4)communities uneducated in ecology and biodiversity values are not active in protection and sustainable development; (5)professionals, especially scientists and university lecturers, of ex–socialist countries need to be acquainted with modern theories of conservation biology and taxonomy in order to help them to integrate their previous and new knowledge and experience and use it with maximum effectiveness; (6)profit–seeking private firms and other similar organizations are not interested in promoting sustainable use of natural resources; (7)proclamation of some territories as protected areas without strengthening the educated manpower and adequate facilities will fail to conserve biodiversity and ecosystems.

Based on these findings, Mongolia is seeking solutions. At present, Mongolia has completed the preparatory stage that can be termed as "the legislation period." The majority of old laws have been updated and some urgent new laws have been developed and approved. In the next century the implementation period should start. Mongolia should make efforts to solve, decrease and if possible, eradicate the socioeconomic roots of the environmental deterioration and loss of biodiversity. This will help the country to make significant achievements in conservation of its unique biodiversity and use of natural resources in a sustainable manner. However, international expertise, experience and financial support will be needed for achieving these objectives.

6. ACKNOWLEDGMENT

We would like to express our gratitude to Mr. H–P. Knegt, Programme Officer of the UNDP–Mongolia, for supplying us with project documents.

7. REFERENCES

Allen, M.E., Friedmann, E.I., McKay, C.P., Meyer, M.A., Druk, A.Y., Kappen, L. and Tsitsigma, Y. 1996. Cyanobacterial primary productivity in extreme arid regions of the Gobi desert. N–31 In: Abstracts of the 96th General Meeting of the American Society for Microbiology. 19–23 May 1996. Ernest N. Morial Convention Center. New Orleans, Louisiana.

Bannikov, A.G. 1954. The Mammals of the Mongolian People's Republic. Publishing House of the USSR's Academy of Sciences, Moscow. 671 pages (in Russian).

Bayartogtokh, B. 1995. Ticks of the Genera *Acariformes* and *Oribatei* in Soils of Mongolia. The Conspectus of the PhD Thesis. Moscow. 37 pages (in Russian).

Bold, A., and Dulamtseren, S. 1981. Rare wild animals and birds in Mongolia. Journal of Science and Life 6: 25–27 (in Russian).

Dorofeyuk, N.I., and Gunin P.D. 2000. A Bibliographic List of Publications on Research Results of the Joint Russian–Mongolian Complex Biological Expedition of the Russian and Mongolian Academies of Sciences, 1967–1995. Biological Resources and Natural Conditions of Mongolia: Proceedings of the Joint Russian–Mongolian Complex Biological Expedition v. 41. Bioinform service, Moscow. 386 pages.

Grubov, V.I. 1955. The Conspectus of Flora of the Mongolian People's Republic. Proceedings of the Mongolian Commission of the USSR's Academy of Sciences, issue 7. Publishing House of the USSR's Academy of Sciences, Moscow–Leningrad. 307 pages (in Russian).

Gunin, P.D. 1990. Ecology of Desertification Processes in Arid Ecosystems. Publishing House of the All–Union Academy of Agricultural Sciences by V.I. Lenin, Moscow. 354 pages (in Russian).

Gunin, P.D., and Zolotokrylin, A.N. 1986. General characteristics of climate. Pages 27–29 In: Deserts of the Trans–Altai Gobi: Natural Conditions, Ecosystems and Regioning. Nauka Publishing House, Moscow (in Russian).

Gunin, P.D., Dugarjav, Ch., Kuzikov, I.V., Krasnoshokov, Yu.N., Rudnev, N.I., and Erdenejav, G. 1996. The principles of preservation of species diversity in specially protected areas (on example of the Orhon–Selenga part of the Lake Baikal basin). Pages 14–15 In: Korsunov, V. M. (Editor) The Conservation of Biological Diversity in the Baikal Region: Problems, Approaches and Practices. Proceedings of the 1st Regional Conference, 14–16 May, 1996. Buryatia Research Centre, Siberian Branch of the Russian Academy of Sciences, Ulan–Ude, v. 1 (in Russian).

Gunin, P.D., Vostokova, E.A. and Matyushkin, E.N. 1998. Protection of Ecosystems of Inner Asia. Proceedings of the Russian-Mongolian Biological Expedition, v. XL. Nauka Publishing House, Moscow. 219 pages (in Russian).

Kazantseva, T.I. 1988. Productivity of the dominant plants of deserts of the Trans-Altai Gobi. Pages 188–201 In: Deserts of the Trans-Altai Gobi. Nauka Publishing House, Leningrad. (in Russian).

Mongolian Statistical Yearbook. 1999. Ulaanbaatar. 222 pages (in Mongolian).

Munkhbayar, Kh. and Borkin, L.Ya. 1990. On taxonomic position of lizards of Gobi's Tian-Shan and Mongolia. Pages 143–174 In: Proceedings of the Zoological Institute of the USSR's Academy of Sciences. "Nauka" Publishing House, Leningrad (in Russian).

Novichkova–Ivanova, L.N. 1980. Soil algae of phytocoenoses of the Sahara-Gobi desert region. "Nauka" Publishing House, the Leningrad branch, Leningrad. 255 pages (in Russian).

Novichkova–Ivanova, L.N. 1988. Epilithic algocenoses of the Trans–Altai Gobi Desert. Pages 276–282 In: Gubanov, I.A., Dorofeyuk, N.I. and Neronov, V.M. (Editors) Natural Conditions, Vegetation Cover and Fauna of Mongolia. Collection of Papers. Puschino (in Russian).

Oyungerel, B. 2001. Ecological and Geographical Problems of Protected Areas of Mongolia. The Conspectus of the PhD Thesis. Ulaanbaatar. 40 pages (in Russian).

Red Book of Mongolia. 1997. Second edition. ADMON Printing House, Ulaanbaatar. 389 pages (in Mongolian).

Rogovin, K.A., Kulikov, V.F., Surov, A.V. and Vasil'eva, N.F. 1986. Small mammals. Pages 62–65 In: Deserts of the Trans-Altai Gobi. Nauka Publishing House, Moscow (in Russian).

Schlegel, H.G. 1993. General Microbiology. 7th edition, University Press, Cambridge. 655 pages.

Shiirevdamba, Ts. et al. 1998. Biological Diversity in Mongolia (First national report). Ministry for Nature & the Environment. ADMON Printing House, Ulaanbaatar. 106 pages (in English and Mongolian).

Sinitsin, V.M. 1956. The Trans–Altai Gobi. Publishing House of the USSR's Academy of Sciences, Moscow-Leningrad. 166 pages (in Russian).

Sokolov, V.E., and Shagdarsuren, O. (Editors). 1983. The Complex Characteristics of Desert Ecosystems of the Trans-Altai Gobi (On Example of the Desert Station and the Great Gobi National Park). Collection of Papers. Puschino. 114 pages (in Russian).

Sokolov, V.E. and Jirnov, L.V. 1986. Large mammals. Pages 135–150 In: Deserts of the Trans-Altai Gobi. Nauka Publishing House, Moscow (in Russian).

Sokolov, V.E., Shagdarsuren, O., Kamelin, R.V., and others (Editors). 1989. Techniques for Development of Maps on Ecosystems and Their Anthropogenic Disturbance for Model Ranges and Stationary Sites with Use of Remote Sensing and Satellite Data (Methodical Recommendations). Ulaanbaatar. 31 pages (in Russian).

Sokolov, V.E., Davaajamts, Ts. and Kamelin, R.V. 1990a. Methodological Problems of Assessment of Conditions of the MPR's Environments. Theses of Reports. Puschino. 133 pages (in Russian).

Sokolov, V.E., Shagdarsuren, O., Kamelin, R.V., and others (Editors). 1990b. Methods of Assessment and Cartography of Modern Status of Desert Ecosystems. Ulaanbaatar. 52 pages (in Russian).

Sokolov, V.E., Orlov, V.N., and Lobachev, V.S. 1996. Class Mammalia. Pages 121–182 In: Rare Animals of Mongolia (Vertebrates). The Institute of Problems of Ecology and Evolution by Severtsov A.N., Moscow.

Stedman-Edwards, P. 1998. Root Causes of Biodiversity Loss. An Analytical Approach. WWF for Nature, April 1998. 86 pages.

Sumya, Ts. 1995. Notes on study of birds in some oases of the Great Gobi National Park. Pages 76–86 In: The Natural Conditions and Biological Resources of the Great Gobi National Park. Ulaanbaatar (in Mongolian).

Tsetseg, B., Daram, D., Onhor, G., and Puntsag, T. 1994. Distribution of streptomycetes in Mongolian Gobi desert soils and their biological activity. Page 229 In: Theses of the 9th

International Symposium on the Biology of Actinomycetes, 10–15 July 1994. All–Union Research Institute of Genetics and Selection of Industrial Organisms. Moscow.

Tsetseg, B. 2003. Unique Microbiota of Mongolia. Pages 196–221 In: Dendeviin, B., Zilinskas, R.A. and Balint, P.J. (Editors) Mongolia Today: Science, Culture, Environment and Development. Routledge Curzon, London and New York.

Tsetsegmaa, D. 1986. Some results of study on algae of the MPR. Page 116 In: Natural Conditions and Biological Resources of the Mongolian People's Republic. Abstracts of Reports of the International Conference, October 1986, Moscow. Nauka Publishing House, Moscow (in Russian).

Ulipkhan, K. 1981. Soil invertebrates. Pages 113–119 In: Desert–Steppes and Northern Deserts of the MPR. Nauka Publishing House, Leningrad (in Russian).

Ulziikhutag, N. 1989. Synopsis of the Mongolian Flora. Ulaanbaatar. 205 pages (in Mongolian).

UNDP–GEF Project "Conservation of the Great Gobi Desert and it's Keystone Species". 2000. Ulaanbaatar.

Vostokova, E.A., Gunin, P.D., Rachkovskaya, E.I. 1995. Ecosystems of Mongolia: Distribution and Present Condition. Biological Resources and Natural Conditions of Mongolia. Proceedings of the Joint Russian–Mongolian Complex Biological Expedition, vol. 39. Nauka Publishing House, Moscow. 223 pages (in Russian).

Watson, R.T, Heywood, V.H., Baste, Y., Dias, B., Gamez, R., Janetos, T., Reid, W., and Ruark, G. 1995. Global Biodiversity Assessment. Summary for Policy–Makers. UNEP. Cambridge University Press.

Yanushev, V.V. and Druk, A.Ya. 1983. Soil animals. Pages 71–72 In: Sokolov, V.E. and Shagdarsuren, O. (Editors) The Complex Characteristics of Desert Ecosystems of the Trans–Altai Gobi (On Example of the Desert Station and the Great Gobi National Park). Collection of Papers. Puschino (in Russian).

Chapter 12

ECOLOGICAL ASSESSMENT OF DEGRADATION PROCESSES IN THE MONGOLIAN PART OF BAIKAL BASIN

P.D. Gunin and S.N. Bazha
A.N. Severtsov Institute of Ecology and Evolution, Russian Academy of Sciences, Russia; email pgunin@online.ru

1. INTRODUCTION

Despite the declaration of Lake Baikal as a UNESCO World Heritage Site in 1996, the environmental quality of the lake and its surrounding ecosystems continues to decline. The southern boundary of the Baikal territory is the state border between Russia and Mongolia. However, this is a political boundary between the two nations and environmental quality of the Lake Baikal area can only be achieved under international cooperation (RFL 2002).

About 52 percent of the Baikal watershed area and over two thirds of the main artery of the Selenga river is in the central and northern part of Mongolia. This region of Mongolia is one its most densely populated and industrialized, and this creates conflicts between economic development of the region and enhanced conservation of resources. Mitigation of the conflicts requires better understanding of ecological conditions in the Mongolian part of the Lake Baikal basin.

In our case study, we describe the results of an inventory of the ecosystems of the southern (Mongolian) part of the Lake Baikal basin at the level of mesoecosystems; we have classified the ecosystems according to their economic use and level of modification by humans. We also describe the spatial distribution of ecosystem degradation processes and the levels of their impacts on the biota of both the Lake Baikal territory within Mongolia as well as in contiguous areas in Russia. The data can be used in the monitoring of anthropogenic pressures on the ecosystems of the Lake Baikal basin. For control purposes, we set aside plots for establishing a transborder reserve in the central part of the Selenga River.

Our study had three objectives: (1)to conduct an inventory of the ecosystems of the southern part of the Lake Baikal basin and classify them according to economic utilization and modification by humans; (2)to analyze processes affecting the Lake Baikal basin's ecosystems and components such as vegetation, soil, topographical relief, etc.; and (3)to assess integrated zoning of the southern part of the basin to help mitigate environmental degradation.

2. ECONOMIC CONDITIONS AND SPATIAL STRUCTURE OF THE ECOSYSTEMS OF THE SOUTHERN LAKE BAIKAL BASIN

2.1 Physiographical Characteristics and Natural Conditions

The basin's total area is 545 thousand km^2 of which 303.9 thousand km^2 resides within Mongolia (National Atlas of Mongolia 1990). In terms of physiographical zoning, the Mongolian part of the Lake Baikal basin is within the Siberian provinces of Khentei, Muren and Eastern Khubsugul, and within the Central Asian provinces of Orkhon, Selenga and Khangai (Murzayev 1952, Preobrazhenskii et al. 1984). Most of our study area is situated in the center of the Asian mainland between 46^0 and 52^{0}N, and 97^0 and 109^{0}E. In the north and northwest the study's boundary coincides with the state border of Russia, and in the remaining portion it passes along the main watershed of Asia dividing the watershed basins of the Arctic and Pacific Oceans, and also the region of the intramainland runoff of the Central Asia. In the south, the study area is bounded by the Khangai mountain range; and in the east, by the Khentei mountain range. The basin's climate is on the whole sharply continental, with a long cold winter, arid and windy spring, hot and relatively wet summer and cool and dry autumn. A characteristic feature of the area's climate is sharp fluctuations of precipitation from year to year (in some places from 50 mm/g in dry years to 500 mm/g in wet years). The basin contains some of Eurasia's highest levels of continental and summer aridity (Vostokova et al. 1995).

The Lake Baikal basin contains several types of soils from sod taiga to chestnut and meadow–chernozem which vary with latitudinal and vertical zonality (Ubugunov et al. 2000).

The floristic composition and plant cover of the southern part of the Lake Baikal basin is characterized by high diversity and ecological uniqueness. It comprises about 2500 species of vascular plants of 600 genera and 100 families (Boikov 1999, Ganbold 1999), which include about 180 endemic species. There are 380 species of mosses and about 450 species of lichens. Geobotanically, the southern part of the Lake Baikal basin is part of the Daurian–Mongolian mountain and forest steppe and the Central Asia steppe subprovinces (Yunatov 1950, Lavrenko et al. 1991).

Natural factors responsible for the high fire hazard of the Lake Baikal basin include extremes of climate and impacts of pest outbreaks. The protracted period of sunlight and high intensity of direct radiation, low precipitation and their irregular seasonal distribution desiccates the soil and dries plant cover. In dry years, the incidence of fires sharply increases, and in the course of the year their greatest number occurs in spring and autumn. The rate of the spreading of fires is a function of wind velocity which typically attains 17–20 m/sec. During dry years evaporation increses and the wind reduces the moisture of the soil and air and promotes the drying of the dead wood and the litter (which function as fuel). The nature of plant cover increases fire hazards since the proportion of flammable plants in the plant communities is high (e.g., coniferous trees like pine and larch, and in the steppes the majority of herbs, mosses and lichens). The probability of fire on forest plots is augmented due to insect pests. On the one hand, the drying and death of pest–affected trees brings about accumulation of fire–hazardous materials, and, on the other hand, changes in microclimate (reduction in relative moisture, increase in wind velocity, desiccation of soil surface horizons). On such plots, the risk of fire sharply increases.

Erosion hazards characteristic of the majority of the landscapes of the Lake Baikal basin are influenced by the properties of orography, climate, soils and plant cover. The strongly dissected relief and heavy precipitation are responsible for the wide distribution of steep slopes which promotes active processes of water erosion. In addition, the valleys bounded by parallel ranges represent natural "wind tunnels" capable of the development of wind erosion and deflation foci. Other natural features influencing wind erosion properties include wind–shifted sandy soils with sparse plant cover and which are relatively less resistant to modification by human activities. A substantial impact on the development of deleterious processes is exerted by thin snow cover in depressions, which causes little moisture accumulation in the soil and does not protect the surface from removal of melkozem and nutrients. Further, the inversion nature of air masses in the basin's depressions in winter creates favorable conditions for stagnation of air masses and accumulation of high concentrations of pollutants in near ground layers.

2.2 Characteristics of Economic Utilization of Natural Resources

Administratively, the Lake Baikal basin comprises: The Metropolitan Region with 4 districts and 12 aimags including 116 somons, 28 of which are separated by the boundary of the watershed basin into unequal parts. Currently, the region's population density is approximately 2.9 people/sq. km and the population is over one million people, accounting for about 60 percent of the total population of Mongolia. The region produces 90 percent of the gross economic output of Mongolia (Erdenzhav 2000).

The Lake Baikal basin area is extensively forested (over 25 percent). The bulk of the forest stocks are spruce forests which account for about 90 percent of the area and 94 percent of the forest timber stock. Concentrated in this region is about 80 percent of all cultivated area of the country, and about 70 percent of cropland. Over 785 thousand ha is concentrated in 5 aimags: Selenge (23.4 percent); Central Tov (3.3 percent); Bulgan (9 percent); Khentei (7 percent) and Zavkhan (6.2 percent). According to agro climate zoning these aimags are situated in agricultural zones with relatively few agriculture policies and lands where overall the cropland area declined 5 to 10–fold and in 2000 it was only 20–30 percent of what it was in the 1970s and 1980s.

During late 1995, Mongolia had approximately 28.6 million head of livestock, including about 2.6 million horses, 3.3 million cattle, 13.7 million sheep, and 8.5 million goats (Mongolian Economy 1996, Graivoronsky 1997). With an area accounting for 19 percent of the country's territory the Selenga River basin had approximately 7.3 million head, which exceeds 25.5 percent of the entire country's livestock.

Analysis of statistical data collected over the administrative subdivisions of the Lake Baikal basin (e.g., aimags, somons) reveals a growth of total livestock from 6.1 million head in 1985 to 8.7 million head in 2000; a particularly significant increase was in goat number (from 672.2 thousand head in 1985 to 2181 thousand in 2000). The estimated number of stock in the basin aimags and their grazing pressure on the rangelands was also accompanied by an increase of livestock heterogeneity in particular aimags.

2.3 Spatial Structure of the Region's Ecosystems and Methodologies for Assessing Modifications of the Lake Baikal Basin

The investigation of the modification of the basin's environment involved three approaches based on: (1)formal principles, (2)principles of administrative division, and (3)landscape and ecological principles.

In the first approach, the boundaries of topographic divisions or grids serve as the object of study. This approach is common in remote study methods; in particular, in the interpretation of quantitative knowledge of the reflection coefficients of the soil–plant cover and its state (Degradation of the Dry Lands of Asia 1999). Taking into account the technical parameters of the data obtained from satellites and stations (e.g., NOAA/AVHRR) the size of each segment affecting the value of the vegetation index is normally equal to 8 km^2; and in case of a higher resolution up to 1 km^2.

Administrative principles are used in the second approach when the objects of assessment are units of the region's political and administrative divisions, i.e., regions and districts for Russia and aimags and somons for Mongolia. According to the administrative division of the Mongolian part of the Selenga River, the number of somons is 116. Administrative principles are normally used in generalizing data on the state of the environment or when preparing annual reports and statistical directories (The Natural Environment of Russia 1995, Nature and Environment in Mongolia 1996, 2000).

In the third approach, landscape or ecological principles are used when the goal of assessment is to try to differentiate the biosphere into small and/or large scale mapping units (e.g., ecosystem types). The studies conducted by the Russian–Mongolian Integrated Biological Expedition, the Russian Academy of Sciences, and the Academy of Medical Sciences, in the territory of Central Asia have been effective where the soil and plant cover has been well studied and when specialized and integrated maps of the regions have been available. The availability of such maps on the area of Mongolia (Ecosystems of Mongolia 1995) has permitted using units of landscape and ecological mapping in determining spatial differentiation of the soil and plant cover in order to indicate the state of ecosystems corresponding to elementary, local and landscape levels of chorological classification of their structure (see, e.g., Vinogradov 2984, Vostokova et al. 1995). As a basic unit we used the level of mesoecosystems, which was formerly used as a basic unit in compiling the map "Ecosystems of Mongolia" (Gunin et al. 1995). This has permitted a more complete inventory of the ecosystems of the southern part of the Lake Baikal basin and their classification according to their natural characteristics and types of economic utilization (Table 1).

In our analysis, 4033 areas at the level of zonal mesoecosystems have been identified which account for 88.4 percent of the basin's area; intrazonal ecosystems, including hydromorphic and river ecosystems account for 10.6 percent, and lake ecosystems, about 1.0 percent, respectively. Among zonal ecosystem types, mountain–forest and mountain–forest–steppe account for 39.2 percent, steppe ecosystems 38.4 percent, and high–mountains 7.4 percent of the basin area. The dominance of mountain–forest, forest–steppe, forest–steppe, and plain–steppe ecosystems demonstrates that the area of the Mongolian part of the Baikal basin is comprised of forest–steppe ecosystems and is characteristically ecotone. The pattern of the distribution of ecosystem types essentially follows zonal regularities and is manifested in a southward increase of the diversity of steppe ecosystems with a reduction of forest and forest–steppe ecosystems. However, the latitudinal distribution of ecosystem types is more complex, which primarily can be accounted for by the effect of orographic factors. For example, forest communities associated with the mountain systems of Khangai and Khentei increase their diversity at latitudes of 48–49° in coordinates corresponding to those mountain structures.

Table 1. Areas and Relationships of the Ecosystems of the Mongolian Part of the Lake Baikal Basin

Ecosystem Types and Indices	**Area (km^2)**	**Percent of Ecosystems Total Area**
Natural and Natural–Anthropogenic		
Glacial–nival and subnival	74	0.02
Mountain–tundra	10,115	3.33
Mountain–meadow–steppe	12,307	4.05
Mountain–forest		
Mountain–forest	87,372	28.75
Forest–steppe	31,691	10.43
Steppe		
Mountain–steppe	59,660	19.63
Moderately–dry	52,230	17.27
Dry	4,357	1.44
Desertified	256	0.08
Hydromorphic		
Valley–river	32,163	10.60
Lake	3,140	1.02
Anthropogenic		
Plough and fallow lands	7,343	2.42
Technogenic	2,912	0.96
Basin Total Area	303,900	100.00

The maximum number of ecosystem species (64) estimated according to the indices of richness according to the Shannon Index was observed within a trapezium of 48–49°N and 100–101°E in the middle and lower reaches of the Chulutyn–gol River east of the Tarbagatai spurs, and the minimum number of species (34) within 48–49° N and 105–106°E, east of the lower reaches of the Tola River. The maximum number of types (26) is localized within 47–48°N and 101–102°E (the upper reaches of the Khoit–Tamyr–Gol and Urd–Tamyr–Gol, and the minimum number of types (10) within 48–49°N and 104–105°E (lower reaches of the Tola River). The maximum fragmentation of the ecosystems into contours (173) is observed at 48–49°N and 98–99°E in the middle part of the Ider valley between the ranges Tarbagatai and Bulgan–Nuru, and the minimum (65) at 47–48°N and 106–107°E south and southwest of Ulan–Bator.

3. ANTHROPOGENIC IMPACTS ON THE ECOSYSTEMS OF THE SOUTHERN PART OF THE LAKE BAIKAL BASIN

3.1 Eco–Biological Assessment of the Ecosystem

Assessment of the current status and anthropogenic modification of the ecosystems of the region was based on methods developed by specialists of the Russian–Mongolian Biological Expedition for the major and intrazonal ecosystems of Mongolia (Methodology 1993, Gunin et al. 1999); the specialists came from the Russian Academy of Sciences, and the Mongolian Academy of Sciences, respectively. The techniques used are based on quantitative indices of changes in particular ecosystem components (vegetation, soil cover, relief) and their differentiation in terms of anthropogenic modification into major groups: absent, slight, moderate, heavy, very heavy, and transition between these categories (Table 2).

Vegetation and its parameters that act as reliable criteria of the assessment of anthropogenic impacts on ecosystems were used to assess the state of ecosystems/communities (Viktorov 1984, Golubeva, 1999). We used the following indices: projection, structure of grass stand, changes in floristic composition compared with communities of undisturbed plots, height of grass stands, sod thickness, introduction of pest species, and abundance and viability of plant species.

For the assessment of the state of ecosystems in terms of soil, we took into account both morphological characteristics of the soil profile and agrochemical indices of the upper soil horizons as thickness of the soil profile, humus horizon and also granulometric composition and humus content in the upper soil horizon. Determination of the level of change in the forms of relief, registration of the qualitative characteristics of the parameters of relief microforms, and, all the more so, mesoforms, may determine the highest level of anthropogenic modification of the ecosystems. This kind of systemic approach requires interpreting the relationship of some individual components in the ecosystem in terms of the extent of their anthropogenic modification, including biological components (animals and vegetation), biotic components (soils), and abiotic components (micro– and mesorelief). The approach has allowed us to update the integral scale of the ecosystem anthropogenic modification level formerly compiled by (Titov et al. 1989) (Figure 1).

Table 2. Qualitative Scales of Anthropogenic Modification of the State of Ecosystems

Modification level	State of ecosystems
Absent (I)	Soil–plant cover is slightly modified; natural regeneration of slightly modified plots is possible.
Slight (II)	Satisfactory, the modification of ecosystems is observed where cattle are grazed; frequently, natural regeneration of the majority of modified plots is possible.
Moderate (III)	Moderate, there are more modified plots than at the preceding stage; natural regeneration is possible but difficult.
From moderate to heavy	Moderate, occasionally poor at forest plots; forest regeneration feasible with temporary bans on grazing, regulation of tree felling.
Heavy (IV)	Poor, soil–plant cover is occasionally modified irreversibly and natural regeneration is extremely difficult.
From heavy to very heavy	Poor, occasionally very poor; natural ecosystems have been virtually destroyed and natural regeneration is frequently impossible.
Very heavy (V)	Very poor, natural ecosystems have been replaced by man–made, or the plots represent an industrial badland (spoils, dirt piles, quarries, etc.); natural regeneration is impossible.

Figure 1. Integrated Scale of Ecosystem Modification and Individual Components

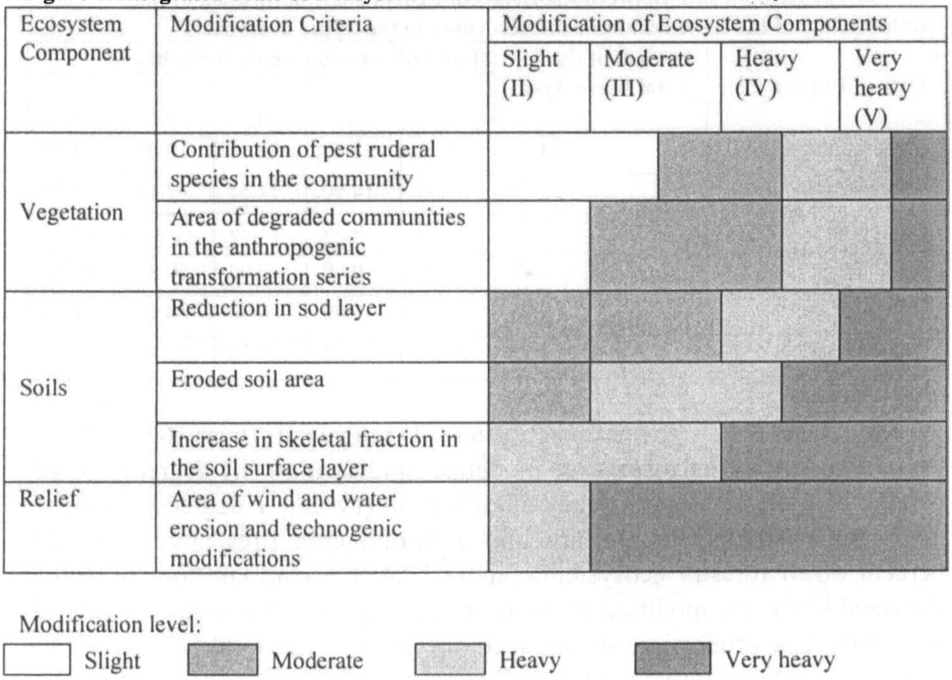

3.2 The Anthropogenic Impacts on the Ecosystems of the Southern Part of the Lake Baikal Basin

We have analyzed ecosystems that have become modified at levels III, IV, and V due to anthropogenic factors. The factors can be divided into two groups: (1)those leading to destruction of natural vegetation, disruption of natural links, and initial ecotype formation; and (2)transformation (sharp or occurring in the course of successional replacement) of indigenous vegetation and the ecotype which largely are recovered after the anthropogenic disturbances are discontinued. The pattern of spatial distribution of disturbances is shown in Table 3. In terms of the types of economic transformation, the basin's ecosystems were classified into: (1)anthropogenic, (2)anthropogenic–natural, and (3)natural–anthropogenic.

Table 3. The Relationship Between the Levels of Ecosystem Modification of the Southern Part of the Lake Baikal Basin According to the Types of Impacts

Type of Impact	Percent of the Total Basin Area	Level of Anthropogenic Disturbance				
		I	II	III	IV	V
Grazing	54.71	7.70	24.42	54.64	13.06	0.18
Felling	25.79	53.70	14.90	11.55	7.12	12.73
Forest fires	7.18	56.59	9.39	11.79	1.08	21.15
Plugging	2.42	0.00	0.00	0.00	50.86	49.14
Urbanized areas	0.96	0.00	0.00	0.00	88.66	11.34
Integrated effect	7.92	21.07	22.52	38.32	12.76	5.33

The largest area in the basin ecosystems is rangelands (over 166 thousand km^2); 54.6 percent of these have moderate modification (III degree) and 15 percent have heavy to very heavy modification (IV and V degrees). Forestry ecosystems exposed to cuttings and anthropogenic fires cover over 20 percent of all forestry ecosystems (about 20.4 thousand km^2 out of 100.5 thousand km^2); the modified areas are concentrated in the northern parts of the basin. The proportion of ploughed–up lands is negligible (2.4 percent) and is currently represented by heavily degraded fallow lands (IVth and Vth degree). The remaining ecosystems (hydromorphic with an integrated impact, technogenic, etc.) account for no more than 10–12 percent of all lands, about half of which are in heavily modified areas.

The major factors that result in changes and disruption of the structural and functional organization of the ecosystems are both locally and widely distributed. These factors often differ not only in their distribution but also with respect to their impacts on ecosystems. For example, widespread livestock grazing can result in slight or moderate impact while localized industries can result in heavy or very heavy impacts which can be irreversible. It also should be noted that direct anthropogenic impacts are the most widespread in the Baikal basin, and are concentrated along the Ulan–Ude–Ulaanbaatar meridian transect in the 104–107 E section.

4. DIAGNOSIS OF DEGRADATION PROCESSES IN SUCCESSIONAL CHANGES OF PLANT COVER

Analyses of data of both of our surveys performed over the last five years as well as the findings of other researchers who previously conducted geobotanical surveys in Mongolia have provided different temporal descriptions of the vegetation of the same plots. These have occurred largely in areas where semi–stationary integrated observations were conducted in the selected testing grounds of the meridian transect Ulan–Ude–Ulaanbaatar crossing the basin of Baikal from north to south. Analyses of the different data and findings has made it possible to identify and assess various impacts on the plant cover. The utilization of satellite images performed in 1989 and in 1999 also has permitted us to model vegetation dynamics.

Based on our assessment of degradation processes responsible for heavy and very heavy anthropogenic modification of the natural environment we have distinguished four groups of hazardous degradation processes: (1)rangeland overgrowth with shrubs, (2)deforestation of forest–steppe ecosystems, (3)depletion of ecosystems with sand soils, and (4)narcotization of agrocenoses in modified ecosystems. We also have differentiated the impacts on the basis of their significance to ecosystems, the economic systems, or social systems, respectively.

In terms of economy, degradation processes include deforestation of forest–steppe ecosystems and the processes of rangeland depletion in hydromorphic landscapes. As discussed previously, deforestation of forest massifs is most intensive in the forest–steppe zone where forest ecosystems are "insular" because of their exposure to maximum phytocenotic pressure on the part of steppe vegetation. Thus, when a tree stand is replaced by a shrub or grass stand the impact often is significant because of a reduction not only of biological diversity but also because of a reduction in forest carrying capacity.

Degradation processes that are significant to social systems stem from the invasion of synanthropous and anthropofilic species of plants, particularly those containing alkaloids and narcotics. Despite the heterogeneous nature of the processes they have a common feature of affecting human health.

4.1 Overgrowth of Rangeland Steppe Ecosystems with Shrubs

In contrast to the European–Black Sea steppes, an important feature of Asian steppes, both of the Trans–Volga, Kazakhstan, and Durian–Mongolian sectors of the Eurasian Province is their overgrowth with various types of shrubs, and semi shrubs of the genera *Caragana, Artemisia, Spiraea, Armeniaca, Amygdalus, Dasiphora* and others. This provides a way to distinguish a special group of shrub steppes (Yunatov 1950) and the classification of shrub steppes into a special vegetation type (Karamysheva 1961). (Neither Yunatov nor Karamysheva regards shrub steppes as a result of pasqual digression under the effect of wild and domestic animals.). Moreover, on changes in the vegetation of steppe reserves (Central–Chernozem, Askania–Nova, Ukrainian Steppe) Abaturov (2001) concludes that an exclusion of steppe rangelands from use resultant in considerable accumulation of dead fall determines mesophytization of plant communities and incorporation of shrub and tree species in them. According to the latest data such species in meadow steppes are *Amygdalus nana, Chamaecyticus ruthenicus* (Bobrovskaya et al. 2000), and in dry steppes species of black and white sagebrush and mock cypress (Parshutina 2000). It is generally thought that the nomadic economy and the conditions in the majority of Mongolian rangelands are not conducive to a wide distribution of pasqual digression. Gorshkova (1977), who was the first to demonstrate the existence of intensive degradation processes in the Trans–Baikal steppes and thus heavily modified rangelands, restricts the distribution of those processes to the steppes of Southern Siberia, i.e., excepting the steppes of Mongolia. Chognii (2000) has demonstrated the contribution of rangeland vegetation, including such species of semi shrubs as *Artemisia frigida* and *A. hangaica* to succession. The data we have obtained have made it possible to specify the pattern of a refined scheme of succession stages of the region under study. According to this scheme, as rangeland pressures increase so too does the role of shrubs and semi–shrubs in the structure of plant communities, which we record as the final (IVth and Vth) stages of pasqual digression.

The IVth stage shows a heavy impact from grazing. Many rangelands situated near watering places, along highways and railroads and livestock driveways are in this stage. The floristic richness in the IVth stage is reduced three or fourfold and the height of the bulk of grass stands by two to two and one–half times. *Cleistogenes squarrosa* and *Arex duriuscula* remain dominant in the grass stands, the role of aboveground production of the shrub *Caragana microphylla* and the semi shrub *Artemisia frigida* sharply increases. In a number of cases their phytomass attains 80 percent of the entire aboveground production. In the central part of the basin, the grass stands are dominated by *Stipa sibiric, Cleistogenes squarrosa,* and *Carex argunensis.*

The Vth stage shows a very heavy impact from grazing and includes rangelands located near wintering places and built up areas; however, the area of such rangelands is negligible. Dominant species in these rangelands are *Artemisia frigida, Arex duriuscula, Cleistogenes squarrosa, Caragana microphylla,* and in some stands the semi shrub *Thymus gobica.*

The ecosystems of the region under study also reveal a separate group of plant species that positively respond to grazing: the shrub *Caragana microphylla*, semi shrub *Artemisia frigida*, small–sod herbs *Cleistogenes squarrosa, Carex duriuscula*, long root *Leymus chinensis* (Gunin et al. 2002). The contribution of shrubs and semi shrub to the communities of the Eastern–Asian sector of the steppe increases with an increase in rangeland pressure. Several species play significant roles on degraded rangelands, including the shrubs *Caragana microphylla, Caragana pygmaea* and semi shrubs *Artemisia frigida, Thymus gobica* and *Potentilla bifurca*; their abundance are directly proportional to the degree of rangeland modification.

4.2 Deforestation of Larch and Pine Forest Ecosystems

To investigate the process of deforestation of larch and pine forests complete geobotanical and taxonomic descriptions were obtained in forest and shrub plant communities on sample sites of 100 to 400 m^2. Taxation measurements included the height and trunk diameter for each tree species; for shrubs, the number of trunks, height and diameter of the shrubs and also the measurements of aboveground phytomass.

The complete succession series is represented by 5 major stages: I–indigenous forests *(Pinus silvestris, Larix sibirica);* II–small leaved forests with a mixture of indigenous species (*Betula platyphylla, Populus tremula, Pinus silvestris*); III–small leaved forests without contribution of indigenous species, including undergrowth or thickets of shrubs with drying trees, or without undergrowth of arborous species (*Betula platyphylla, B. fusca, Populus tremula, Cotoneaster melanocarpa);* IV–thickets of shrubs without involvement of any trees (*Cotoneaster melanocarpa, Spiraea aquilegifolia, Amygdalus pedunculata, Armeniaca sibirica, Betula fusca*, and *Salix* sp.*);* V–steppe plots (*Carex korshinskyi, C. pediformis* and *Stipa baicalensis).*

The plant cover of the IIIrd to stage IV of the post–cutting successions shows a reduction of floristic diversity 1.5–fold. At stage IV trees and their undergrowth are absent, and the species diversity in the herb layer declines 1.5–fold. On the other hand, in the communities of stages III–IV some similarity is observed: 50 percent in the shrub layer, and 25.8 percent of the herbaceous layer. In those layers, similarity is accounted for by dominant and auxiliary species. In that case the cenotic significance of species in communities of stages III and IV does not change. The conditions of development of coniferous forest are not very favorable as early as stage III of the succession series. The forest is weakened and is drying: the proportion of dry birches accounts for 25 percent, that of dry aspens 20 percent. At stage IV the conditions are unfavorable for regeneration and development of any forest, not only coniferous, but also small leaved trees. Spontaneous regeneration of pine forests after clear cutting may take a long time, since in the undergrowth of young small leaved forests, pine undergrowth is normally absent, or its role is negligible. Without forest regeneration work, i.e., plantation of pine undergrowth, the small leaved forest stage lasts for an indefinitely long time.

Thus, the succession series shows replacement of indigenous forests by small leaved forests, shrub communities, and subsequently by steppes.

Succession replacements of coniferous forests at the southern border of their distribution can be attributed to irreversible vegetation changes. This is confirmed by: (1)absence of coniferous trees and their undergrowth at all stages, beginning in stage III; (2)absence of undergrowth of any arborous trees and viability of trees at all stages, beginning in stage III; (3)increase in the level of grass infestation of the soil at the final stages; and (4)augmentation of the processes of soil drying at the final stages which prevent development of tree and undergrowth seedlings.

4.3 Desertification of Ecosystems with Soils of Light Granulometric Composition

One problem of desertification in Mongolia concerns the natural preconditions of the origin of processes of degradation and increasing ranges of anthropogenic impacts on the environment (Gunin 1991, Vostokova et al. 1995). Of special importance for the development of desertification in the Selenga River basin is degradation of the soil–plant cover of the ecosystems on light sandy soils.

According to our definition of desertification, it includes only the processes that bring about penetration of more arid elements into degraded ecosystems. These include: (1)incorporation into modified phytocenoses of dry steppe species typical of deserts, (2)the formation of barkhan forms of relief on modified plots with sand soils, and (3)development of solonchaks in forest–steppe as a result of water erosion of salt–containing waste piles. The desertification criteria we use are development of fragments of desert landscapes or manifestation of some particular components of arid ecosystems. These are to be distinguished from the: (1)development of the elements of arid mesorelief, (2)development of arid biota types, and (3)development of ecosystems characteristic of arid landscapes.

The desertification processes we have described are observed in different types of soil–forming rocks, under various relief conditions and in various natural zones ranging from steppes to forest–steppes and coniferous broad–leaved sub zones of the forest zone. Today, desertification foci are recorded in several regions of the basin, but their area is so far negligible.

Identification of desertification processes is based largely on revealing their indicators, i.e., visually identifiable features or objects. These include elements of plant cover which are different for steppe and semi–deserts, where against the background vegetation, desert plant species are readily identifiable. The first species to advance through disturbed habitats into neighboring zones are pest anthropophillic species (*Corispermum mongolicum, Agriophyllum pungens, Peganum harmala, P. nigellastrum, etc.*), characteristic of sandy or salinized habitats.

4.4 Depletion of Rangelands of Hydromorphic Landscape

The rangelands of the hydromorphic landscapes are common in river valleys and lake depressions, where the major anthropogenic factors of long–term dynamics of plant communities are constant rangeland pressures, concentration of the human population and livestock, and unregulated human traffic. Virtually throughout the entire area of hydromorphic ecosystems there occur digressive replacements of phytocenoses in the course of succession development of meadow vegetation (Titov et al. 1999, Ubugunov et al. 2000).

Such vegetation replacements have been studied in greatest detail in the northern part of the basin in the floodplains and on the terraces of the rivers of the steppe and forest–steppe zones. These and other observations in different regions of the basin support the conclusion that the pathways of development of plant communities converge in floodplain ecosystems. For example, succession series are characterized by an emergence of pest and unpalatable species even under moderate rangeland pressures. The phytocenoses of digression stages III and IV in meadow steppe sod soils are represented by meso– and xeromorphic species *Carex duriuscula, Artemisia frigida, A. commutata, A. laciniata, and A. adamsii.* At the final stage of succession series, the initially different dominant species that occur virtually everywhere (e.g., on terraces and floodplains of meadow communities) are the edifier perennial iris species (*Iris lactea, I. Bungei*). Studies have revealed an increase in the values of productive phytomass up to 25.0–30.0 centers /ha due to increase in the proportion of irises in the communities (from 5–10 percent to 67–70 percent).

Another common type of succession change is when *Achnaterum splendens* becomes dominant in the course of use of rangelands in hydromorphic communities. According to Vostokova and Kazantseva (1995) the role of the latter in the total phytomass of *Achnaterum splendens* approaches 70 percent, which indicates its edifying role in the above community type.

On the whole, anthropogenic successions of hydromorphic vegetation of the floodplains are digressive, since they bring about development of highly productive groupings of herbaceous perennials in terms of phytomass of little value as fodder. Regenerative successions of hydromorphic vegetation of the floodplain virtually do not occur due to recurrent anthropogenic pressure. Only when degraded iris communities in hydromorphic landscapes are removed from rangeland rotation, there occurs restoration of graminean–sedge–forb meadows, where the role of irises can be reduced to minimum values.

4.5 Narcotization of Agrocenoses in Fallow Lands and Heavily Modified Ecosystems

We regard the current distribution of the *Cannabis* species in such a vast region as the Baikal basin as a manifestation of syngenetic succession. According to Rabotnov (1978) this succession reflects not only the process of population into vacant areas, but also incorporation into established cenoses. The distribution of *Cannabis* is primarily promoted firstly by the biological properties of this species, and, secondly, by the presence of conditions providing the germination of seeds and seedlings.

According to the classification by Grime (1979), *Cannabis sativa* is a species with ruderal strategy, characterized by rapid growth, which annually forms a large number of seeds. The aggressiveness of *Cannabis sp.* may also be determined by the fact that compared with other pest species, it appeared in the basdin region relatively a longer time ago. Voronov (1973) believes that of the pest species the most aggressive is a more ancient species, i.e., the one that settled down earlier. For the Mongolian territory this is supported by findings of *Cannabis* in the area of ancient settlements.

Cannabis also shows a high environment forming capacity, successfully competing with species forming phytocenoses into which it is incorporated and where frequently monocenoses are formed from tall (3.0 m and over) above ground organs with large aboveground mass providing up to 2.5 to 3.0 tons/ha of biomass in ruderal plant communities. As it forms settlements, *Cannabis* acts as an anthropochore, anemochore and hydrochore. When transferred, cannabis acts as a psammophil, being associated with sandy soils, or as nitrophil, settling down in soils with a high content of organic matter. The investigations conducted in the Baikal basin have reveled the dynamics of competitive capacity of *Cannabis* in different zones if its distribution ranges: (1)in the incorporation zone, (2)in the zone of maximum phytocenotic role, (3)in the zone of declining phytocenotic contribution.

In the population–forming zone, *Cannabis* uses various population mechanisms for forming colonies. It forms primary and secondary colonies whose time of origin is difficult to determine. When the primary *Cannabis* colonies are formed, *Cannabis* acts as an anthropochore, anemochore and hydrochore. Its seeds are transferred by transport and by the wind along transportation routes, ditches and roadside banks, and in that case *Cannabis* acts as a psammophil associated with sandy soils. The secondary colonies are formed by means of water streams, where *Cannabis* budlets from transportation roads are transferred to erosion hollows, fertilizer and manure dumps and to garbage habitats.

In the flourishing population zone (maximum phytocenotic and environment forming role) *Cannabis* mostly is deposited by winds stretching along croplands to form sediments of dust–like sandy particles in adjacent areas and also in croplands and fallow lands where it gets as an admixture to the seeds. In that zone *Cannabis* also populates settling basins of industrial water and channel dumps. *Cannabis* populates some particular areas. The area occupied by *Cannabis* in the interfluves may range from 1–2 to 100–120 ha. We recorded such considerable areas in the somons Jargalant, Altan–Bulag, Dzun–Kharaa and the vicinity of the town of Darkhan. A characteristic feature of that zone is a great density of *Cannabis* population, including considerable height of stems and maximum biomass.

In the zone of reduced phytocenotic contribution, or a stage of declining population succession, *Cannabis* accounts for isolated plots with individual sizes of not more than 100–110 cm. This area includes zones of the most ancient *Cannabis* distribution (an area of the most ancient human burial sites). A characteristic feature of that zone is a decline of morphological parameters and *Cannabis* production down to 10–15 centers/ha, which is associated with depletion of the needed resources. Although the viability of *Cannabis* individuals has been depressed, it still forms monocenoses.

5. ECOLOGICAL ZONING OF THE SOUTHERN (MONGOLIAN) PART OF THE LAKE BAIKAL BASIN

In solving conservation problems at regional levels, it is practically important to use the results of eco–biological assessments to ascertain the threats and degradation of areas for planning and zoning purposes. The ecological zoning that we conducted for the southern part of the Baikal basin has made it possible for us to clearly differentiate its territory in terms of different relationships between ecosystems with different degrees of modification into three categories: high, moderate and slight.

Two regions in the central part of the basin have high levels of threats and degradation over about 50 to 75 percent of their area. A region with moderate ecological risk and degradation were the Cis–Khubsugul and Orkhon–Tol districts, where the proportion of ecosystems with heavy and very heavy modification does not exceed 50 percent; among the groups of degradation processes are combinations of rangelands overgrown with shrubs, deforestation of forest–steppe ecosystems, and desertification of hydromorphic ecosystems.

Districts with a low level of ecological threats and degradation are situated in the southern part of the basin and occupy about one third of the area under study. They are associated with the upper parts of the water catchment's areas of Khangai and Khentei. The above regions are characterized by minor modification (10–25 percent) and absence or negligible development of degradation processes. A presence in those areas of a considerable network of specially protected zones and nature parks (about 10 percent of the entire basin) appears to promote conservation of the regional ecosystems. The proposed zoning may improve environmental management by distinguishing the regions that require conservation or other protection measures.

6. CONCLUSION

The inventory of natural, natural–anthropogenic and anthropogenic ecosystems and the eco–biological assessment of their status has allowed us to assess the quantitative relationships of the major types of ecosystems and also their differentiation with respect to economic utilization, development of degradation processes and the level of anthropogenic modification; in turn, this has contributed to the efficacy of zoning of the Mongolian part of the Selenga River basin for environmental decisionmaking.

The structures of the ecosystem cover of the southern part of the Baikal basin are 4033 areas at the level of zonal and intrazonal mesoecosystems, which account for 88.4 percent of zonal ecosystems in the basin's area. Intrazonal ecosystems including hydromorphic and river ecosystems account for 10.6 percent; and lake ecosystems account for no more than 1.0 percent. The dominance of mountain–forest and plain–steppe ecosystems is indicative that the Mongolian part of the Baikal basin should be classified as a region with a predominant distribution of mountain–forest–steppe ecosystems, which confirms its ecotone nature.

The eco–biological assessment of the state of ecosystems led us to conclude that the detrimental anthropogenic impacts on the structure and functioning of the ecosystems are manifested at the physiognomic level when the modification is at the IIIrd degree. Under the Vth degree of modification the ecosystems are classified as completely transformed anthropogenic. Analysis of modification as determined by various anthropogenic factors reveals that overgrazing has to date heavily and very heavily modified 13.2 percent of ecosystems' area, and rangeland ecosystems that have been moderately modified account for 54.6 percent of their total area. Felling and fires have modified forestry ecosystems heavily and very heavily in over 20 percent of the forest covered territory. Virtually all of the agroecosystems of the region have been anthropogenically modified to the IIIrd to Vth degree.

Diagnosing the degradation processes responsible for heavy and very heavy anthropogenic modification of the natural environment and determination of their differentiated impacts on the ecosystems, economy and human health has led us to identify hazards on the basis of their economic and social significance. The level of hazard (from hazardous to very hazardous) was determined by the nature of their impacts in one, two or three spheres of human activity, respectively: (1)ecologically hazardous degradation processes are those of desertification and rangeland overgrowth with shrubs, (2)economically hazardous degradation processes are deforestation of forest–steppe ecosystems and processes of rangeland depletion in hydromorphic landscapes, and (3)socially hazardous degradation processes are those of invasions of narcotic containing plants into modified ecosystems.

Data of spatial distribution of ecosystems heavily and very heavily modified anthropogenically has revealed regions with moderate, high and very high levels of ecological threats and degradation. Among such regions were those where the total area of heavily modified and very heavily modified ecosystems is 10–25 percent, 25–50 percent, and 50–75 percent of the region's area, respectively. In terms of geographical distribution, the ecologically threatened regions are associated with the landscapes of the Orkhon–Tol and Orkhon–Selenga river systems and the Selenga mid–mountain area which also coincides with the regions which are best developed agriculturally and industrially.

Conservation and sustainable use of biodiversity in the Lake Baikal basin require management based on ecosystem approaches appropriate to the entire basin and detailed knowledge of the extent of threats and degradation to the different ecosystem types. Minimally, management should: (1)ensure that the laws of Russia and Mongolia conform with one another in their regulation of economic activity to conserve the basin; (2)establish a network of strictly protected reserves in ecologically important areas of the basin such as the central part of the Selenga River and which might also include a transborder Russian–Mongolian Biosphere Reserve with several clusters in the middle reaches of the Selenga River; (3)monitoring the dynamics of aquatic and solid runoff, including accumulation of pollutants in the water and bottom sediments, so that baseline monitoring stations in the mouths of the tributaries of first and second orders of the Selenga River (the Orkhon River, the Tola River, the Khara–Gol River, the Yero–Gol River and the Egiin–Gol River) can be established.

7. ACKNOWLEDGEMENTS

The study was performed under the framework of the Russian–Mongolian Integrated Biological Expedition, Russian Academy of Sciences and Academy of Sciences of Mongolia. The authors extend their sincere thanks to professors E.A. Vostokova and Ch. Dugarjav, and Drs. N.N. Slemnev, A.V. Prischepa, I.M. Miklyaeva for constant support and participation in fieldwork when investigating the anthropogenic modification processes in Lake Baikal ecosystems.

8. REFERENCES

Abaturov, B.D. 2000. The ecological consequences of ungulates pasture for semi–desert ecosystems. Pages 57–83 In: Aliakrinskaya I.O. (Editor) The Reports on XIX Annual Readings of Memory of the Academician V.N. Sukachev, Ecological Processes in Arid Biocenoses, Scientific Advice on Ecology the Biological Systems of RAS, Moscow. (In Russian)

Boikov, T.G. 1999. Rare plants and phytocenoses of Transbaikalia. Science publ., Novosibirsk. 253 pages. (In Russian)

Bobrovskaja, N.I., Kazantseva, T.I., et al. 2000. Pages 40–47 In: Restoration of vegetation of meadow steppes in Central–Chernozeme zone. Arid Ecosystems: Moscow 6: 11–12. (In Russian)

Chognii, O. 2000. History of used pasture by Mongolia. Pages 39–41 In: Dugarjav, Ch. and Gunin, P.D. (Editors) Central Asian Ecosystems. Proceedings of Abstracts. Ulaanbaatar. (In English)

Danilov–Daniljan, V.G. (Editor). 1995. Environmental Nature of Russia. ECOS. Moscow. 118 pages. (In Russian)

Erdenezhav, G. 2000. Natural conditions and biological resources of Orkhon–Selenga basin's ecosystems. Manaj Mongol Journal 3: 34–53. (In Mongolian)

Ganbold, E. 2000. Vegetation of Northern Mongolia, Report for the Doctor of Biological Sciences Degree. St. Petersburg. 67 pages. (In Russian)

Golubeva, E.I. 1999. Methods of Diagnostics of an Anthropogenic Disturbed Ecosystems Condition. Geogr. Faculty of MSU. Moscow. 68 pages. (In Russian)

Gorshkova, A.A. 1977. Ecology and Pastural Digression of Steppe Communities of Transbaikalia. Novosibirsk, 192 pages. (In Russian)

Grime, J.P. 1979. Plant Strategies and Vegetation Processes. Chichester; New York: John Wiley and Sons. 222 pages.

Graivoronski, V.V. 1997. Modern nomadism of Mongolia: Social problems of the transitive period (1980–1995). Moscow 2: 5. (In Russian)

Gunin, P.D. 1991. Natural processes of desertification in arid ecosystems. Report for Doctor of Biological Sciences Degree. Moscow. 82 pages. (In Russian)

Gunin, P.D., Miklyaeva, I.M., Bazha, S.N. 2002. Succession dynamics of rangeland ecosystems of Mongolia. Pages 122–140 In: Chuluun, T., and Ojima, D. (Editors). Fundamental Issues Affecting Sustainability of the Mongolian Steppe. Interpress. Mongolia, Ulaanbaatar.

Gunin, P.D., Vostokova, E.A., Dorofeyuk, N.I., Tarasov, P.E., Black, C. (Editors). 1999. Vegetation Dynamics of Mongolia. Kluwer Academic Publishers, Dordrecht. 238 pages.

Gunin, P.D., and Vostokova, E.A. (Editors). 1993. Methodology of Modern State Estimation and Ecosystems Mapping Under Extreme Natural Conditions. Puschino Centre for Science of RAS. 203 pages. (In Russian)

Gunin, P.D., and Vostokova, E.A. (Editors). 1995. The map "Ecosystems of Mongolia", scale 1: 1000 000. 15 sheets. Moscow.

Karamisheva, Z.V. 1961. Vegetation of a northeast part Kazakhstan steppes. Botanical Institute Issues AS USSR. Sulfurs. 3–Geobotany 13: 464–486.

Kharin, N.G., Tateishi, R., and Harahsheh, H.H. 1999. Degradation of the Drylands of Asia. CERES, Chiba University, Japan. 81 pages.

Lavrenko, E.M., Karamisheva, Z.V., and Nikulina R.I. 1991. Steppes of Eurasia. Science publ. Leningrad. 145 pages. (In Russian)

Mongolian Economy and Society in 1995. Statistical Yearbook. 1996. Ulaanbaatar. 127 pages.

Murzayev, E.M. 1952. Mongolian People's Republic. Geographical description. Geographiz, Moscow. 472 pages. (In Russian)

Nature and Environment of Russia. 1995. ECOS, Moscow. 118 pages. (In Russian)

Nature and Environment in Mongolia. 1996. Ulaanbaatar. 130 pages. (In English)

Nature and Environment in Mongolia. 2000. Ulaanbaatar. 95 pages. (In English)

Parshutina, L.P. 2000. Steppes of the Rostov area and their modern condition. Arid Ecosystems: 6, N 11–12: 57–65. (In Russian)

Preobrazhenskii, V.S., Fadeeva, N.V., Kelner, Yu.G., Tsegmid, Sh., and Tulgaa, H. 1984. The experience of landscape–typological zoning of Mongolian territory basing on remote sensing. Izvestiya of the Academy of Sciences of the USSR. Ser. Geography 4: 5–14. (In Russian)

Rabotnov, T.A. 1978. Phytocenology. MSU Publ. Moscow. 384 pages. (In Russian)

The National Atlas of the Mongolian Peoples Republic. 1990. GUGK USSR & GSK MPR, Ulaanbaatar–Moscow. 44 pages. (In Russian)

(RFL 2000). The Law of Russian Federation "On Protection of Baikal Lake". ECOS, Moscow. 25 pages. (In Russian)

Titov, U.V., Ogar, N.P., et al. 1990. Methods of an Assessment of a Modern Condition of River Valleys in Mongolia. Ulaanbaatar. 41 pages. (In Russian)

Ubugunov, L.A., Ubugunova, V.M., Korsunov, V.M. 2000. The Soils of Gidromorphic Ecosystems of Central Asia. BNC–Publ., Ulan–Ude. 215 pages. (In Russian)

Vinogradov, B.V. 1984. The Ecosystems Space Monitoring. Science publ., Moscow. 320 pages. (In Russian)

Voronov, A.G. 1973. Geobotany. High School Press. Moscow. 384 pages. (In Russian)

Vostokova, E.A., Gunin, P.D., Rachkovskaja E.I., et al. 1995. Ecosystems of Mongolia: Distribution and Modern Condition. Science publ., Moscow. 223 pages. (In Russian)

Vostokova, E.A., and Kazantseva T.I. 1995. Distribution, ecology and productivity of communities Achnatherum splendens (Trin) nevski in Mongolia. Vegetation Resources 4: 77–87. (In Russian)

Chapter 13

PLANT DIVERSITY AND SUCCESSION OF ARTIFICIAL VEGETATION TYPES AND ENVIRONMENT IN AN ARID DESERT REGION OF CHINA

Li Xin–Rong, Zhang Jing–Guang, Liu Li–Chao, Chen Huai–Shun, Shi Qing–Hui

Shapotou Desert Experiment and Research Station, Cold and Arid Regions Environmental and Engineering Research Institute, Chinese Academy of Sciences, Lanzhou 73000, China; email Lxinrong@ns.lzb.ac.cn

1. INTRODUCTION

Nearly 400 million of people in arid and semiarid regions in China are affected by desertification. Desertification is not only a major obstacle to the country's regional economic development but also poses a serious threat to the biodiversity, which constitutes a material basis for long–term human well–being. Consequently, mitigation of desertification is important to promoting sustainable development. This study describes the establishment and development of restorative sand–fixing vegetation in China's Shapotou region where railways have been affected by increasing problems of desertification, and further provides a theoretical basis for greater understanding of the ecological mechanisms of restoring biodiversity decertified regions.

2. DESCRIPTION OF THE STUDY AREA

The study area is located in the Shapotou region in Zhongwei county of Ningxia at the southeast fringe of Tengger Desert at 37°27′N and 104°57′ E.

It is part of a steppified desert zone in a transitional region between desert and oasis. Natural vegetation in the study area is dominated by *Hedysarum scoparium* and *Agriophylum squarrosum*, with a vegetation cover of one percent or so. This area is 1500m above sea level and covered with high and dense reticulate barchanic chains. Soil substrate in the area is loose and impoverished shifting sand with stable moisture content ranging from two to three percent. According to over 40 years of observations, annual mean air temperature in the area is 9.6°C, extreme minimum temperature –25.1°C, extreme maximum temperature 38.1°C, annual sunshine hours 3264.7h, annual mean precipitation 181.6mm, annual mean wind velocity $2.9 m \cdot s^{-1}$ and the annual number of dust storm days is 59.

Since 1956, to ensure safe operation of the Baotou–Lanzhou railway the Chinese Academy of Sciences and related institutions began establishing protective systems to prevent sand encroachment on the railway, emphasizing sand fixation in combination with block measures (Zhao Xingliang 1988). A first step was to erect lm × lm straw checkerboard on shifting sand surfaces and then plant sand binders inside each square as sand surface becomes stable. The sand fixation belt of the protective system is 500m wide to the north side of the railway and 200m wide to its south side, with a total length of 16km. Over the 40 years of restoration the plant succession in the area has significantly improved and the fixation of sand surface has created a better condition for the recolonization of many annual plants and has led to the formation of microbiotic crust (Li Xinrong et al. 2002). Approximately nine fungi species, over 300 insect species and many other microorganisms now occur in the area. Species restored include *Caragana korshinskii*, *Hedysarum scoparium*, *Caragana intermedia*, *Atraphaxis bracteata*, *Artemisia sphaerocephala*, *Artemisia ordosica* and *Calligonum arborescens*; invaded indigenous species include *Bassia dasyphylla*, *Eragrostis poaeoides*, *Setaria viridis*, *Sonchus brachyotus*, *Salsola ruthenica*, *Artemisia capillaris*, *Chloris virgata*, *Aristida adscensionis*, *Cleistogenes squarrosa*, *Allim mongolicum*, *Scorzonera divaricata*, *Corispermum declinatum*, *Euphorbia humifusa*, *Echinops gmelinii*, *Agriophyllum squarrosum*, *Artemisia sphaerocephala*, *Inula salsoloides*, *Stellaria dichotoma*, *Stipa glareosa* and *Artemisia ordosica*. In addition, approximately 65 bird species entered the area, of which 60 are brood birds; other colonized animals numbered over 30 species, nine of which are rodents. The colonization and propagation of large number of species changed a former moving dune–dominated landscape into a complex artificial–natural desert ecosystem.

3. STUDY METHODS

3.1 Selection and Survey of Sample Plots

Ten quadrants (10m × 10m) were arranged in each regular observation planted sand binders in 1956, 1964, 1973, 1981 and 1987 in the 16km–long and 500m–wide artificial vegetation area to the north of the railway; in total 50 quadrants were arranged (Table 1). (Because the plot planted sand binders in 1973 to the north side of the railway was disturbed by grazing, the same year's plot to the south side of the railway was selected.). The survey area of shrubs was 10m × 10m, while that of herbs was 1m × 1m. The composition of plant species and quantitative characteristics of plant communities in the sand fixation areas of different years were observed. Samples were taken for analysis and soil moisture and nutrient data were collected.

3.2 Measurement of Biodiversity

In this study Simpson index D (Simpson 1949), Shannon–Wiener index H' (Peet 1974) and Pielou (Pielou 1975) evenness index J are used to measure α diversity,

$$D = 1-\Sigma p_i^2;\quad H' = -\Sigma p_i \ln p_i;\quad J = (-\Sigma p_i \ln p_i)/\ln S$$

where p_i is the relatively importance value of species i (relative height + relative coverage), S is the total number of species i in the quadrant, i.e., an abundance index.

Beta (β) diversity refers to the species replacement degree with the environmental gradient changes, some scholars refer to this as species turnover rate, species replacement rate or rate of biotic change. Its measurement can be classified as binary attribute data and quantitative data measurement methods. In this study the well–defined binary attribute data Sorenson index and two widely used quantitative data C_N and C_{MH} were used for calculations (Whittaker 1972, Ma Keping et al. 1995). The Sorenson index,

$$C_j = j/(a+b)$$

where j is the common species number of two communities, a and b are the species number of community A and community B respectively. And, the Bray–Curtis index (Bray et al. 1957),

$$C_N = 2jN/(aN+bN)$$

where aN is the species number of sample plot A; bN is the species number of sample plot B; jN is the sum of fewer individual number of common species in sample plot A (jaN) and B (jbN). In the calculation the significant values of species were used to replace individual numbers according to the Morisita–Horn index (Li Xinrong 2001)

$$C_{MH} = 2\Sigma(an_i \cdot bn_i)/[(da+db)aN \cdot bN]$$

where aN and bN are the same as in above formula; an_i and bn_i are the individual number of species in sample plot A and B respectively (replaced by significant values), and

$$da = \Sigma an_i^2/aN^2;\ db = \Sigma bn_i^2/bN^2$$

Therefore,

$$C_{MH} = 2\Sigma(an_i \cdot bn_i) \cdot aNbN/[bN^2\Sigma an_i^2 + aN^2\Sigma bn_i^2]$$

4. RESULTS

4.1 Changes in Vegetation Succession Processes

Shifting sand control in Shapotou region is a successful model in desertification mitigation in an arid region of China. Prior to the establishment of artificial protective systems the most striking feature of the desert environment was its growing sand surface. The erection of straw checkerboard stabilized sand surface created a condition for the establishment of sand–fixing plants (Xiao Honglang et al. 1996). As vegetation cover reached ten percent, wind velocity became reduced by twenty to forty percent. Wind velocity at 20cm above ground surface in the plant restored area reached $2.9 m \cdot s^{-1}$, the corresponding figure in the shifting sand area was about $5.9\ m \cdot s^{-1}$, or twice that of the former (Shi Qinghui 1993). Stable physical environments created necessary preconditions for the formation of sand surface crust and soil development (Duan Zhenghu et al. 1996).

The features of habitat changed as a function of time after plant restoration. With time, significant changes took place in vegetation composition; community structure gradually changed from single shrub and semi–shrub species into a complex annual herb–dominated structure; shrub gradually became less abundant in the restored community; and finally total vegetation cover and biomass decreased with succession time. With extension of sand fixation time, soil crust thickened, soil moisture content (mean content in 0–300cm soil layer) decreased, soil microorganisms increased and soil nutritional condition improved. Meanwhile, the number of species in the vegetation composition increased. In addition, more than 28 years after plant restoration the probability of new or invading species in the vegetation composition became less and the species composition tended toward a balanced state (species number varied between 12–14). (*Artemisia ordisica* regeneration was high, followed by some annual plants such as *Eragrostis poaeoides, Bassia dasyphylla, Corispermum declinatum, Salsola ruthenica* and *Aristida adscensionis.*). It is worth noting that the occurrence of *Stipa glareosa* in the sand fixation area and *Stipa bungeana* in the old fixed sand land on the terrace of Yellow River exhibited an increasingly intensified steppification feature of artificial vegetation.

Some of the specific changes which occurred with sand fixation in different years included:

- 1956– vegetative cover (percent) 25.2; dominant plants: *Artemisia ordosica, Bassia dasyphylla, Eragrostis poaeoides*; community biomass (g/100m^2) 17,052; soil crust thickness (cm) 2.1; soil moisture (percent) 1.16; soil bacteria (10/gm) 32,200; soil actinomyces (10/gm) 48.3; soil fungi (10/gm) 4.34; soil N (g/kg) 0.02; soil P_2O_5 (g/kg) 1.59; soil K_2O (g/kg) 60.0.
- 1964– vegetative cover (percent) 27.6; dominant plants: *Artemisia ordosica, Bassia dasyphylla, Caragana korshinskii*; community biomass (g/100m^2) 13.180; soil crust thickness (cm) 1.0; soil moisture (percent) 1.17; soil bacteria (10/gm) 63,752; soil actinomyces (10/gm) 35.8; soil fungi (10/gm) 6.23; soil N (g/kg) 0.74; soil P_2O_5 (g/kg) 1.50; soil K_2O (g/kg) 60.0.
- 1973– vegetative cover (percent) 30.2; dominant species: *Bassia dasyphylla, Salsola ruthenica, Caragana intermedia, Artemisia ordosico*; community biomass (g/100m^2) 29,800; soil crust thickness (cm) 0.6; soil moisture (percent) 1.40; soil N (g/kg) 0.68; soil P_2O_5 (g/kg) 1.52; soil K_2O (g/kg) 60.0.
- 1981– vegetative cover (percent) 36.3; dominant plants: *Agriophyllum squarrosum, Artemisia ordosico*; community biomass (g/100m^2) 36,412; soil crust thickness (cm) 0.4; soil moisture (percent) 1.51; soil bacteria (10/gm) 20,237; soil actinomyces (10/gm) 44.0; soil fungi (10/gm) 5.22; soil N (g/kg) 0.54; soil P_2O_5 (g/kg) 1.51; soil K_2O (g/kg) 60.0.

- 1987– vegetative cover (percent) 30.0; dominant plants: *Artemisia ordosica, Atraphaxis bracteata*; community biomass (g/100m^2) 39,743; soil crust thickness (cm) 0.3; soil moisture (percent) 2.02; soil N (g/kg) 0.38; soil P_2O_5 (g/kg) 1.04; soil K_2O (g/kg) 60.0.
- Control– vegetative cover (percent) <1.0; dominant plants: *Artemisia squarrosum, Hedysarum scoparium*; community biomass (g/100m^2) 0.0; soil crust thickness (cm) 0.0; soil moisture (percent) 3.03; soil bacteria (10/gm) 15,698; soil actinomyces (10/gm) 1.57; soil fungi (10/gm) 0.11; soil N (g/kg) 0.03; soil P_2O_5 (g/kg) 0.01; soil K_2O (g/kg) 60.0.

4.2 Plant Diversity Changes and Time

In areas where the vegetation has experienced an α succession time of 42 years (or planted in 1956) the plant species have a relatively high diversity based on the Simpson Index (D) average (D = 0.706~0.822 H' = 1.393~1.893), while the vegetation planted ten years ago has a lower diversity (D = 0.501~0.702 H' = 0.819~1.074). However, the diversity indices of the artificial vegetation planted in 1973 to the south side of the railway were lower than those of the artificial vegetation planted in 1981 due to frequent grazing disturbance. Based on the diversity indices, the plant diversity in Shapotou region is still rather low, although the artificial vegetation there has experienced several decades of succession. This suggests that artificial vegetation has simple structure, low community stability and high sensitivity to exotic disturbances. Therefore, without careful management in the use of vegetation (especially grazing) inverted succession of vegetation and the developing or evolving ecosystems may be damaged.

4.3 Beta Diversity Changes

Beta (β) diversity includes the differences of species composition among different communities. The smaller the common species number of different communities or different sites of a certain environmental gradient, the larger the β diversity (Ma Keping et al. 1995). Ecologically speaking, β diversity also indicates the relative extent in which species divide habitats, and its measured value can be used to make comparisons of habitat diversity among different sites. β diversity together with α diversity constitute the overall diversity or biotic heterogeneity of a site. Hence, the determination of β diversity is significant because when it is assessed over time it can contribute to a greater understanding of the structural and functional processes of biotic communities (Ma Keping et al. 1995) (see, e.g., Table 1).

Table 1. Comparison of Plant Diversity and Calculation of Beta in Sand Fixation Areas

	1956	1964	1973	1981	1987
Simpson Index (D) (Average)	0.706–0.822 0.767	0.595–0.856 0.752 0.7	0.627–0.777 0.696	0.631–0.788 0.711	0.501–0.702 0.539
Shannon–Wiener Index (H) (Average)	1.393–1.893 1.642	1.232–1.814 1.51	1.247–1.633 1.385	1.171–1.69 1.390	0.819–1.074 0.859
Pielou Evenness (J) (Average)	0.638–0.961 0.701	0.661–0.862 0.775	0.554–0.743 0.646	0.658–0.877 0.745	0.524–0.712 0.534
Sorenson Index (C_N)		0.801	0.657	0.538	0.819
Bray–Curtis Index (C_N)		0.216	0.189	0.034	0.332
Morisita–Horn Index (C_{MH})		0.845	0.307	0.242	0.935

As shown in Table 1, the Sorenson index, Bray–Curtis index, and Morisita–Horm index were used to measure the β vegetation diversity over five different time periods. Although the Sorenson index is a formula of binary data and the last two indices are the formulae of quantitative data, all three calculated results are very similar. Based on calculated β diversity values, a comparison of plant communities of five different periods was made. The β diversities of 1964 vegetation and 1981 vegetation were calculated (C_{MH} = 0.328, C_N = 0.024, C_j = 0.412). The comparative results showed that the difference in species composition between 1981 vegetation and 1987 vegetation was largest (C_{MH} = 0.935, C_N = 0.332, C_j = 0.819), suggesting that the turnover rate of species is greatest, followed by those of the 1964 and 1956 vegetation. From this it follows that after over 40 years of succession the turnover rates of restored plant species were largest in two time intervals, namely a 10 to 20–year interval and 30 to 40–year interval. In the first time interval a large number of annual herbaceous species invaded and the community changed from single shrub layer into a multilayer structure. In the second time interval high shrub species retreated and restored planted *Artemisia ordosica* began to decline. However, large amounts of seed were reproduced successfully, community structure became more complex and patchy, and herb and algae–lichen layers were apparent with herb as the dominant layer patch (its coverage is larger than that of shrub layer and moss layer). These results indicate that temporally the dynamical changes in plant diversity to a certain degree can reflect the feature of vegetation succession processes.

(The β diversity is calculated according to the communities before and after the planting years, hence no β diversity value of the first planting year exists.).

4.4 Relation Between Biodiversity and Habitat

After the restored plant species were established in shifting sand lands, the stability of the physical environment increased and the formation of sand surface crust was closely related to the diversity of species. The crust thickness in the range 0.32–2cm is positively correlated with the diversity of plant species. However, with increase in crust thickness, its resistance to rain water infiltration increased and this helped increase the chances of deep–rooted plants such as *Hedysarum scoparium* and *Caragana korshinskii* dying out because of lack of precipitation reaching the deep sand layer. On the other hand, because sand surface crust has better water retention, it creates a condition for the germination of many annual herbaceous plants which often develop and constitute the main body of community diversity.

The increase in soil nutrient content also is positively correlated to the increase of biodiversity, and that the decrease in sandy soil moisture content has less significant influence on plant diversity. This decrease is because the ephemeral annual plants that are the main components formed as a result of the development of the community diversity often complete their life cycles under conditions of lower rainfall. Therefore, the accumulation of soil organic matter and nutrients helps increase the plant diversity. The decrease in soil moisture content obviously affects the survival of shrubs.

This is also an important reason why the vegetation shows such a succession pattern from the dominant community consisting of shrubs and semi–shrubs to the community consisting of annual herbaceous species and semi–shrubs. From this it follows that the dynamical changes in plant diversity are closely related to the tendency of vegetation successional strategies and trajectories.

5. CONCLUSION

The establishment of the protective system of artificial vegetation along the railway in Shapotou region improved the environmental condition of the desert region as vegetation succession progressed; it also created conditions for the formation of plant diversity dominated by annual herbaceous plants. As succession progressed the abundance of (some) plant species increased; the percentage of annual herbaceous plants increased; the abundance of others such as restored (artificial) planted shrubs gradually became less abundant in the overall plant diversity composition. The species composition of restored plant species gradually showed a dynamically balanced state which was related to a thickened sand surface crust, reduced moisture content in the sand layer, and limited reproduction of deep—rooted shrubs and perennial herbaceous plants.

Diversity indexes of restored plant species reflect the common features of simple structure of desert vegetation and their sensitivity to various disturbances. Further detailed study is required to regulate and maintain the stability of the vegetation and to allow scientists to better ascertain the role of restored plant species in protection against desertification.

Based on our results, the determination and analysis of β diversity associated with time since restoration seems to be a good indicator of vegetative succession. The different β diversity indices show consistent results insofar as the plant successional processes experienced two important and relatively rapid turnover or replacement stages of species. These stages have important significance to the ecological restoration, oasis construction and management in arid desert regions.

The improvement of soil substrate helps to restore and increase the diversity of artificial vegetation in desert regions; limited precipitation has less significant influence on the plant diversity. However, with further succession the interspecific and intraspecific competition resulted from increased species diversity may inhibit the increase of species diversity.

6. ACKNOWLEDGEMENTS

This research was funded by the National Natural Science Foundation of China (project 90202015), the innovation project of CAREERI, Chinese Academy of Sciences (210031) and the foundation of open Shapotou Station (9703). We also gratefully acknowledge Dr. John Lemons for valuable comments on the manuscript.

7. REFERENCES

Bray, J.R., and Curis, J.D. 1957. An ordination of the upland forest communities of southern Wisconsin. Ecological Monographs 27: 325–349.

Duan, Z. H., and Liu, X.M. 1996. A study on the nutrient elements cycling of artificial vegetation area in Shapotou. Journal of Desert Research 16: 76–79.

Li, X.R. 2001. Study on shrub community diversity of Ordos plateau, Inner Mongolia, Northern China. Journal of Arid Environments 47: 271–279.

Li, X.R., Wang, X.P., Li, T., and Zhang, J.G. 2002. Microbiotic soil crust and its effect on vegetation and habitat on artificially stabilized desert dunes in Tengger Desert, North China. Biology and Fertility of Soils 35: 147–154.

Ma, K.P., Liu, C.R., and Liu, Y. 1995. Measurement of biotic community diversity: II β–diversity. Chinese Biodiversity 3: 38–43.

Peet, R.H. 1974. The measurement of species diversity. Annual Review of Ecological Systems 5: 285–307.

Pielou, E.C. 1975. Ecological diversity. John Wiley & Sons Inc., New York.

Shi Q. H. 1993. Annual Report of Shapotou Desert Research and Experiment Station, Chinese Academy of Sciences. Lanzhou: Gansu Science and Technology Press. Pages 89–107.

Simpson, E.H. 1949. Measurement of diversity. Nature 163:688.

Whittaker, R.H. 1972. Evolution and measurement of species diversity. Taxon 21: 213–251.

Xiao, H.L., Zhang, J.X., and Li, J.G. 1996. Development of fertility of aeolian soil in the processes of shifting sand fixation in the southeastern Tengger Desert. Journal of Desert Research 16: 64–69.

Zhao, X. L. 1988. A study on the control shifting sand dunes of Shapotou region in the edge of southeastern Tengger Desert. Ningxia People's Press, Yinchuan. 506 pages.

Chapter 14

PLANT PRODUCTION AND DIVERSITY AT DESERTIFICATION STAGES IN HORQIN SAND GRASSLAND REGION, CHINA

Chang Xueli [1] and Zhao Halin [2]

[1] *Department of Geography, Yantai Normal College, Shandong Yantai 264025, China; email xlchang@eyou.com, changxueli@fm365.com;* [2]*Cold and Arid Region Environmental and Engineering Reserarch Institute, Chinese Academy of Sciences, Lanzhou 73000*

1. INTRODUCTION

The selection of biodiversity indices has importance implications on the conclusions drawn from such studies (Cousins 1991). As simple and useful indices, species richness and the Shannon–Wiener (SW) indices have been widely used relative other indices for studying species diversity (Gough et al. 1994, Guo 1995). These two indices are used by many ecologists to describe species composition and function attributes of plant communities and to identify relationships between productivity and diversity (Hooper et al. 1997, Tilman et al. 1997). Due to the complexity, variation, and scale dependence of the traits of vegetation, there are many inherent uncertainties in the measurements and ultimate calculations of species richness and SW indices (Palmer 1988, Hegazy et al. 1998).

In this case study, we employed species richness and the SW indices to describe the diversity of the Horqin sandy grassland and analyze the impacts of diversity upon productivity, species, composition and functional diversity. The major arm of this study is to illustrate the function and effect of diversity in the Horqin sandy grassland ecosystem and provide decision makers with enhanced knowledge for restoration and rehabilitation of the vegetation in this region.

2. STUDY AREA

The study area is located at Naiman Banner, Inner Mongolia (42°14'10"–43°32'20"N, 120°19'40"–121°31'44"E) which borders the south part of the Horqin sandy grassland. It has a temperate continental semi–arid climate. Mean annual temperature is 6.1–6.4 °C and accumulative temperature (≥10 °C) is 3161.1 °C. Mean annual precipitation is 366 mm, 67.4 percent of which falls in May–July. Mean annual evaporation is 1935 mm. The zonal soil is sandy chestnut soil, which has been degraded into sandy soil due to wind erosion and human activities. Geomorphologically, the study area is characterized by sand dunes alternating with gently undulating interdunal lowlands. Dominant plant species include some annuals, perennials, and shrubs such as *Artemisia halodendron*, *Artemisia scoparia*, *Artemisia frigida*, *Caragana microphylla*, *Mellissitus ruthenicus*, *Cleistogenes squarrosa*, and *Setaria viridis*.

3. DEFINITION OF DESERTIFICATION STAGES

According to Zhu and Liu's classification criteria for desertification degree (1981), the desertification progression in the Horqin sandy grassland can be divided into four stages: (1)latent desertification stage, (2)slight desertification stage, (3)moderate desertification stage, and (4)severe desertification stage. Alternatively, these stages are also named, (1)fixed sand dune, (2)semi–fixed sand dune, (3)semi–moving sand dune, and (4)moving sand dune. There are different dominant species and vegetation coverage in the different desertification stages (Table 1).

Table 1. Dominant species and vegetation coverage at different desertification stages in the Horqin sand grassland

	Latent	Slight	Moderate	Severe
Dom. species	*Caragana microphylla, Cleistogenes squarrosa, Atemisia frigida, Lespedeza davurica*	*Atemisia Halodendro, Mellissitus Ruthenicus, Artemisia scoparia*	*Atemisia Halodendron, Pennisetum centrasiaticum*	*Agriophyllum squarrosum*
% Cover	≥50	30–50	10–30	≤10

4. MEASUREMENTS OF PRODUCTIVITY AND DIVERSITY

In August 1999, we randomly chose 72 quadrants (1m × 1m) at fixed, semi–fixed, semi–moving and moving sand dunes, corresponding to four desertification stages (each had 18 quadrants) to investigate vegetation composition, individual numbers, height, biomass, and emergence frequency of plant species. Twenty–nine species were identified (Table 2) and were divided into composition and functional groups. The composition group included three categories: annuals, perennials, and shrubs; the functional group included legumes, C4 plants, and C3 plants (Hooper et al. 1997, Tilman et al. 1997). Above ground biomass of plant (dry weight) was used to represent the productivity in this study (Gough et al. 1994). The SW index H was used to measure α diversity,

(1) $H = -\sum P_1 \times \ln P_1 \qquad (P_1 = n_1/N)$

(2) $P_1 = (ID/ID + H/H + F_1/F) \times 100$

where, H represents the Shannon–Wiener index; n_i, importance values of ith species (or categories); N, summation of the importance values of all species (or categories) in a quadrant; P_i, occurrence probability of ith species (or categories). ID_i, H_i and F_i are respectively individual number, mean height and occurrence frequency of ith species (or categories) around the quadrant; ID, H and F, are respectively the total species (or categories) individual numbers, the summation of height, and the frequency in the same quadrant. Where P_i is the relative importance value of species I (relative height + relative coverage + relative frequency).

Table 2. Species composition at the different desertification stages in the Horqin sand grassland (V denotes presence)

Desertification stages	Latent	Slight	Moderate	Severe
Plant species				
Caravan microphylla	V	V		
Atemisia frigida	V			
Atemisia halodendron	V	V	V	V
Mellissitus ruthenicus	V	V		
Cleistogenes squarrosa	V			
Lespedeza davurica	V	V		
Pennisetum centrasiaticum			V	
Phragmites australis		V	V	
Staria viridis	V	V	V	V

Table 2 (continued)

Desertification stages	Latent	Slight	Moderate	Severe
Eragrostis pilosa	V	V		
Digitaria sanguinalis	V	V		
Chenopodium acuminatum	V	V		
Euphobia humifusa	V	V	V	
Corispermum didutum	V	V	V	
Artemisia scoparia	V	V		
Salsola collina	V	V		
Ferula bungeana	V			
Ixeris chinensis	V	V	V	
Cynanchum thesioides	V			
Convolvulus arvensis	V	V		
Euphobia esula		V		
Chenopodium aristatum	V	V		
Aillum bidentatum	V			
Gueldenstaedtia stenophylla	V			
Erodium stephanianum	V	V		
Agriophyllum squarrosum			V	V
Leymu secalinus	V			
Echinops gmelini		V		
Bassia dasyphylla		V	V	

5. RESULTS

5.1 Characteristics of Diversity Composition

In the Horqin sandy grassland, the composition of life form groups and the functional groups of the vegetation varied at the different desertification stages (Table 3). For the life form composition group, the annual category had the highest species richness at the different desertification stages; its importance values were also highest at all the stages except the latent desertification stage. One primary feature was that at the latent desertification stage annual species exhibited the lowest importance (34.6) and the highest species richness (14). On the contrary, at the severe desertification stage, they displayed the highest importance value (62.0) and the lowest species richness (2). This implies that the annual category had different impacts on the desertification progression (from latent to severe desertification stage), and that it played a dominant role in enriched species composition of sandy grassland vegetation at the latent desertification stage. At the severe desertification stage, the annuals, as a constructive category, played an importance role to halting desertification progression. The species richness and the importance value of the perennial category were the lowest among the three life form composition categories. The highest species richness value and the highest importance value for the perennial's were only 6 and 22.2 respectively at the latent desertification stage. Both the species richness and importance values were reduced to 0.0 at the severe desertification stage. The importance value of shrub category varied little form 27.4 to 43.2, demonstrating that the shrubs might exert a buffer effect in the desertification process.

Table 3. Species richness and importance values of composition diversity groups at the desertification stages in the Horqin Sand Grassland. (IP, importance value; SR, species richness)

	Annual		Perennial		Shrub	
	IP	SR	IP	SR	IP	SR
Latent stage	34.6	14	22.2	6	43.2	3
Slight stage	56.7	14	10.0	4	33.4	2
Moderate stage	61.6	6	10.6	2	27.4	1
Severe stage	62.0	2	0	0	38.4	1

Table 4. Species richness and importance values of functional diversity groups at desertification stages in the Horqin sand Grassland. (IP, importance value; SR, species richness)

	Legume		C_4 plant		C_3 plant	
	IP	SR	IP	SR	IP	SR
Latent stage	26.4	4	35.6	8	37.9	11
Slight stage	22.9	3	39.8	8	37.3	9
Moderate stage	0.0	0.0	54.9	4	45.1	5
Severe stage	0.0	0.0	61.6	2	38.4	1

The composition analysis of the functional groups shows that the legume category only existed at the latent and slight desertification stages. Their importance value and species richness were lower than C3 and C4 plant categories in the Horqin sandy grassland (Table 4). Comparison of C3 and C4 plant categories indicates that, with the desertification progression, the importance value increased from 35.6 to 61.6 for C4 plant category while it varied between 37.3 and 45.1 for C3 plant category. In addition, the species richness of the C4 plant category varied less than that of the C3 plant category during the desertification process (e.g., 2–8 and 1–11, respectively.

5.2 Relation of Plant Productivity and Diversity

Correlation analysis shows that the relationship between productivity and life form composition diversity was closest (absolute value) at the latent desertification stage; the correlative degree was –0.534. The correlative degrees between productivity and the species diversity and between productivity and the functional diversity were lower than that in this stage (Table 5). Analysis data indicate that the values of life form composition diversity varied from 0.78 to 1.23; the productivity varied form 316g/m^2 to 645 g/m^2. Regression analysis shows that the relation between the life form composition diversity and the productivity was a power function as $y = 589.89\ x^{-0.1696}$ ($r = 0.72$) (y, productivity; x, life form composition diversity index), suggesting that the productivity decreased with the increase in the life form composition diversity. At the slight desertification stage, the relation between the productivity and the functional diversity was closest. The correlative degree was 0.722. On the other hand, the relation between the productivity and the life form composition diversity was lower at this stage than it was in the latent desertification stage. The functional diversity indices varied from 0.86 to 1.18, the productivity ranged from 189g/m^2 to 412 g/m^2. The relation could be expressed as a linear regression equation: $y = 8.1476x + 206.1$ ($r = 0.74$). Suggesting that the productivity increased with an increase of the functional diversity.

The relationships between the productivity and the life form composition diversity were closest at the moderate and severe desertification stages and the correlative degrees were –0.574 and –0.431, respectively. The life form composition diversity varied from 0.56 to 1.06, the productivity varied from 130g/m^2 to 319 g/m^2 at the moderate desertification stage. The relation could be expressed as a logarithmic function: $y = -45.53 \ln(x) + 307.34$ ($r = 0.61$). At the severe desertification stage, the life form composition diversity varied from 0.42 to 0.91 and the productivity varied from 31g/m^2 to 136 g/m^2. The relation between the productivity and the life form composition diversity was the most complex at all the stages, it showed multinomial regression function as $y = 0.0266x^3 - 0.3772x^2 - 4.7132x + 126.69$ ($r = 0.47$).

Table 5. The correlative degrees between the productivity and diversity indices at the different desertification stages in the Horqin sand grassland

Desertification stage	Latent	Slight	Moderate	Severe
Species diversity	0.129	0.119	–0.195	–0.300
Life form composition diversity	–0.534	–0.240	–0.574	–0.431
Functional diversity	–0.238	0.722	–0.399	0.318

The regression relation between the productivity and the diversity indices obviously changed at the different desertification stages. The functional diversity had the closest relation to the productivity at the slight desertification stage; the life form composition diversity had closest relations to the productivity at the other stages. According to the regression analysis, the regression relations between the productivity and the life form composition diversity were most significant at the latent desertification stages 0.72 > $P(|r| > r_{0.001})$ – 0.679; significant at the moderate desertification stage 0.61> $P(|r| > r_{0.01}) = 0.561$, and less significant at the serve desertification stage 0.47 > $P(|r| > r_{0.05}) = 0.444$. It was also obviously to figure out relation between the productivity and the functional diversity at the slight desertification stage, relation between them was also most significant 0.74 > $P(|r| > r_{0.001})$ – 0.679.

6. DISCUSSION

The implication of the relationship between productivity and diversity in plant ecological study has been recently emphasized by many ecologists (Hooper et al. 1997, Tilman et al. 1997). The measurements of life form composition, functional diversity, and productivity can not only identify the relationships between these parameters but also facilitate the rehabilitation and restoration of vegetation because the knowledge of the relationships can help decision makers to distinctly preserve the stability and the sustainable use of vegetation ecosystems; in this case, in the Horqin sandy grassland (Li 1993, Chang et al. 2000). In addition, many ecologists pay more attention to plant functional groups in ecological studies. Plant functional groups not only have close relations with energy flow and mass cycle in ecosystems, but also influences ecosystem stability and adaptive strategies of plant species (Chapin et al. 1997, Grime et al. 1997, Hooper et al. 1998, Meissner et al. 1999). Grime (1993) suggests that the plant functional group is a foundation in community and ecosystem ecology study, and can be used in building ecological modeling, and adjusting and predicting community and ecosystem responses.

In the Horqin sandy grassland, the desertification process distinctly affects the diversity of vegetation. It not only causes changes in species richness, but also influences the composition of life forms and the function groups. During the desertification process, the species diversity was reduced from 2.63 to 0.64; the life form composition diversity from 1.06 to 0.67; the functional diversity from 1.09 to 0.67. Legume plant category disappears after the moderate desertification stages; C3 and C4 plant categories exist at all the stages. However, the proportion of C4 plants category trends to increase with the desertification progression as indicated by an increase of its importance value. On the other hand, annual and shrub categories also exist at all the desertification stages and have more obvious effects than perennial category. Especially, annual category plays an importance role in the desertification process because its importance value increases with the desertification progression.

Understanding the relations between plant productivity and diversity depends on how to select diversity indices to describe vegetation at the different desertification stages. In general, productivity is related most closely to the life form composition diversity, moderately to the functional diversity, and weakly to the species diversity. Although the productivity is closely related to the life form composition diversity, they are negatively correlated at all the stages, indicating that the community composed of simple plant life forms would obtain higher productivity than that composed of complex plant life forms. The functional diversity has a closest relation to and a positive correlation with the productivity at the slight desertification stage, indicating that the higher evenness is the result of the higher productivity at this stage in the sandy grassland. This result differs with Tilman (1997) who obtained in the experimental plots with artificially controlled plants species richness and diversity components with Savanna annual plants. We also have not found that the productivity curves has single peaks with increasing diversity indices; our results therefore differ with the result of Gough et al (1994) because he employed merely species richness to analyze the relation between the productivity and diversity.

7. ACKNOWLEDGEMENTS

We thank Prof. S. R. Liu, Department of Biological Sciences of Inner Mongolia University, for providing us a list of C4 plants in the Hoeqin sandy grassland. This project was supported by the National Natural Sciences

Foundation of China (49601020) and The Innovation Project of CAREERI, CAS (210066).

8. REFERENCES

Chang, X.L., Yang, C., Zhao, H.L., and Xu, J. 2000. Influence upon the productivity of sandy grassland by plant species diversity in Horqin region. Acta of Applied Ecology 11(3): 395–398. (in Chinese).

Chapin, F.S., Walker, B.H., Hobbs R.J., Hooper D.U., Lawton, J.H., Sala, O.E. and Tilman, D. 1997. Biotic control over the functioning of ecosystems. Science 277: 500–509.

Cousins, S. H. 1991. Species diversity measurement: Choosing the right index. Tree 6: 190–192.

Gough, L., Grace, J.B., and Taylor, K.L. 1994. The relationship between species richness and community biomass: The importance of environmental variables. Oikos 70: 271–279.

Grime, J.P. 1993. Ecology sana frontieres. Oikos 68: 385–392.

Grime, J.P, Thompson, K., and Hunt R. 1997. Integrated screening validates primary axes of specialisation in plants. Oikos 79: 259–281.

Guo, Q.F. 1995. Advances and trends in species diversity studies. Pages 89–107 In: Li Bo (eds.) Lectures of Modern Ecology. Chinese Science Press, Beijing. (in Chinese).

Hegazy, A.K., El–Demerdash, M.A., and Hosni, H.A. 1998. Vegetation, species diversity and floristic relations along an altitudinal gradient in south–west Saudi Arabia. Journal of Arid Environment 38: 3–13.

Hooper, D.U. and Vitousek, P.M. 1997. The effects of plant composition and diversity on ecosystem processes. Sciences 277: 1302–1304.

Hooper, D.U. and Vitousek, P.M. 1998. Effects of plant composition and diversity on nutrient cycling. Ecological Monograph 68: 121–149.

Li, Y.H. 1993. Grazing dynamics of the species diversity in Aenurolepidium Chinese steppe and Stipa Grandis steppe. Acta Botanica Sinica 35: 877–884. (in Chinese).

Meissner, R.A. and Facelli, J.M. 1999. Effects of sheep exclusion on the soil seed bank and annual vegetation in chenopod shrublands of South Australia. Journal of Arid Environments 42: 117–128.

Palmer, M.W. 1988. Fractal geometry: A tool for describing spatial patterns of plant communities. Vegetatio 75:91–102.

Tilman, D., Knops, J., Wedin, D., Reich P. 1997. The influence of functional diversity and composition on ecosystem processes. Science 277: 1300–1302.

Zhu, Z.D. and Liu, S. 1981. Desertification process and its rehabilitation division in north China. Chinese Forestry Science Press, Beijing. 23 pages. (in Chinese).

Chapter 15

AQUATIC BIODIVERSITY IN ARID AND SEMI–ARID ZONES OF ASIA AND WATER MANAGEMENT

Brij Gopal
School of Environmental Sciences, Jawaharlal Nehru University, New Delhi 110067, India; email brij00@123india.com

1. INTRODUCTION

Contrary to the general impression, and paradoxical as it may appear, arid and semiarid regions are often quite rich in water resources and have many aquatic habitats. Some of the world's large rivers such as Nile, Indus and Murray–Darling, pass through the arid regions though their source of water lies far outside the arid zone (Williams 2000). Many large and deep lakes, mostly brackish or saline, also occur within the arid regions; for example Caspian Sea, Aral Sea, Dead Sea, Lake Chad, and the Great Salt Lake. There are hundreds of seasonal and ephemeral streams and thousands of shallow, freshwater and saline lakes (now better known as wetlands) of varying size and depth, many of which are permanent.

Yet, assessments of and conservation strategies for biodiversity in arid and semiarid regions tend to overlook the aquatic habitats. Our knowledge of the arid and semiarid zone ecosystems has been synthesized in several relatively recent books (Goodall and Perry 1979; West 1983; Evenari et al. 1985, 1986). These publications rarely refer to aquatic habitats or their biota. Only during the past few years, the richness and peculiarities of biodiversity in aquatic habitats of arid regions of Australia, Africa and North America have started receiving some attention (Robarts and Bothwell 1992, King et al. 1996; Witham et al. 1998; Williams 1998ab, 2000; Brendonck and Williams 2000), but the aquatic ecosystems of arid and semiarid regions of Asia (Figure 1) remain almost totally ignored. A recent contribution on wetlands of the world's arid zones to the First Conference of Parties of the UN Convention to Combat Desertification made no reference to any wetland in Asia though the list of 39 references included four from Asia (Kingsford 1997). However, a more recent paper by Mirabzadeh (1999) discussed in some detail the nature, distribution, values, threats and future of wetlands in the entire arid region of Asia except South Asia, parts of China and Mongolia. In another recent publication, valuable information on fisheries, plankton and benthos in many lakes of the Central Asian region has been summarized (Petr 1999a).

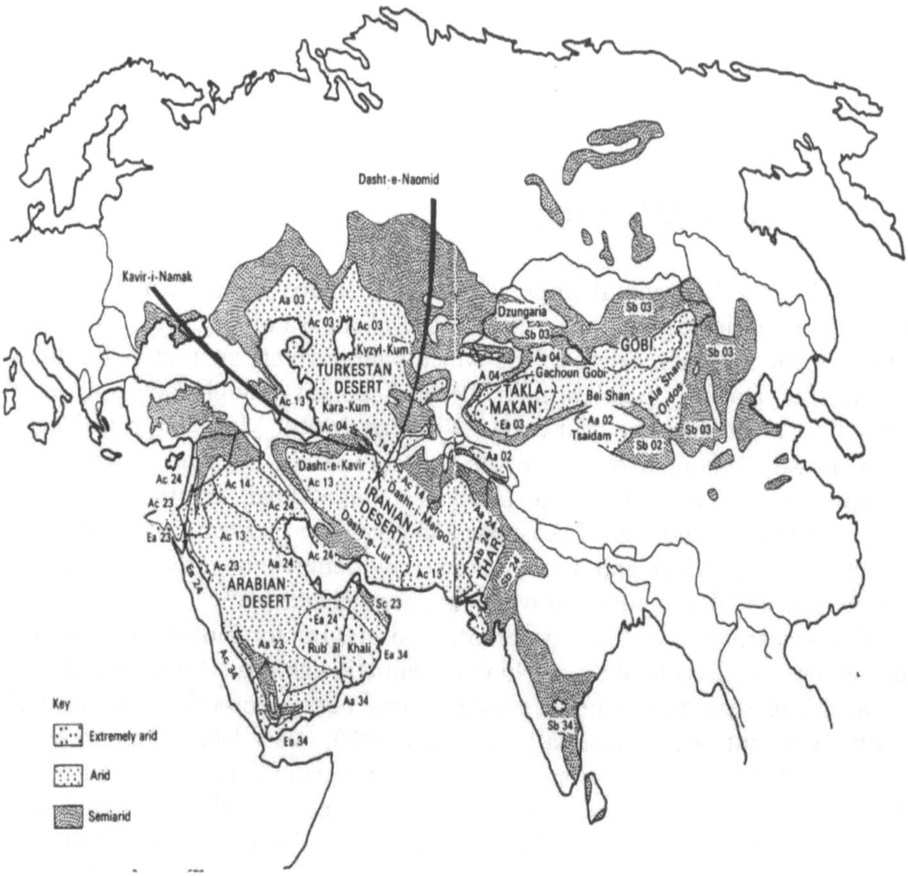

Figure 1. Arid regions of Asia according to Meigs (1953). E = extremely arid, A = arid, S = Semiarid; a = no distinct season of rain, b = summer rain, c = winter rain; first digit represents mean temperature of the coldest month and second digit represents mean temperature of the hottest month: 0 = <0°C, 1 = 0–10°C, 2 = 10–20°C, 3 = 20–30°C, 4 = >30°C. Reproduced from Shimida (1985).

This paper, therefore, attempts to review the major characteristics of aquatic habitats and their biodiversity in Asia's arid and semiarid regions. Further, threats to aquatic biodiversity are discussed particularly in the context of water resource management practices and their impacts. Though aquatic habitats throughout the arid and semiarid zones of Asia from the Middle East to Central Asia and Mongolia are examined briefly, examples are used mainly from the Indian subcontinent to discuss the issues of conservation and water resource management.

2. ARID AND SEMIARID REGIONS OF ASIA

Whereas the arid or semiarid regions account for nearly two–thirds of the African and Australian continents, they cover only about one third of Asia (Figure 1). Whereas Africa has the largest hyper–arid area, the Asian continent has more arid and semiarid area (13.3 million km^2) than Africa. However, most of the Middle East (about 84 percent), Central Asia, Mongolia, and western China (much of Qinghai, Xinjiang and Xizang provinces are also arid or semiarid (see Petr 1999a). In South Asia, most of Afghanistan and Pakistan are arid or semiarid (Nazneen 1995, Adil 2000). There is some disagreement on the limits of arid and semiarid regions in India. While Bharucha (1955) set an upper limit of precipitation for semiarid regions at 25 cm, others consider 50 cm as appropriate (Dyer 1955). Meher–Homji (1965, 1967, 1972) has reviewed various methods of defining limits of arid and semiarid regions. Various criteria result in somewhat different boundaries at the periphery but the core area remains unchanged. Parts of Rajasthan, west of Aravalli ranges and Gujarat, covering about 12 percent of India's total land area, are arid. Semiarid region extends to most of Rajasthan and parts of Haryana. Parts of peninsular India are also categorized as semiarid (Meher–Homji 1965). Some areas of even Sri Lanka, which lies close to Equator, are arid (Costa and de Silva 1995).

The arid regions of Asia are of greater significance beyond their biodiversity or aquatic habitats. The valleys of Indus, Tigris and Euphrates rivers are well known for the beginning of agriculture in their floodplains. Irrigation was also developed first in the Indus valley. The arid and semiarid regions of Asia have a higher density of human population and livestock, and are also known for their cultural diversity that is closely linked with biological diversity in the region.

Further, Asia's arid and semiarid regions are remarkable for the great contrasts and extremes of landscape features and climate. High mountain ranges with snow covered peaks form a nearly contiguous chain from east to west through the arid regions. Besides vast sandy plains with sand dunes, at the other extreme, there are large depressions well below the sea level. Thus, the region includes the highest peaks and lowest places on the earth (Dead Sea, 394 m below sea level). The world's hottest and driest places also lie within Asia's arid zone. The temperature extremes lie between more than 50° C during summer in hot deserts and –30° C in winters at high altitudes. Similarly, the diurnal variations in temperature are also very large. Annual precipitation exhibits large temporal–seasonal, inter–annual and long–term variability. Some years receive heavy rainfall that equals the average precipitation of 10 or 20 years in one day while there is no rain in several consecutive years. The timing of precipitation within the summer and winter seasons is also highly variable, and rainfall is generally patchy (Evenari 1985).

3. AQUATIC ECOSYSTEMS OF ARID AND SEMIARID ASIA

Large hydrological variability coupled with geomorphological variability ensures a large diversity of aquatic habitats varying in size, depth and duration of water. Salinity is also commonly associated with aridity, and therefore, saline lakes with various degrees of salinity are widespread.

Water bodies in arid and semiarid region owe their origin to the runoff from areas outside them and to snow melt. Cold deserts of Asia are surrounded by glacier–covered mountain peaks and hence, are often richer in water bodies than the hot deserts such as the Arabian Desert. Most standing waters are brackish or saline. Despite low amounts of precipitation, water bodies develop also due to poor drainage. Soils in hot arid regions are generally rich in calcium and therefore, impervious subsurface layers are formed by precipitation of calcareous pans that readily cause waterlogging. Precipitation readily accumulates in depressions for sufficiently long periods to allow many organisms to inhabit these pools. Most aquatic habitats are, therefore, temporary and seasonal.

Several perennial streams, which originate invariably in areas with high precipitation pass through the arid regions and form extensive floodplains due to large seasonal changes in their discharge. There are numerous seasonal and ephemeral streams. However, the landscape is dotted with innumerable human–made water bodies. Large spatial and temporal variability in rainfall has necessitated storage of seasonal runoff and transport of water through canals for long distances.

Only the Arabian Peninsula is without significantly large aquatic habitats, but extensive intertidal mudflats, mangroves and lagoons occur along the coasts of the Arabian Gulf, Gulf of Oman and the Arabian Sea. Mountains and highlands of Saudi Arabia, Oman and Yemen have a large number of natural springs that are harnessed into extensive networks of human–made channels referred to as "aflaj." This region also has a large number of retention reservoirs some of which are permanent and recharge reservoirs that are temporary.

In Syria, aquatic habitats are confined mostly to riverine marshes along the course of Euphrates and a shallow salt lake Sabkat al Jabboul. Several shallow water bodies are formed also after winter rains. Lebanon, Israel and Jordan have several water bodies. Azraq oasis in eastern Jordan is a large (about 100 km^2) complex of marshes fed largely by springs. Lake Tiberias (freshwater) and the Dead Sea (salt) are well known lakes in Israel. Apart from the seasonal marshes of Ammik Lebanon has two important human–made lakes, Tanayel and Qaroun.

In Iraq, extensive network of lakes and marshes occurs in the floodplains of Tigris and Euphrates rivers which flood over 15000 km^2 in their lower reaches (Mesopotamia) during winter and spring after winter rains in headwaters (mostly in Turkey). Rzoska (1980) provided a detailed account of these two rivers and associated habitats. Some of the temporary or permanent shallow water bodies, called hawr or hor, formed by inundation of lowlands spread over hundreds of square km area; e.g., Tharthar (2700 km^2), Habbaniya (430 km^2), Hawr al Hammar (2500 km^2) and Dibbis (1985 km^2). Other important areas are: Hawr Dalmaj, Hawr Ibn Najim, Hawr al Suwayqiyah, Hawr as Saniyah and Hawr Hawizeh. Several large reservoirs constructed on these rivers, mainly for irrigation and flood control, include Samarra, Kut, Ramadi, Hindiya, Yao, Meshkab, Hafar and Akaika.

Iran is equally rich in aquatic habitats particularly in its northern part where a 700 km long network of lakes, lagoons, and marshes occurs south of the Caspian Sea. Anjali wetlands (Anjali Murdab) are among the most significant. In northern Azarbayjan province lies a large (4830 sq km) hypersaline lake Orumiyeh (Urmieh) with large areas of important marshes such as Ghara Gheshlaq, Dorgeh Sanghi, Gerde Gheet and Kobi. Three major rivers, Karun, Dez and Kharkeh flood the southwestern plain and form extensive seasonal wetlands. Shadegan marshes alone cover over 2900 sq km area. Further south along the Gulf coast, floodplain and deltaic wetlands occur in Helleh river basin. Numerous saline, brackish (Perishan, Maharloo, Bakhtegan, Tashk) and freshwater lakes (Dasht–e–Arjan and Haft Barm lakes) occur in the central Fars province. The Neiriz (Niriz or Neyriz) lake alone, when flooded during periods of heavy rainfall, covers some 1800 sq km but gets fragmented into many smaller shallow water bodies/marshes during the dry period (Löffler 1953, 1956). The Seistan basin, contiguous with Afghanistan on the east, has another complex of freshwater lakes of which Hamoun–i–Puzak, Hamoun–i–Sabari and Hamoun–i–Helmand are prominent. Mangroves and mudflats along the Gulf and Arabian Sea coast are other major aquatic habitats.

Afghanistan has four major river systems namely Kabul, Amu, Helmand and Harirud. Jamoun–i–puzak, Lake Hashmat (near Kabul), Dasht–e–Nawar and Abi–Estada are important among lakes (Petr 1999b). A few small lakes occur along river Kabul and in Bande Amir Valley in Hazarajat Mountains.

Among Central Asian Republics, Kyrgyzstan has 1923 lakes covering about 684 km^2 whereas Tajikistan has more than 700 km^2 of wetlands. Important lakes in the region are Kourgaldzhin, Tengiz, Zaysan and Alakol in Kazakhstan, Issyk–kul, Chatyrkel and Sankel in Kyrgyzstan, Arnassy, Shurkul and Sartigamish in Uzbekistan, and Sevan in Armenia (Mirabzadeh 1999, Mitrofanov and Petr 1999, Savvaitova and Petr 1999 ab). A large area of Mongolia forms a part of the vast Central Asian closed drainage basin with more than 4000 rivers and streams traversing the territory (Finch 1996). Several large rivers such as Selenge, Shishkhed, Bulgan, Onon, Kherlen and Khalkhin Gol drain towards the Arctic and Pacific. There are about 3500 lakes of which more than 85 percent are less than 100 ha in area and another 13 percent range between 100 and 1000 ha. Some of the very large lakes include Uvs Nuur (3350 km^2), Urureg (238 km^2), Achit (290 km^2) and Tolbo in Altai region, Khovsgol Nuur (2620 km^2) in Hangayn region, Khar Us Nuur (1852 km^2), Khar Nuur (575 km^2), Durgun (305 km^2) and Khyargas Nuur (140 km^2) in the Great Lakes Valley and Boon–tsagaan (252 km^2) and Orog (140 km^2) in Gobi valley (Finch 1996, Dulmaa 1999).

Although most of the western China is arid to semiarid (Walker and Yang 1999), many large rivers (Brahmaputra, Mekong, Yangtse, Huangho) pass through it. Besides, there are hundreds of lakes, mostly saline (Wang 1987, Williams 1991).

In India, parts of Rajasthan, west of Aravalli ranges and Gujarat are arid, though the semiarid region extends to most of Rajasthan and even parts of Haryana. There is some disagreement on the limits of arid and semiarid regions in India. While Bharucha (1955) set an upper limit of precipitation for semiarid regions at 25 cm, others consider 50 cm as appropriate (Dyer 1955). Meher–Homji (1962, 1965, 1972) has reviewed various methods of defining limits of arid and semiarid regions. Various criteria result is somewhat different boundaries at the periphery, but the core area remains unchanged. Also parts of peninsular India are categorized as semiarid (Meher–Homji 1965).

Many rivers, mostly ephemeral, drain these areas. Seasonal rivers, such as Banas, form narrow floodplains. Natural water bodies are few and some examples are, salt lakes at Sambhar, Deedwana, and Pachpadra. In Gujarat, flooding occurs naturally during rains in the Kutch peninsula, which has a complex of saline marshes. All along the Gujarat coast mangroves occur in small patches. Nal Sarovar is probably the only natural shallow water area. Thousands of large and small reservoirs are, however, the dominant feature of Gujarat and Rajasthan as well as other semiarid parts of peninsular India.

In Pakistan, Indus is the main river, which forms extensive floodplains and a large delta with mangroves and intertidal mudflats. The river has been regulated at several places to divert water into an extensive canal system. Besides many natural water bodies, there are numerous man–made reservoirs constructed mainly for irrigation, domestic water supply and aquaculture (Nazneen 1995).

4. BIODIVERSITY OF AQUATIC HABITATS

An assessment of the biodiversity of aquatic habitats in the arid and semiarid regions can be made readily from the enormous size of waterfowl populations that use them seasonally. Arid and semiarid zone wetlands of Asia lie along the flyways of numerous species of waterfowl, which migrate annually from northern latitudes southwards and therefore, constitute very important habitats for staging and over–wintering. It was indeed the biodiversity of these arid zone wetlands that led to the adoption of the Convention on Wetlands of International Importance, especially for the Waterfowl (better known as the Ramsar Convention) at Ramsar (Iran) in February 1971. It is surprising that aquatic habitats of arid and semiarid regions have yet been considered as poor in biodiversity (Wiggins et al. 1980, Maltby 1991).

Waterfowl diversity has generally been well documented and frequently monitored. Whereas some habitats support very large populations of one or few species (e.g. flamingoes in Rann of Kutch marshes and Sambhar lake vide Hussain and De Roy (1993), others support very large number of species (e.g., 370 species in Keoladeo National Park, Rajasthan vide Chauhan and Gopal (2001) and 200 species in Nal Sarovar Wildlife Sanctuary in Gujarat). Many species of waterfowl in Asia's arid region are rare and threatened. These include Dalmatian pelican (*Pelecanus crispus*), great white pelican (*Pelecanus onocrotalus*), pygmy cormorant (*Phalacrocorax pygmeus*), northern bald ibis (*Geronticus eremitta*), marbled teal (*Anas angustirostris*), red breasted goose (*Branta ruficollis*), lesser white fronted goose (*Anser eruthropus*), Siberian crane (*Grus leucogeranus*), white headed duck (*Oxyura leucocephala*), white eyed gull (*Larus leucophthalmus*) and Audouin's gull (*Larus audounii*).

Fish are also relatively well known from the arid regions of Asia. Though fish diversity in cold deserts is not as impressive as in other tropical regions, it is quite significant in view of the restricted distribution of most of the species. Cyprinids and cobitids are common and many species are endemic. There are 64 species, representing 11 families of fish in Mongolia most of which are confined to one or the other drainage basin (Dulmaa 1999). Banister (1980) listed 84 species including 60 cyprinids and 10 cobitids from Euphrates and Tigris river basins in Iraq. More than 100 species of fish, mostly cyprinids, cobitids and siluriformes have been recorded from rivers in Afghanistan (Petr 1999b). *Schizothorax* and *Noemacheilus* are most dominant genera. It is interesting that several species recorded earlier from Seistan basin such as *Adiposia macmahoni*, *Adiposia rhadinaea* (both endemic), *Discognatus adiscus*, *Discognathus phryne* and others (Annandale 1919–1921) have not been reported in later surveys (Coad 1981). Only two species of *Noemacheilus* had been reported from the high altitude lakes of Ladakh (Gopal et al. 2002) though River Indus has a richer fish fauna. Recent surveys conducted by the Wildlife Institute of India have found 24 species of fish in the region (BC Choudhary, personal communication). Various rivers, including Indus, flowing through the arid and semiarid parts of Pakistan have diverse and characteristic fauna that differs appreciably between different drainage basins (Nazneen 1995). Fish diversity is considerably high also in some reservoirs of hot semiarid regions of India (Durve 1976, Mahajan et al. 1984, Ajithkumar and Vijayan 1988) but in most cases it is due to stocking for aquaculture.

Biodiversity of higher plants is generally low. Grasses (particularly Phragmites), sedges (species of *Cyperus*, *Scirpus* and *Carex*), and cattails (*Typha* species) are common throughout the arid regions. Among submerged macrophytes, species of *Potamogeton*, *Zannichelia*, *Najas*, *Chara*, etc., are very common. Species richness declines greatly with increasing salinity. Typical halophytes occur in saline marshes whereas mangroves along the coast are relatively poor in species, usually dominated by *Avincennia*.

Assessments of biodiversity generally tend to focus on larger organisms and particularly those of direct benefit to mankind. In fact, aquatic biodiversity in arid and semiarid regions is not too small, but it comprises of less conspicuous organisms (Williams 1998b, 2000). Arthropods constitute a large proportion of the earth's total biodiversity, and majority of the arthropods like crustaceans live in aquatic habitats, while many species of insects, depend on these habitats for their larval life. A significant proportion of this fauna live in arid/semiarid regions and play an important role in the food web of waterfowl.

All major groups of fauna occur in arid and semiarid water bodies, and many taxa (e.g., Notostraca, Conchostraca, Anostraca) are restricted to temporary waters. Ostracoda is another crustacean group common in temporary waters–fresh, saline or marine (Keeley and Zedler 1998). Citing examples from Australia, Williams (2000) pointed out that the aquatic faunal diversity in the temporary waters of arid and semiarid regions might be as high or even higher than that of permanent waters. Endemic, rare and endangered species in a region are important components of biodiversity. On continental and regional scales, many species in several groups of aquatic organisms are endemic to arid and semiarid zones.

Aquatic habitats of Asian arid regions are only poorly explored and an inventory of biodiversity of smaller organisms other than waterfowl and fish has rarely been made. A few examples will illustrate the biodiversity of different arid regions of Asia. Annandale (1919–1921) collected from various aquatic habitats in Seistan (Afghanistan) 10 molluscs, 19 insects, 4 ostracods, 6 cladocerans, 3 copepods, 5 annelids, 4 polyzoans, 3 sponges and several organisms of other groups. The biodiversity of Lake Issyk–kul includes 299 species of phytoplankton, 117 species of zooplankton (mostly rotifers) and 224 zoobenthic species (Savvaitova and Petr 1999a). Lake Sevan (Armenia), on the other hand, has quite low diversity that includes 26 phytoplankton, 15 zooplankton and zoobenthos dominated by a few chironomids and leeches (Savvaitova and Petr 1999b). Tso Moriri, a high altitude lake in the cold desert of Ladakh, provides an example of unexplored biodiversity. Maheshwari (unpublished) recorded more than 40 new species of chironomids during a single visit to the lake in August 2000 (Gopal et al. 2002). Highly saline water bodies are characterized by the occurrence of salt–tolerant species such as *Dunaliella* and *Artemia* which contribute significantly to the food webs.

Brief mention must be made here of periodic fluctuations of biodiversity in arid and semiarid regions. As pointed out earlier, the hydrology of arid zones is characterized by very large temporal variability in precipitation, besides its scarcity. The arid zone organisms are adapted to prolonged dry periods. Both plants and invertebrates perennate through seeds, eggs and resting stages from which they emerge rapidly upon availability of water and often complete their life cycle within a short period. These adaptations are relatively well known (Williams 1987, Bonis et al. 1995, Brendock et al. 1998), although studies from the Asian arid regions are rare (Middleton et al. 1991, Chatterjee and Gopal 1998). However, even small changes in the depth, duration or timing of water availability result in significant differences in species composition of the communities in different years or adjacent habitats (Keeley and Zedler 1998, Simovich 1998). Recently, Gopal and Chauhan (2001) have discussed these fluctuations in relation to monsoonal variability in South Asia (see also Chauhan and Gopal 2001 for an example from semiarid India).

5. VALUES AND FUNCTIONS OF BIODIVERSITY

Besides providing food and habitats to huge populations of migratory waterfowl and several other migratory species, arid and semiarid zone wetlands such as seasonally flooded marshes are major providers of biological resources such as food, fodder, fiber, fuel. Natural water bodies (both rivers and lakes) provide fairly large amounts of fish. Higher fish yields have been obtained through aquaculture in numerous reservoirs with exotic species (Nazneen 1995, Sugunan 1995, Petr 1999a). Waterfowl have also been widely hunted for food. Plants like reeds and cattails are important as fuel, thatch for hutments, and also as fodder. Various seasonal wetlands are valuable grazing grounds for a variety of animals in both hot and cold climates. Cattle and buffaloes in arid and semiarid parts of Gujarat and Rajasthan depend greatly on seasonal wetlands, whereas in Ladakh sheep and goat graze in extensive sedge marshes along water bodies.

The vegetation of these habitats plays also an important role in checking both wind and water erosion. The aquatic habitats in arid regions not only enhance the aesthetic value of the landscape, but also have a moderating influence on the microclimate of the surrounding areas, which are therefore preferred for human settlements.

6. HUMAN IMPACTS

Over 700 million people and an even greater population of grazing animals such as cattle, camel, sheep and goat inhabit arid and semiarid regions of Asia. Along with a steady rise in human population, arid areas are being increasingly urbanized. Both agriculture and industrialization are growing rapidly. This has resulted in manifold increase in demand for water, which is met by excessive exploitation of surface and groundwater and inter–basin transfer through canals. The arid regions of Rajasthan and Gujarat in India are excellent examples. The extraction of groundwater has resulted in a rapid fall of groundwater table. The recharge of aquifers is slow, if not totally lacking, due to the formation of hard pans below the soil surface.

Habitat destruction and degradation are widely regarded as most important factors responsible for the loss of biodiversity. In arid and semiarid zones, destruction or degradation of aquatic habitats occurs with any hydrological change that disrupts natural hydrological regimes. The management of water resources therefore becomes the prime agent of habitat degradation and loss. The utilization of meager surface water resources readily turns into "over–exploitation" and high evaporation rates in hot arid regions contribute to accelerated drying of the water bodies. The shortening of the hydroperiod is followed by direct disturbance as the land is often brought under various land uses including irrigated agriculture. The consequent loss of habitat means the loss of its biodiversity.

An example of changes in biodiversity following water extraction is offered by Lake Sevan (Armenia). After withdrawal of water started in 1938 for hydropower and irrigation, the water level fluctuated for some period, but declined considerably during 1950s and 1960s, exposing large littoral areas. The resultant loss of spawning areas (shallow littoral zone) and its food (gammarids) caused a steep decline of the endemic Sevan trout fisheries. Low water levels coupled with increasing pollution also resulted in several other changes in biotic communities. The appearance of blue green algal blooms replacing diatoms, abundance of rotifers in place of copepods, and decline of *Chara*–dominated macrophyte beds are some notable effects (Savvaitova and Petr 1999b).

Construction of dams and diversion of stream flows seriously affect the downstream habitats (e.g., River Indus in Pakistan and R. Banas in Rajasthan). Reduced freshwater discharges to coastal waters also result in increased salinity and decline of mangroves (e.g. Saurashtra in Gujarat).

Paucity of water and extended dry periods lead to overexploitation of wetland resources (e.g., overgrazing) and consequently, habitat degradation. It may, however, be pointed out that the arid zone ecosystems, both terrestrial and aquatic, have coevolved with grazing animals over many centuries. Therefore, complete exclusion of grazing and forage removal is also detrimental to biodiversity. The accumulated organic matter accelerates the filling up of the water bodies and alters the hydrological regime. Undesirable weedy species increase as competition from the grazed species decreases. Further, habitat changes also occur and result in a decline in biodiversity (Gopal and Chauhan 2001).

Creation of water bodies and transporting water from long distances have since historical times been a major human response to nearly unpredictable availability of water in arid and semiarid regions. Despite the fact that these human–made aquatic habitats have become important habitats for numerous species, particularly migratory waterfowl, disruption of natural hydrological regimes by the habitats with longer, often permanent, hydroperiods have caused many problems for biodiversity of both aquatic and terrestrial habitats. Waterlogging has been a common phenomenon in large areas of arid regions in India due to seepage from canals and excessive use of irrigation water (NIH 1992–93). In Rajasthan Canal (now Indira Gandhi canal) command area alone, about 8000 ha land is permanently inundated and another 1000 ha have been converted to marsh. The changes in biodiversity, particularly loss of native species and invasion by exotics are well documented in the arid parts of Rajasthan receiving water from Rajasthan Canal. The highly nutritious grass *Lasiurus sindicus*, which requires less than 150 mm annual rainfall, has almost disappeared in the command area giving way to less nutritious species. Similarly, more than 17000 ha in Mahri and Ukai–Kakrapar command areas have become waterlogged in Gujarat due to canals (Khandelwal et al. 1996, IRMA 2001).

Increased water availability in arid regions has caused serious socioeconomic and other impacts. Not only the agricultural production has increased and hence, the economy of the farmers and others associated with agricultural trade has improved, the newfound affluence has resulted in the displacement of local nomadic communities whose traditional grazing grounds have been brought under cultivation (Mathur 1984, Sinha 1996). Further, it has also caused a sharp rise in criminal activities and social conflicts. Another major consequence has been the emergence of malarial epidemic and other diseases unknown to hot arid regions (Shiva and Shiva 2000).

Enhanced availability of water leads to increased pollution of both surface and ground waters. Besides domestic and industrial effluents, agrochemicals are becoming a major source of pollution. Growing salinity and fluoride pollution are also consequences of altered hydrology of arid regions. However, there is very little specific information available on the impact of various pollutants on wetland biota in Asia.

Another anthropogenic impact of increasingly greater importance comes from intentional or accidental introduction of exotic species some of which are highly invasive. Fish introductions have been widespread in Central Asian lakes tote fisheries. Many introduced species have established well. Introduction of Sevan salmon (*Salmo ischchan gegarkuni*) a native of Armenia, caused rapid decline of native fisheries in Lake Issyk–kul in Kyrgyzstan though Sevan salmon itself was saved from extinction. Even unintentional introductions have succeeded (e.g., *Pseudorasbora parva* in Irtysh river basin in Kazakhstan). In other countries, carps have been extensively introduced for increasing fish yields at the cost of native fishes.

In tropical arid and semiarid regions, several exotic macrophytes have become a serious problem. Water hyacinth is among the most common invasive species, which accelerates drying of the water body through increased evapotranspiration and also causes rapid siltation through excessive inputs of organic matter. Another exotic, *Ipomoea fistulosa* is fast becoming a serious threat to aquatic habitats especially in the arid tropical regions.

7. CONSERVATION OF BIODIVERSITY

For centuries, humans have recognized the importance of aquatic habitats in arid and semiarid regions, and have taken steps to protect them for various purposes such as grazing, water supply or simply recreational hunting. Long before conservation became a buzzword, steps were taken by the former rulers in some countries to conserve natural aquatic habitats and create new ones. Enhancing the habitat as a duck shooting reserve more than a century ago created Keoladeo National Park (India) itself. Countries from arid regions were among the first to join the Ramsar Convention, which promotes conservation of aquatic habitats (wetlands) worldwide. Until February 2002, 16 Asian countries with whole or a large part of their territory as arid or semiarid, had joined the Ramsar Convention and designated 67 sites as wetlands of international importance.

In most countries, however, there is no specific policy aimed at wetland conservation. The wetlands designated under the Ramsar Convention represent only a small fraction of total wetlands in these regions. A significant aspect of arid zone wetlands is the large variability in species richness and composition of communities at the same time over short distances due to differences in hydrological regimes (Timms 1993). Thus, it is very important that the countries in arid and semiarid regions develop a national policy for conservation of wetlands in the context of their river basins.

8. CONCLUSION

Conservation of aquatic biodiversity is directly related to the management of water resources. This is more significant in arid and semiarid regions where extraction of water as well as increased supply through water storage and irrigation projects is decimating native biodiversity. Whereas natural habitat such as floodplains, marshes and mangroves have been eliminated by changes in land use, numerous man–made reservoirs and waterlogged areas provide no substitute though they often attract large populations of waterfowl. Both natural and man–made aquatic habitats are also getting degraded due to pollution, overexploitation of natural resources, increased salinization and introduction of exotic species. Conservation of aquatic biodiversity in arid zones therefore requires sustainable management of water resources.

9. REFERENCES

Annandale, A. 1919–1921. Report on the aquatic fauna of Seistan, with subsidiary studies. Parts I–V. Records of the Indian Museum 18: 1–253.

Adil, A.W. 2000. National biodiversity strategies and action plans (NBSAP) for Afghanistan. Paper presented at the First Workshop on National Biodiversity Strategies & Action Plans in Northeast and East Central Asia: Experiences and Lessons (Beijing, 26–28 April 2000). 10 pages.

Ajithkuma, R., and Vijayan, V.S. 1988. On the fish of Keoladeo National Park, Bharatpur (Rajasthan). Journal of the Bombay Natural History Society 85: 44–49.

Banister, K.E. 1980. The fishes of the Euphrates and Tigris. Pages 95–108 In: Rzoska, J. (Editor). Euphrates and Tigris: Mesopotamian Ecology and Destiny. Monographiae Biologicae 38. Dr W. Junk Publishers, The Hague. 132 pages.

Bharucha, F.R. 1955. Structural and physiological features of Rajasthan desert. Pages 34–37 In: Proceedings of the Montpellier Symposium on Plant Ecology. UNESCO, Paris.

Bonis, A., Lepart, J., and Grillas, P. 1995. Seed bank dynamics and coexistence of annual macrophytes in a temporary pool and variable habitat. Oikos 74: 81–92.

Brendonck, L., Riddoch, B.J., Van de Weghe, V., and Van Dooren, T. 1998. The maintenance of egg banks in very short–lived pools–a case study with anostracans (Branchiopoda). Archiv für Hydrobiologie, Special Issues 52: 141–161.

Brendonck, L. and Williams, W.D. 2000. Biodiversity in wetlands of dry regions (drylands). Pages 181–194 In: Gopal, B., Junk, W.J. & Davis, T.A. (Editors) Biodiversity in Wetlands: Assessment, Function and Conservation. Vol. 1. Backhuys Publishers, Leiden. 353 pages.

Chatterjee, K., and Gopal, B. 1998. Experimental study of emergence of zooplankton in temporary waterbodies in relation to dry periods. Verhandlungen der internationale Verieinigung fur Limnologie 26: 1309–1315.

Chauhan, M., and Gopal, B. 2001. Biodiversity and management of Keoladeo National park (India)–a wetland of international importance. Pages 217–256 In: Gopal, B., Junk. W.J., & Davis, J.A. (Editors) Biodiversity in Wetlands: Assessment, Functions and Conservation, Volume 2. Backhuys Publishers, Leiden. 311 pages.

Coad, B.W. 1981. Fishes of Afghanistan, an annotated checklist. Publications in Zoology 14. National Museum of Canada, Ottawa. 26 pages.

Costa, H.H., and de Silva, P.K. 1995. Limnological research and training in Sri Lanka: state of the art and future needs. Pages 63–104 In: Gopal, B. and Wetzel, R.G. (Editors) Limnology in Developing Countries, Vol. 1. International Scientific Publications, New Delhi. 230 pages.

Dulmaa, A. 1979. Hydrobiological outline of the Mongolian lakes. Internationale Revue der gesamten Hydrobiologie 64: 709–736.

Dulmaa, A. 1999. Fish and fisheries in Mongolia. Pages 187–236 In: Petr, T. (Editor) Fish and Fisheries at Higher Altitudes: Asia. FAO Fisheries Technical Paper 385. Food and Agricultural Organization of the United Nations, Rome.

Durve, V.S. 1976. Fisheries of the lake Jaisamandh, Rajasthan. Journal of the Inland Fisheries Society of India 8: 19–23.

Evenari, M. 1985. The desert environment. Pages 1–22 In: Evenari, M., Noy–Meir, I. and Goodall, D.W. (Editors). Hot Deserts and Arid Shrublands, A. Ecosystems of the World 12A. Elsevier, Amsterdam.

Evenari, M., Noy–Meir, I., and Goodall, D.W. (Editors). 1985. Hot Deserts and Arid Shrublands, A. Ecosystems of the World 12A. Elsevier, Amsterdam. 365 pages.

Evenari, M., Noy–Meir, I., and Goodall, D.W. (Editors). 1986. Hot Deserts and Arid Shrublands, B. Ecosystems of the World 12B. Elsevier, Amsterdam. 451 pages.

Finch, C. (Editor). 1996. Mongolia's Wild Heritage. Ministry of Nature and Environment, Ulan Baatar, Mongolia, and Avery Press, Boulder, CO, USA. (see online version at http://www.un-mongolia.mn/wildher/online-intro.htm).

Goodall, D.W., and Perry, R.A. (Editors). 1979. Arid Land Ecosystems. Volume 1. IBP series 16. Cambridge University Press, Cambridge. 881 pages.

Gopal, B. ,and Chauhan, M. 2001. South Asian wetlands and their biodiversity: The role of monsoons. Pages 257–276 In: Gopal, B., Junk, W.J. & Davis, T.A. (Editors) Biodiversity in Wetlands: Assessment, Function and Conservation. Vol. 2. Backhuys Publishers, Leiden. 311 pages.

Gopal, B., Chauhan, M., and Zutshi, D.P. 2002. High altitude lakes of Ladakh: Tso Moriri and Tso Kar. WWF–India, New Delhi. (in press).

Hussain, S.A., and De Roy, R. 1993. Directory of Indian Wetlands. WWF–India, New Delhi. 263 pages.

IRMA (Institute of Rural Management, Anand). 2001. White Paper on Water in Gujarat. Report to the Government of Gujarat, Gandhinagar. Institute of Rural Management, Anand, Gujarat.

Keeley, J.E., and Zedler, P.H. 1998. Characterization and global distribution of vernal pools. Pages 1–14 In: Witham, Carol W., Bauder, E., Belk, D., Ferren, W. and Ornduff, R. (Editors) Ecology, Conservation and Management of Vernal Pool Ecosystems. Proceedings from a 1996 Conference. California Native Plant Society, Sacramento, CA, U.S.A. 296 pages.

Khandelwal, M.K., Gupta, S.K., and Tyagi, N.K. 1996. Mismatch between canal water supply and demand in Ukai–Kakrapar Irrigation Command. In: Gupta, S.K. and Tyagi, N.K. (Editors) Waterlogging and Soil Salinity in Ukai Kakrapar Command: Causes and Remedial Measures. Central Soil Salinity Research Institute, Karnal.

King, J.L., Simovich, M.A., and Brusca, R.C. 1996. Species richness, endemism and ecology and crustacean assemblages in northern California vernal pools. Hydrobiologia 328: 85–116.

Kingsford, R.T. 1997. Wetlands of the world's arid zones. Briefing paper at the First Conference of Parties of the UN Convention to Combat Desertification, Rome, 29 September to 10 October 1997.

Löffler, H. 1953. Limnologische Ergebnisse der Osterreichischen Iran expedition 1949–50. Naturwissen–schaftliche Rundschau 6: 64–68.

Löffler, H. 1956. Limnologische Untersuchungen an Iranischen Binnengewässern. Hydrobiologia 8: 201–278.

Mahajan, C.L., Narayana, H.S., Ratnam, B.V., and Jain, B.L. 1984. Final Report of Pilot Project on Reservoir Fisheries Management in Rajasthan. University of Rajasthan, Jaipur. 112 pages.

Maltby, E. 1991. Wetlands and their values. Pages 8–26 In: Finlayson, C.M. and Moser, M. (Editors) Wetlands. Facts on File, Oxford. 224 pages.

Mathur, P.C. 1984. Social Aspects of Development and Management of Natural Resources: A Survey of Literature on Irrigation and Water Management in India, with Special Reference to Rajasthan Canal Project, 1984. WP 13. Institute of Development Studies, Jaipur. 76 pages.

Meher–Homji,V.M. 1965. Aridity and semi–aridity: A phytoclimatic consideration with reference to India. Annals of Arid zone 4: 152–158.

Meher–Homji, V.M. 1967. On delimiting arid and semi–arid climates of India. Indian Geography Journal 42: 1–6.

Meher–Homji,V.M. 1972. An approach to the problem of defining aridity in the Indo–Pakistan subcontinent. Geographical Reviews of India 34: 205–215.

Meigs, P. 1953. World distribution of arid and semi–arid homoclimates. Pages 203–210 In: Anonymous (Editor) Reviews of Research on Arid Zone Hydrology (Arid Zone Research Series 1). UNESCO, Paris. 364 pages.

Middleton, B.A., van der Valk, A.G., Davis, C.B., Mason, D.H., and Williams, R.L. 1991. Vegetation dynamics and seed banks of a monsoonal wetland overgrown with *Paspalum distichum* in northern India. Aquatic Botany 40: 239–259.

Mirabzadeh, Parastu A. 1999. Wetlands in western Asia. Background paper for 'Conservation and Wise Use of Wetlands in Western Asia. Ramsar Bureau, Gland, Switzerland. 18 pages.

Mitrofanov, V.P. ,and Petr, T. 1999. Fish and fisheries in the Altai, northern Tien Shan and Lake Balkhash (Kazakhstan). Pages 149–167 In: Petr, T. (Editor) Fish and Fisheries at Higher Altitudes: Asia. FAO Fisheries Technical Paper 385. Food and Agricultural Organization of the United Nations, Rome. 304 pages.

Nazneen, S. 1995. State of limnology in Pakistan. Pages 191–230 In: Gopal, B. and Wetzel, R.G. (Editors) Limnology in Developing Countries, Vol. 1. International Scientific Publications, New Delhi. 230 pages.

NIH (National Institute of Hydrology). 1992–93. Status of Waterlogging, Soil Salinity and Alkalinity in India. Technical Report TR 143. National Institute of Hydrology, Roorkee.

Petr, T. (Editor). 1999a. Fish and Fisheries at Higher Altitudes: Asia. FAO Fisheries Technical Paper 385. Food and Agricultural Organization of the United Nations, Rome. 304 pages.

Petr, T. 1999b. Coldwater fish and fisheries in Afghanistan. Pages 138–148 In: Petr, T. (Editor) Fish and Fisheries at Higher Altitudes: Asia. FAO Fisheries Technical Paper 385. Food and Agricultural Organization of the United Nations, Rome. 304 pages.

Robarts, R.D., and Bothwell, M.L. (Editors). 1992. Aquatic Ecosystems in Semi–arid Regions: Implications for Resource Management. National Hydrology Research Institute Symposium Series 7, Environment Canada, Saskatoon. 404 pages.

Savvaitova, K.A., and Petr, T. 1999a. Fish and fisheries in Lake Issyk–kul (Tien Shan), River Chu and Pamir lakes. Pages 168–186 In: Petr, T. (Editor) Fish and Fisheries at Higher Altitudes: Asia. FAO Fisheries Technical Paper 385. Food and Agricultural Organization of the United Nations, Rome. 304 pages.

Savvaitova, K.A., and Petr, T. 1999b. Fish and fisheries in Lake Sevan, Armenia, and in some other high altitude lakes of Caucasus. Pages 279–304 In: Petr, T. (Editor) Fish and Fisheries at Higher Altitudes: Asia. FAO Fisheries Technical Paper 385. Food and Agricultural Organization of the United Nations, Rome 304 pages.

Shimida, A. 1985. Biogeography of the desert flora. Pages 23–78 In: Evenari, M., Noy–Meir, I. and Goodall, D.W. (Editors). Hot Deserts and Arid Shrublands, Ecosystems of the World 12A. Elsevier, Amsterdam.

Shiva, M., and Shiva, V. 2000. The Political Ecology of the Resurgence of Malaria in India. Third World Network, Penang, Malaysia.(http://www.twnside.org.sg/title/india-cn.htm)

Simovich, M.A. 1998. Crustacean biodiversity and endemism in California's ephemeral wetlands. Pages 107–118 In: Witham, C.W., Bauder, E., Belk, D., Ferren, W. and Ornduff, R. (Editors) Ecology, Conservation and Management of Vernal Pool Ecosystems. Proceedings from a 1996 Conference. California Native Plant Society, Sacramento, CA, U.S.A. 296 pages.

Sinha, S. 1996. Resisting Change? Adaptations by Traditional Pastoralists to the Rajasthan Canal Project. Drylands Issue Paper IP 62. IIED, London. 21 pages.

Sugunan, V.V. 1995. Reservoir Fisheries of India. FAO Fisheries Technical Paper 345. Food and Agricultural Organization of the United Nations, Rome. 423 pages.
Timms, B.V. 1993. Saline lakes of the Paroo, inland New South Wales, Australia. Hydrobiologia 267: 269–289.
Walker, K.F., and Yang, H.Z. 1999. Fish and fisheries in western China. Pages 237–278 In: Petr, T. (Editor) Fish and Fisheries at Higher Altitudes: Asia. FAO Fisheries Technical Paper 385. Food and Agricultural Organization of the United Nations, Rome. 304 pages.
Wang, H. 1987. The water resources of lakes in China. Chinese Journal of Oceanography and Limnology 5: 263–280.
West, N. (Editor). 1983. Temperate Deserts and Semi–Deserts, A. Ecosystems of the World 5. Elsevier, Amsterdam. 522 pages.
Wiggins, G.B., Mackay, R.J., and smith, I.M. 1980. Evolutionary and ecological strategies of animals in annual temporary ponds. Archiv für Hydrobiologie, Supplement 58: 97–206.
Williams, D.D. 1987. The Ecology of Temporary Waters. Croom Helm, London. 205 pages.
Williams, W.D. 1991. Chinese and Mongolian saline lakes: A limnological overview. Hydrobiologia 210: 33–66.
Williams, W.D. (Editor) 1998a. Wetlands in a Dry Land: Understanding for Management. Workshop Proceedings. Environment Australia, Biodiversity Group, Canberra, Australia. 335 pages.
Williams, W.D. 1998b. Dryland wetlands. Pages 33–47 In: McComb, A.J. and Davis, J.A, (Editors) Wetlands for the Future. Gleneagles Publishing, Adelaide. 780 pages.
Williams, W.D. 2000. Biodiversity in temporary wetlands of dryland regions. Verhandlungen der internationale Vereinigung für Limnologie 27: 141–144.
Witham, C.W., Bauder, E., Belk, D., Ferren, W., and Ornduff, R. (Editors) 1998. Ecology, Conservation and Management of Vernal Pool Ecosystems. Proceedings from a 1996 Conference. California Native Plant Society, Sacramento, CA, U.S.A. 296 pages.

Chapter 16

TRADITIONAL MANAGEMENT OF BIODIVERSITY IN INDIA'S COLD DESERT

K. Chandrasekhar[1], R. Gavali , K.S. Rao[2], R.K. Maikhuri[3], and K.G. Saxena[1]

[1]School of Environmental Sciences, Jawaharlal Nehru University, New Delhi 110067, India, eamil saxena2002in@yahoo.com; [2]Sustainable Development and Rural Ecosystems Programme, G.B. Pant Institute of Himalayan Environment and Development, Kosi–Katarmal, Almora 263643, India; [3]Garhwal Unit, G.B. Pant Institute of Himalyan Environment and Development, P.O. Box 92, Srinagar (Garhwal), India

1. INTRODUCTION

India's arid region can be divided into a hot arid zone in the west, inhabited by the Aryans, and a cold arid region in the north India, inhabited by indigenous tribal communities. This article deals with biodiversity issues in India's cold desert region.

Traditional resource uses and management demonstrate that biodiversity has been a key organizing principle in traditional landscape management. The goal of biodiversity conservation and its sustainable utilization can be better achieved if conservation development policies build on strengths and weaknesses of traditional knowledge and institutions.

2. COLD DESERT

2.1 Location, Terrain, Climate and People

The Himalayas, the planet's youngest mountain system, comprises two major biogeographic zones: trans–Himalaya and main Himalaya. The trans–Himalaya zone spreads over 3 million square kilometers in Afghanistan, Pakistan, India and the Tibetan plateau in China (Anonymous

1994). In India, this ecoregion covers the Ladakh region with elevations varying from 3000 meters (m) to more than 6500 m. The climate is arid: Average annual rainfall is 100 millimeters. The region is inhabited by tribes with Mongoloid affinity. People are Buddhists and thus have a religious appreciation for wildlife. The highest permanent settlements are at about 4000 m, the highest cultivation at 4300 m, and the highest temporary grazing settlements at 4750 m. The region, one of the most thinly populated in south Asia, remained isolated until 1974 when it was opened to tourism.

2.2 Flora

A 1994 survey sampled 985 species belonging to 353 genera and 79 families. Asteraceae is the most dominant family followed by Poaceae and Brassicaceae. The flora comprise four major elements: boreo–alpine, Euro–Siberian or Trans Palaearctic, southern European or Mediterranean elements, and Siberian–Mongolian. The area is an important repository of medicinal plants and many medicinal species have been designated rare and endangered species.

2.3 Fauna

Among large mammals, four species–snow leopard (*Panthera uncia*), wolf (*Canis lupus*), ibex (*Capra ibex*) and bharal (*Pseudois nayaur*) are widely distributed across the region. Conversely, Ladakh urial (*Ovis vignei*), Tibetan argali (*Ovis ammon*), wild ass (*Equus kiang*), Tibetan gazelle (*Procapra picticaudata*), Tibetan antelope (*Pantholops hodgsoni*), wild yak (*Bos grunniens*), wild dogs (*Cuon alpinus*), lynx (*Felis lynx*) and brown bear (*Ursus arctos*) are confined to isolated pockets (Mallon 1991, Fox et al. 1991). Different agencies give different conservation status to different species. For example, the International Union for the Conservation of Nature and Natural Resources' Red Book lists snow leopard and wild yak as endangered species and Tibetan wolf and wild ass as vulnerable species (Anonymous 1990a). The Convention on International Trade in Endangered Species of Fauna and Flora cites Tibetan argali, Tibetan antelope and urial as "threatened with extinction" (Anonymous 1990b).

2.4 Land Use–Cover

The regional landscape can be divided into categories. One, is settled crop–livestock mixed farming landscapes characterized by moderately dissected terrain, gentle slopes, and lower altitudes (less than 4300 m). These village–scale landscapes, which vary from 1 to 5 square kilometers in area, are highly dispersed. Two, are almost pristine areas, including permanent snow areas and adjoining alpine vegetation, that are practically impossible to access for any consumptive resource use. Three, are remaining areas characterized by a highly dissected terrain and steep slopes that nomadic tribes (Gaddis) use for summer grazing.

2.5 Village Landscapes

Throughout the region, villages are divided into large landholders (more than 1 hectare) and small landholders (less than 1 hectare). Village landscapes may be geographically stratified into riverside village landscapes and hillside village landscapes. Riverside villages are smaller in size and located at lower altitudes. They are characterized by a larger cultivated land to pasture land ratios than hillside villages. Pastures in riverside villages are too small to satisfy local requirements; hillside villages, meanwhile, have a surplus of forage. Meadows in hillside villages can be divided into two subregions: (1)areas nearby dwellings where households have equal resource use rights; and (2)areas away from dwellings, called Doksa, that can be used only by small landholders who assume the role of Dogpa and Tarjia. Dogpa take milking dzomos (a cross between yaks and cows) and cows to the Doksa for summer grazing (July to September). Tarjia do the same with horses. In return, Dogpa traditionally are entitled to 50 percent of milk products; Tarjia, meanwhile, are paid in the form of grain or cash.

3. TRADITIONAL FARMING

Hillside and riverside village communities differ more in their composition of livestock holdings and cropping patterns than in their total land and livestock holdings. The proportion of sheep, yaks and donkeys is substantially higher in hillside villages than in riverside villages that have a higher proportion of goats.

All agricultural land is privately owned. Decisions on crops to be grown and ploughing, manuring, mulching and weeding intensity are taken at the household level. However, religious heads (Lamas) fix the schedule for sowing and harvesting and community consensus determines the amount of water that each household receives for irrigation.

Only one crop can be harvested each year. Barley (*Hordeum vulgare*) is the main staple food crop complemented by wheat (*Triticum aestivum*). The latter is grown at altitudes of less than 3600 m. Covered barley (*Hordeum winlense*) and black pea (*Pisum arvense*) are grown mainly as fodder crops. The latter's seeds have human food value but are bitter in taste and eaten only during food shortages. Cover barley and wheat residues are mixed with meadow hay for livestock stall feeding in winter. Cultivation of green pea (*Pisum sativum*) as a cash crop, which recently began in the Spiti catchment, is rapidly replacing traditional crops. Similarly, large scale potato cultivation has marginalized traditional staple crops in Lahul region, northwest of Spiti catchment (Jina 1995).

Throughout the region, agriculture is organic and requires irrigation. Livestock excreta generated during winter months of stall feeding is incorporated during ploughing. Above ground biomass of *Artemisia salsoloides*, *Polygonum tortuosum* and *Thermopsis inflata* is turned into mulch. Subsurface water flow is used for irrigation in higher altitudes. Snow melt flows are used for irrigation in lower altitudes where subsurface water sources are inadequate. Water is stored in small ponds and diverted through downslope channels. Ponds are shared and managed jointly by 3 to 6 families. Manure, mulch and irrigation inputs vary depending on the crop and altitude. About 10 percent of cultivated land is left fallow each year. Agricultural expansion no longer takes place in the cold desert region due to shortages in irrigation water.

Unlike some mountain regions where policies encourage the cultivation of cash crops to boost the economy (Singh et al. 1997, Ives and He 1998, Renaud et al. 1998), expansion of potato and pea cultivation in cold desert region is driven by sociocultural changes from a subsistence to market economy that is facilitated by improved accessibility and supplies of staple food grains at prices subsidized by the government. Because sweet pea and potato require less irrigation water, their cultivation is feasible via traditional irrigation systems. Marginalization of such traditional crops as *Hordeum winlense* and *Pisum arvense* and increasing stress on cash crops (Chu–Lin et al. 1999) has led to a narrowing of the traditional food base and has created additional risks for local livelihoods (Maikhuri et al. 1996, Giampietro 1997).

4. LIVESTOCK AND CROP DEPREDATION

Snow leopards prefer a rocky terrain and areas of low human land use. In contrast, wolves prefer open valley areas close to settlements. There is a strong similarity in wild herbivores predated by these two carnivores as well as partial overlap of their habitats (Fox and Nurbu 1990).

Local people view wolves as more serious threats than snow leopards because the former is a year round livestock predator while the latter attacks mostly in winter. Estimates of livestock killing by snow leopard fall within a range of 1 to 12 percent of total stockholding (Schaller *et al.* 1987, Fox et al.

1988, Jackson 1991, Mallon 1991, Oli et al. 1994, Mishra 1997). Assessment of killings by wolves has not received as much attention as those by snow leopard. A survey of three villages in and near Kibber wildlife sanctuary (Mishra 1997) shows that almost all livestock killings over an 18 month period were caused by snow leopards and none by wolves. However, temporal changes and alterations in ecological conditions also could be responsible for the variation in the frequency of attacks by different predators. A comparison of a typical riverside and a hillside village in Spiti catchment shows that killings caused by snow leopard were more frequent in the former and by wolf in the latter.

Attacks by snow leopards inside stock pens at night cause severe losses due to multiple killings. It is not clear if livestock predation derives from shortages of wild prey of snow leopards or from their opportunistic behavior. Precise estimation of predation losses is difficult because predation as a cause of livestock death is confounded by such factors as disease, poor nutrition, weather and accident. Assessments on livestock depredation, moreover, are based on killings reported by farmers who often deliberately inflate the figures to increase compensation grants from government or who fail to distinguish between proximate and ultimate causes of livestock deaths (Wagner 1988).

Livestock have less ability to escape than wild herbivores and therefore are easily predated (Nowell and Jackson 1996). Local people traditionally accept losses due to snow leopards, an infrequent hazard. Precautions taken to minimize this loss include: (1)grazing by goats, sheep, donkeys and cows is kept close to permanent settlements; (2)robust males are selected and kept as community animals to enhance the ability of domesticated yaks and cross–breeds to escape; (3)only yaks, dzomos and horses are taken to the Doksa; (4)a few non–milking dzomos are kept outside encampments at night to curb attacks on more valuable animals.

Retaliatory killings of snow leopards by local people are likely only when an animal causes frequent and large–scale damage and when leopards are found guarding kills of large domestic animals (Fox and Chandawat 1988, Fox and Nurbu 1990, Oli et al. 1994).

Blue sheep and snow cocks are traditional delicacies but local people rarely hunt these herbivores, partly because of religious and sociocultural norms discouraging wildlife killings (Fox and Nurbu 1990, Mallon 1991) and partly because of the ability of these herbivores to escape. Whenever a blue sheep is captured, its meat is shared by all families of the village. People also understand that reduction in population of wild herbivores may increase the frequency of attacks on livestock. Thus there seems no immediate threat to the snow leopards' or wolves' main natural preys.

Crop depredation by wildlife is as serious a problem as livestock depredation both in subhumid and humid cold areas (Maikhuri et al. 2000) and hot desert regions (Shekhar 1998). But it is not as serious a problem in arid and semiarid cold regions. Not one case of crop depredation was observed or reported in the Spiti catchment of the cold desert region, possibly because of limited crop husbandry and abundance of wild forage.

5. GRAZING PRACTICES

Village meadows are divided into hay making areas near dwellings and grazing areas away from dwellings. Each shows a moisture gradient, with moist habitats in depressions on one extreme and highly dry habitats on steep slopes on the other. Grazing areas are foraged by all village livestock (excluding those in charge of Dogpa) for 2 to 3 months during the peak growing season. In the hay areas, grazing takes place during May and June, and the following September through November after harvesting of hay and mulch material. Livestock is stall fed from December to April. The hay area is divided into plots distributed equitably among households. The community fixes the period of hay harvesting (usually 12 days in August). When the hay area is grazed, the entire area is considered common land. Private ownership of grazing/hay making land (Zhao and Zhou 1999, Singh and Saxena 1998) is not observed in Indian cold desert. Grazing in hay areas tend to favor growth of hay/mulch species such as *Cicer microphyllum*, *Cousinia thomsoni*, *Seseli trilobum*, *Polygonum tortuosum* and *Poa calliopsis*. A family violating the community norms is penalized by burning the hay harvested during non–permissible periods and forcing offenders to contribute to the community fund.

Of the four most dominant species in a community/habitat type, *Caragana versicolor* and *Cousinia thomsoni* are common to all sites on steep and moderate slopes, and *Carex infuscata*, *Carex stenophylla* and *Poa calliopsis* on all marshy habitats in depressions. *Kobresia royleana* and *Seseli trilobum* are the distinguishing codominants on steep slopes at higher altitudes, *Lonicera spinosa* and *Krascheninnikovia ceratoides* on moderate slopes at lower altitudes, *Cicer microphyllum* and *Lindelofia stylosa* on very gentle slopes at higher altitudes, *Halerpestis sarmentosa* in marshy habitats at higher altitudes, and *Plantago tibetica* in marshy habitats at lower altitudes.

Although it is difficult to draw conclusions on impacts of traditional hay making and grazing on biodiversity potential from such field samplings, variable impacts of these resource uses are clearly evident. Hay making seems to promote dominance of *Cousinia thomsoni* and *Cicer microphyllum* on drier slopes. In marshy habitats, grazing seems to facilitate dominance of *Pedicularis rhinanthoides* across the altitude sampled.

While livestock population has increased recently, the grazing area has not. In discussions, none of the respondents identified fodder shortage as limiting livestock productivity or causing a decline in the availability of other locally valued plant resources. Experimental studies are needed to ascertain the carrying capacity of these meadows.

6. FORAGING

Dry steep slopes traditionally are subjected to lower intensity grazing (one day a week) compared to moderate slopes and moist habitats in depressions (six days a week). Lower intensity grazing in dry areas is based on a widespread perception of poor regrowth if subject to high intensity grazing compared to moist sites.

Forage preferences vary with time, community structure and habitat, and livestock type. All livestock forage aboveground except donkeys, which often uproot *Astragalaus zanskarensis* and consume its underground parts. There is considerable similarity in species preferred by goats and sheep and by dzomos and cows. Divergence in foraging behavior is more marked on slopes than in moist depressions. During early stages of vegetation regrowth (June), livestock consume *Carex curta* in the highest quantities in moist habitats. In dry habitats (steep and moderate slopes) *Caragana versicolor* is the preferred forage of sheep and goats, *Festuca valesiaca* of donkeys, and *Agropyron canaliculatum* of cows and dzomos. At the peak stage of vegetation growth (August), *Polygonum tortuosum* is the preferred forage of sheep and donkeys, *Festuca valesiaca* of goats, and *Cousinia thomsoni* of cows and dzomos in dry habitats. In moist sites, *Plantago tibetica* is foraged most by all livestock except sheep which preferred *Carum carvi*. At the end of growing season (September), *Artemisia salsoloides* is the preferred species of sheep, dzomos and cows, *Polygonum aviculare* of goats and *Astragalus zanskarensis* of donkeys in dry sites. In moist sites, *Poa calliopsis* is most frequently foraged by livestock except donkeys that display a preference for *Carex infuscata*.

Grazing causes environmental heterogeneity by reducing dominance and nutrient enrichment from livestock excreta. Livestock also may help the germination and dispersal of some species. Since people use almost all dung deposits in grazing land as fuel, the contribution of livestock excreta to biodiversity and nutrient cycling may not be significant.

7. CONVENTIONAL CONSERVATION

Crop husbandry, animal husbandry, and wild biodiversity are closely integrated subsystems traditionally managed at the village landscape scale, where resource use and conservation are integrated with the livelihood of local communities. Traditional sociocultural mechanisms of fostering systematic and regulated use of wild plant resources have evolved to optimize economic outputs from domesticated biodiversity. Conventional approaches to conservation such as the establishment of wildlife sanctuaries and national parks have assumed traditional practices are detrimental to the conservation of wild biodiversity and ecosystem function. Such perceptions often lead to conflicts between people and protected area managers.

7.1 Interference with Traditional Grazing Systems

When an area is legally protected, traditional grazing rights are partially (in wildlife sanctuaries) or completely (in national parks) terminated. Such a policy, however, is only partially enforced because of a lack of any physical fencing and limited capacity and commitment of protected area staff. Indeed termination of grazing rights in protected areas may enhance plant diversity and ecosystem function if the grazing pressure is intense. Long–term benefits from such enhancement may encourage local communities to reduce livestock holdings and switch to alternative livelihood strategies (Sharma and Shaw 1993, Fox 1993, Maikhuri et al. 2000). Such local responses are possible only when conservation agencies demonstrate the unsustainability of traditional grazing practices and offer alternatives to the benefits denied due to the termination of grazing rights. No scientific evidence shows conclusively that traditional grazing practices are detrimental to the goal of biodiversity conservation and ecosystem function. If disturbances caused by traditional resource uses are moderate, diversity may decline following abandonment of hay making and livestock grazing and recuperation of lost diversity may be slow and costly (Kotiluoto 1998, Stampfli and Zeiter 1999). A few long–term studies on arid ecosystems suggest herbivores cause species specific changes at site–scale while landscape–scale changes may be linked to short and long–term variation in climatic patterns (Reyerson and Parmenter 2001). Ecological impacts of traditional resource use and management must be scientifically evaluated before such uses are abandoned or modified.

7.2 Linkages Between Local and Outside Communities

Local communities traditionally allow grazing of livestock from outside the region supervised by nomadic tribes (Gaddis) because: (1)the tribes use areas not grazed by local livestock and thus do not threaten local livelihoods; (2)the presence of livestock brought by the tribes reduces the probability of depredation of local livestock; and (3)nomadic tribes barter essential commodities carried from the foothills that are unavailable locally. Since 1950, all uncultivated lands has been declared "public forest and wasteland." The Forest Department, the custodian of these public lands, began granting grazing permits to individuals in all areas except those that were protected (Chakravarti–Kaul 1998). In the process, department officials ignored traditional institutions favoring reciprocity between the nomadic Gaddis and local communities. Local communities protested in some parts of the Himalayas against policies that threatened their livelihood, forcing government to restore, at least partially, traditional institutions (Price and Thomson 1997). Such responses thus far have not taken place in cold desert regions but are likely in the future if conservation is promoted at the cost of local economic development.

7.3 Sustainable Land Management

Increasing influence of market forces is leading to a replacement of traditional fodder yielding crops by non–fodder (e.g., potato) and poor fodder (e.g., sweet pea) crops. Increasing stress on such cash crops, in turn, implies reduction in fodder availability from private farm land and more pressure on meadows to meet livestock feed needs. Such trends may curb the productivity of meadows, which could threaten local livelihood as well as biodiversity. Cultivation of fodder, fuelwood and medicinal species harvested from the wild in agricultural fallow land (10 percent of presently cultivated area) could reduce pressure on protected areas (Rao et al. 1999, Saxena et al. 2001). *Cousinia thomsoni* and *Cicer microphyllum* among fodder species, *Caragana versicolor*, *Astragalus zanskarensis* and *Krascheninnikovia ceratoides* among fuelwood species, and *Allium humile* and *Dactylorhiza hatagirea* among medicinal species are some of the potential species that could be cultivated. Enhancement of local capacity and knowledge with appropriate scientific and institutional inputs is needed for successful cultivation of these species. Farmers traditionally tend to maximize manure and mulch inputs. They remain unaware that it is the synchrony between nutrient release from manure/mulch and crop requirement that determines crop yields and not the quantity of these inputs. Improvement in traditional manuring and mulching practices could also help reduce pressure on meadows without any economic loss to local communities. Irrigating some meadows near settlements and enhancing local skills could further improve the local economy while limiting pressure on wild resources.

7.4 Wildlife Killing

Penalties for killing wildlife imposed by protected area managers is another key change that follows enforcement of protected areas. Local people often dislike this policy because it treats outsiders hunting wildlife for game or economic gains on par with local people who resort to killing only when a wild animal turns extremely hostile. Involvement of local people for surveillance of game or commercial hunting, together with interventions that keep the predators away from settlements and their resource catchments, could help resolve such conflicts.

7.5 Compensation for Livestock Depredation

Protected area management provides cash compensation for livestock killed by wildlife. However, funds are too low to compensate for the losses, and the procedure is too complicated for the community to understand (Mishra 1997, Maikhuri et al. 2000). Compensation policies, moreover, contribute to the erosion of such traditional institutions as Dogpa and Tarjia that have been instrumental in reducing risks posed by depredation and overgrazing.

8. CONCLUSION

Conventional conservation strategies have sought to achieve their goals by enforcing legal protection rather than by demonstrating that conservation, apart from its global benefits, is in the interest of local communities. This has engendered local skepticism and opposition. High levels of biodiversity within protected areas derive, partly if not fully, from the sustainability of traditional resource use and management over centuries. Exclusion of local communities from protected areas and the promotion of conservation divorced from the development aspirations of local people cannot be justified either ethically or scientifically. Nevertheless, traditional practices, which survived in the past in a subsistence economy, cannot be fully effective in the emerging market economy. As a result, conservation objectives could be better achieved by capitalizing on the strength of indigenous knowledge systems, overcoming the weaknesses of traditional resource use and management practices through scientific and institutional inputs, and involving local communities in the management planning, implementation and monitoring of protected areas.

9. ACKNOWLEDGEMENTS

We are grateful to the Director, G.B. Pant Institute of Himalayan Environment and Development and the Director, Wildlife Institute of India for the facilities, to the Ministry of Environment and Forests and Tropical Soil Biology and Fertility Programme for financial support and to Dr. G.S. Rawat for their suggestions and advice.

10. REFERENCES

Anonymous 1990a. Red list of Threatened Animals. World Conservation Monitoring Centre, IUCN, Cambridge.

Anonymous 1990b. Appendices I and II, Convention on International Trade in Endangered Species of Wild Fauna and Flora. CITES Secretariat, Lausanne.

Anonymous. 1994. Indian Cold Desert: A Status Report on Biodiversity. Wildlife Institute of India, Dehradun.

Chakravarty–Kaul, M. 1998. Transhumance and customary pastoral rights in Himachal Pradesh: Claiming the high pastures for Gaddis. Mountain Research and Development 18: 5–17.

Chu–Lin, L., Fox, J., Xing, L., Lihang, G., Kui, C., and Jieru, W. 1999. State policies, markets, land use practices and common property: 50 years of change in a Yunnan village. Mountain Research and Development 19: 133–139.

Fox, J. 1993. Forest resources in a Nepali village in 1980 and 1990: Positive influence of population growth. Mountain Research and Development 13: 89–98.

Fox, J. ,and Chundawat, R. 1988. Observations of snow leopard stalking, killing and feeding behavior. Mammalia 52: 137–140.

Fox, J.L., and Nurbu, C. 1990. Hemis, a national park for snow leopard in Indiaís trans–Himalaya. Int. Ped. Book of Snow leopards 6: 71–84.

Fox, J.L., Sinha, S.P., Chundawat, R.S., and Das, P.K. 1988. A field survey of snow leopard presence and habitat use in northwestern India. Pages 99–111 In: Freeman, H. (Ed.) Proceedings of the Fifth International Snow Leopard Symposium, International Snow Leopard Trust, Bellevue, Washington, and Wildlife Institute of India, Dehradun, India.

Fox, J.L., Sinha, S.P., Chundawat, R.S., and Das, P.K. 1991. Status of snow leopard Panthera uncia in northwest India. Biological Conservation 55: 283–298.

Giampietro, M. 1997. Socioeconomic constraints to farming with biodiversity. Agriculture, Ecosystems and Environment 62: 145–167.

Ives, J.D., and He, Y., 1996. Environmental and cultural change in the Yulong Xue Shan, Lijiang District, N W Yunnan, China. Pages 1–16 In: Chaing, M. (Ed.) Montane Mainland South East Asia in Transition. Chiang Mai University, Chiang Mai, Thailand.

Jackson, R. 1991. A Wildlife Survey of the Qomolangma Nature Preserve, Tibet Autonomous Region, People's Republic of China. Woodlands Mountain Institute, Franklin, West Virginia.

Jina, P.S. 1995. High Pasturelands of Ladakh Himalaya. Indus Publishing Company, New Delhi.

Kotiluoto, R. 1998. Vegetation changes in restored semi–natural meadows in the Turku Archipelago of SW Finland. Plant Ecology 136: 53–67.

Maikhuri, R.K., Rao, K.S., and Saxena, K.G., 1996. Traditional crop diversity for sustainable development of Central Himalayan agroecosystems. International Journal of Sustainable Development and World Ecology 3: 8–31.

Maikhuri, R.K., Nautiyal, S., Rao, K.S., Chandrasekhar, K., Gavali, R., and Saxena, K.G. 2000. Analysis and resolution of protected area–people conflicts in Nanda Devi Biosphere Reserve, India. Environmental Conservation 27: 43–53.

Mallon, D.P. 1991. Status and conservation of large mammals in Ladakh. Biological Conservation 56: 101–119.

Mishra, C. 1997. Livestock depredation by large carnivores in the Indian trans–Himalaya: Conflict perceptions and conservation prospects. Environmental Conservation 24: 338–343.

Nowell, K., and Jackson, P. 1996. Status and Conservation Action Plan: Wild Cats. IUCN/SSC Cat Specialist Group, IUCN, Gland, Switzerland.

Oli, M.K., Taylor, I.R., and Rogers, M.E. 1994. Snow leopard Panthera uncia predation of livestock: An assessment of local perceptions in the Annapurna Conservation Area, Nepal. Biological Conservation 68: 63–68.

Price, M.F., and Thompson, M. 1997. The complex life: Human land uses in mountain ecosystems. Global Ecology and Biogeography Letters 6: 66–70.

Rao, K.S., Maikhuri, R.K., and Saxena, K.G. 1999. Participatory approach to rehabilitation of degraded forest lands: A case study in a high altitude village of Indian Himalaya. International Tree Crops Journal 10: 1–17.

Renaud, F., Bechstedt, H.D., and Udomachai, N., 1998: Farming system and soil conservation practices in a study area of northern Thailand. Mountain Research and Development 18: 345–356.

Ryerson, D.E. and Parmenter, R.R. 2001. Vegetation change following removal of keystone herbivores from desert grasslands in new Mexico. Journal of Vegetation Science 12: 167–180

Saxena, K.G., Rao, K.S., Sen, K.K., Maikhuri, R.K., and Semwal, R.L. 2001. Integrated natural resource management: Approaches and lessons from the Himalaya. Conservation Ecology 5: 14. [online] URL: www.consecol.org/vol5/iss2/art14

Schaller, G.B., Li Hong, Lu Hua, Ren Jurang, Qiu Mingjiang, and Wang Haibin. 1987. Status of large mammals in the Taxkorgan reserve, Xinjiang, China. Biological Conservation 42: 53–71.

Sekhar, N.U. 1998. Crop and livestock depredation caused by wild animals in protected areas: The case of Sariska Tiger Reserve, Rajasthan, India. Environmental Conservation 25: 160–171.

Sharma, U.R., and Shaw, W.W. 1993. Role of Nepalís Royal Chitwan National Park in meeting the grazing and fodder needs of local people. Environmental Conservation 20: 139–142.

Singh, G.S., Rao, K.S., and Saxena, K.G., 1997: Energy and economic efficiency of the mountain farming system: A case study in the north–western Himalaya. Journal of Sustainable Agriculture 9: 25–49.

Singh, G.S., and Saxena, K.G. 1998. Sacred groves in the rural landscape: A case study of Shekhala village in Rajasthan. Pages 153–161 In: Ramakrishnan, P.S., Saxena, K.G., and Chandrashekara, U.M. (Eds.) Conserving the Sacred for Biodiversity Management, Science Publishers, New Hampshire/Oxford and IBH, New Delhi.

Stamfli, A., and Zeiter, M. 1999. Plant–species decline due to abandonment of meadows cannot be easily reversed by mowing. A case study from the southern Alps. Journal of Vegetation Science 10: 151–164.

Wagner, F.H. 1988. Predator Control and Sheep Industry: The Role of Science in Policy Formation. Regina Books, Claremont, California.

Zhao, Xin–Quan, and Zhou, Xing–Min. 1999. Ecological basis of alpine meadow ecosystem management in Tibet: Heibei alpine meadow ecosystem research station. Ambio 28: 642–647.

Chapter 17

CONSERVING BIODIVERSITY IN ARID REGIONS: EXPERIENCES WITH PROTECTED AREAS IN INDIA

Malavika Chauhan
School of Environmental Sciences, Jawaharlal Nehru University, New Delhi, India; email mach0400@mail.jnu.ac.in

1. INTRODUCTION

In the Indian subcontinent, natural ecosystems have become islands, rich in biological diversity and natural resources in a landscape poor in biomass and species. With India's population over one billion, and still growing, the protection of these areas increasingly becomes more difficult.

The 10th General Assembly of the International Union for the Conservation of Nature (IUCN) in New Delhi in November 1969 recommended that all governments agree to a generalized definition of the term "National Park." It further decided to create different zones within each park, based primarily on function, e.g., wilderness zone and transition zone. The wilderness zone consisted of large areas where no development other than trails and primitive shelters would be allowed. The transition zone would contain roads, the interpretation centre, tourist accommodations and other facilities (IBWL 1970). The agreement on definitions and zoning was formulated with the idea that there would be no human habitation (with the exception of tourists) within national parks.

A landmark in Indian conservation history was the establishment of the Indian Wild Life Protection Act of 1972 (WLPA). The WLPA enabled the establishment of protected areas. Before the enactment of the WLPA some states had their own legislation, e.g., the Hailey National Park Act of U.P. under which the present Corbett National Park was set up as the Hailey

National Park. Officially, four categories of PAs inspired by the IUCN (1964) categorization of PAs were declared under the WLPA:

National Parks: Highest degree of protection, virtually no human activity barring passage, management work, and tourism allowed by law. According to the WLPA: "(i) No alteration of the boundaries of a National Park shall be made except on a resolution passed by the legislation of the state. (ii) No persons shall destroy, exploit, or remove any wildlife from a National Park, or destroy or damage the habitat of any wild animal, or deprive any wild animal of its habitat within such National Park except under and in accordance with a permit granted by the Chief Government, being satisfied that such destruction, exploitation, or removal of wildlife from the National Park is necessary for the improvement and better management of wildlife therein, authorizes the issue of such permit (iii) No grazing of any livestock shall be permitted in a National Park and no livestock shall be allowed to enter therein except where such livestock is used as a vehicle by a person authorized to enter such National Park."

Sanctuary: Somewhat lesser degree of legal protection than national parks. While points (i) and (ii) of the WLPA applicable to national parks are also applicable to sanctuaries, the difference is that in sanctuaries certain types of rights and activities might be permitted. Livestock grazing and collection of forest produce, for instance, may be permitted.

Closed Area: The state government may, by notification, declare any area closed to hunting for such period as may be specified in the notification. No hunting of any wild animal shall be permitted in a closed area during the period specified in the notification. Other activities, however, are permitted to continue.

Game Reserve: The state government may, by notification declare any area to be a game reserve. No hunting of any wild animal shall be permitted in such a reserve except under and in accordance with a license issued under this section by the Chief Wildlife Warden or authorized officer.

The category of "Game Reserve" has been omitted in a 1991 amendment to the WLPA, and that of "Closed Area" has hardly been used in India. Today India has 80 national parks and 441 sanctuaries, covering over 4.5 percent of its area. In addition, two other PA categories not included in the WLPA but with administrative significance in India are:

Tiger Reserves: "Project Tiger" launched in 1973 was a large and ambitious single species conservation project. It was based on the philosophy that a species cannot be preserved in isolation and in particular the tiger, being at the apex of a large and complex biotype, requires that its habitat be protected. As a result nine tiger reserves were established in various parts of the country in 1973–74. The reserves were planned to have a "core" area free of most human activities and a "buffer" where restricted human land use is allowed. By 1996, the number of tiger reserves had increased to 23. These reserves in themselves are not legal entities but contain national parks or sanctuaries as their core and in some cases their buffer zones. Management concentrates on the elimination of human disturbance in the core and restriction of human activities in the buffer, habitat management to the extent of restoring areas disturbed earlier by human activity and intensive research and monitoring of wildlife.

Biosphere Reserves: In 1972 UNESCO, through its Man and Biosphere Programme (MAB) proposed the creation of "Biosphere Reserves" (BR). The concept dealt with the conservation of biodiversity and sustainable use, with emphasis on humans as integral parts of the ecosystem and the need to involve them in conservation activities. It was visualized that the programme would involve natural and social scientists, conservation and development groups, management authorities and local communities to look at how best biodiversity would be conserved, taking into consideration local communities. BRs were planned to have a undisturbed core zone, a well defined buffer zone and a flexible transition area which would contain a variety of activities (agriculture) and where local communities, management agencies, scientists, NGO's and cultural groups can work together (UNESCO 1995). India has seven BRs, though none are registered with UNESCO. Further, this category is not designated under the WLPA and therefore has no legal protection.

The network of PAs in India currently covers an area of 8.1 million ha, encompassing about 14 percent of the country's forest area and 4.61 percent of its land mass. From six national parks and 59 wildlife sanctuaries in 1970, the respective numbers increased to 85 and 462 by 1998 (WII 1998).

Although conservation in many cases does require the imposition of regulations over access to resources, problems arise when such regulations are applied without prerequisite study and detailed research of individual cases. A complete ban on the use of resources from many PAs (national parks) in the country created a situation where local populations living within and around the areas were suddenly deprived of their traditional livelihood. Despite the ban on resource use, a study by the Indian Institute of Public Administration in the 1980s revealed that 69 percent of surveyed PAs had humans living within their boundaries, and that 64 percent of them had rights, leases or concessions to extract fuel and fodder, to graze, or to carry out other activities. It is estimated that there are at least 3 million people living within Indian PAs and several million more people using them from adjacent settlements (Kothari et al. 1989).

The purpose of this paper is threefold. One, to present a case study of Keoladeo National Park. Two, to briefly describe some PAs in Gujarat and their management problems. Three, to present conclusions on lessons learnt from PA management focusing both on the needs for protection and the needs for humans to use resources.

2. KEOLADEO NATIONAL PARK, RAJASTHAN

Keoladeo National Park, Rajasthan, known especially for being the only wintering site for one section of the western population of the Siberian crane (*Grus leucogeranus*), is undoubtedly one of India's oldest and most species–rich parks. Designated as a Ramsar site in 1981 and a World Heritage Site in 1985, it lies 50 km west of the city of Agra. Once part of the floodplain of two seasonal rivers, the park is comprised of a series of human–made impoundments surrounded by scrub–savannah. Much of the wetland habitat has been created over the past century by constructing dykes and earthen mounds, and planting trees and shrubs for nesting. Besides monsoonal rainfall the park receives water from a nearby reservoir via a canal, which is then distributed to different impoundments through a series of sluice gates. The waters bring with them millions of fish fry upon which piscivorous species depend. Extensive habitat management has resulted in a high diversity that comprises about 370 species of avifauna (including 111 migratory birds and 64 winter visitors), 90 wetland and 280 terrestrial plants and 50 fish. There are also many species of mammals, amphibians and reptiles, algae, microorganisms, and invertebrates that have not yet been enumerated.

During the past two decades, Keoladeo National Park has faced several management problems and threats to its biodiversity. Until 1981, villagers had access to the park's natural resources (*Vetiveria* grass, fodder, wood, fuel wood, etc.) and grazed their cattle and buffaloes in the area. Upon establishment as a national park, all forms of utilization were prohibited. The elimination of grazing resulted in rapid, weedy growth of *Paspalum distichum,* an aquatic species which until then had been kept under control by grazing buffaloes. Consequent changes in habitat (less open water) and vegetation (reduction in floating leaves, and submerged and emergents such as *Scirpus tuberosus*) directly affected waterfowl habitat and impacted the Siberian crane, amongst other species. Bans on the removal of deadwood and grass resulted in accelerated growth of coarse, drought tolerant species (*Desmostachya bipinnata, Saccharum spontaneum*) and an increase in the frequency of small fires. With the loss of grazing grounds, surrounding villages have shifted to an intensive agriculture–based economy. The increasing demands of water for agriculture and domestic use in the surrounding villages, coupled with periodic drought have created problems of water management. During drought, the biodiversity declines and shifts occur in species composition; e.g., while herons disappear pelicans arrive in large numbers. The exotic water hyacinth has been a major problem (some of it is removed manually), and another exotic *Ipomoea fistulosa* is spreading (Gopal 1991).

2.1 Management Issues and Practices

The removal of grasses and fodder in Keoladeo National Park is now allowed through a system of permits. In support of this and other resource utilization, efforts have been made to augment the park's water supply by pumping groundwater. Although recommendations have been made to maintain water levels in different parts of wetlands throughout the year and that year–to–year variation should be kept to a minimum, there is a need to more completely understand and manage the park's long–term hydrology in order to protect the park' biodiversity. For example, during the past century, Keoladeo experienced several major floods and droughts that are common phenomena in monsoonal South Asia. Whereas the biota are adapted to monsoonal vagaries, periodic floods and droughts maintain the wetlands in a "successionally" young state. Other factors such as grazing by buffaloes as well as human use of the park has also altered and, in some instances, increased biodiversity. Whereas grazing and harvesting of resources occurs in the park and may be used as tools for the management of vegetation, it is nearly impossible to simulate long–term hydrology. Therefore, the park's biodiversity may change significantly if it does not periodically experience prolonged deep flooding.

It is unlikely that the park will be able to continue to draw the required minimum of 9.5 million m^3 water from the Ajan Bund reservoir, and therefore a suggestion has been made to bring water from River Chambal (Vijayan 1991). Ali and Vijayan (1986) also recommended that the annual variation in the amount of water released into the park should be kept to a minimum, the time of releasing water should be fixed, and that the water level in various areas should be kept relatively constant during successive years. However, these recommendations have not been implemented. It must be stressed that inter– and intra–annual variations in water regimes are normal phenomena in this region, and that the biota are well adapted to them. Periodic droughts and floods are unlikely to cause any major changes in the park's biological diversity. However, a series of droughts and/or a trend of lower water volume over many years without a severe flood can adversely affect the park's biota. A detailed analysis of long–term changes in the monsoon behaviour together with that of the water requirements of the park and the local villages is required to develop appropriate strategies for water management.

Keoladeo National Park has been managed variously since it was created a century ago. Initially, management was confined to the distribution of water in different impoundments. At that time, water availability was probably more than adequate, though severe droughts, like one in 1941 also occurred (Auffenberg 1980). After Keoladeo was declared a bird sanctuary, the management focused on the enhancement of avian habitat by planting trees and shrubs.

High soil salinity during the dry period and flooding during the monsoon were major problems in the establishment of saplings of *Acacia*. To overcome this, saplings were planted over earthen mounds created in some impoundments. However, salts rose to the surface of the mounds and formed a crust on drying, thereby inhibiting the growth of the saplings. Therefore, salt tolerant *Prosopis juliflora* was planted and watered frequently. After the initial planting and tree development in the 1960s and 1970s, the exercise was abandoned. During a severe drought in 1979 four deep bore wells were drilled in the park. However, they could not provide enough water to fill the impoundments to desired depths. Consequently, *Typha* invaded areas of the park and had to be removed by bulldozing the entire area during 1981.

Other management problems include the decline of the population of Siberian cranes, the spread of weeds, and increased amounts of pesticides and other agricultural chemicals.

Factors responsible for the near total decline of this western population of the Siberian cranes are many and complex, including hunting in Afghanistan and Pakistan along their migratory route. Many consider the species to be an indicator of the "health" of Keoladeo's wetlands, and link the decline in population to a degradation of habitat. (This consideration was a main reason for placing Keoladeo on the Montreux Record or Ramsar Convention). Water hyacinth, a nuisance for decades, has been manually removed several times without success. Another exotic, *Ipomoea fistulosa* is spreading rapidly but has not yet attracted attention. Park management has often failed to take timely and appropriate action to control these weeds because of their perceived benefits aspects until they become a nuisance. *Typha* stands, for example, were valued for the large number of nesting weaver birds, without recognizing the habitat needs of other waterfowl, or the loss of biodiversity. Hyacinth was considered useful for species that walk, feed and nest on the mats and later for the removal of heavy metals.

Four Sarus cranes and 40 ring doves were found dead outside the park during 1988 and early 1989, possibly due to pesticide poisoning, highlighting the rapidly growing problem of pesticide use in surrounding agricultural fields. Pesticides have also been considered responsible for the failed nesting of some raptor species in the park. There are other problems in the park, presently ignored. For example, soils are rich in calcium.

Formation of calcium carbonate granules (kankar) occurs and their accumulation below the soil surface forms a hard pan that creates a barrier to water movement through the soil and to the emergence of seedlings and young sprouts. Grazing animals such as buffaloes help in regularly breaking the hard pan. The development of hardpan in the absence of grazing may adversely affect the regeneration of wetland vegetation. The larvae of *Parapoynx diminutalis* (Lepidoptera) is reportedly a serious pest that considerably inhibits the growth of *Nymphoides cristatum* (John and Nanjappa 1988). It is important to monitor such pests and pathogens that may affect the habitat and/or food of waterfowl and to initiate mitigating measures wherever possible.

2.2 Discussion

Prior to Keoladeo's status as a national park, it had been successfully managed for eight decades that saw a steady rise in biodiversity in its human–made wetlands. Problems relating to biodiversity management arose only during the past twenty years, with after its legal conservation status was raised to a national park status, and a ban on use of park resources was issued. At the same time, the water supply from Ajan Bund declined. Though park managers have begun to manage vegetation by issuing harvesting permits, management of the hydrology depends largely on an external water supply, and this remains a problem. The area experiences periodic floods and droughts characteristic of a monsoonal climate. Increasing water demands for intensive agriculture and a rapidly growing human population around the park and in its catchments have greatly reduced water availability. While adequate water is available in years with good rainfall, "destructive" flooding has not occurred since 1974. Periodic droughts together with changes in hydrological regimes and a ban on the removal of dead wood and litter have accelerated terrestrial succession. Further, periodic floods and droughts maintain the wetland in a "successionally" early state, and large seed banks, rich in species, ensure the maintenance of high biodiversity. Prior to the ban on resource use, buffaloes and humans played pivotal roles in Keoladeo's ecology. These roles can be used to manage and sometimes increase biodiversity; however, for this to occur impoundments need to be cleared of dead trees and efforts should be made to replace them with native species such as *Mitragyna* and *Barringtonia* that are adapted to periodic flooding. All exotic weeds must be effectively controlled.

In conclusion, the management of Keoladeo National Park should be based on a better understanding of its long–term hydrology and the successional processes that are influenced by grazing and various forms of human use. The biodiversity of Keoladeo is unlikely to be greatly affected unless its "ecological character" is altered by reduced incidence of severe flooding, lack of regular selective harvest and accumulation of detritus (Chauhan and Gopal 2001).

3. PROTECTED AREAS IN THE STATE OF GUJARAT

Most parts of Gujarat lie in a semi–arid to arid climatic region that receive very little and erratic rainfall, from an average of 250mm to a maximum of 700mm in a few southern hilly areas. Soils, vegetation, topography and geology, which heavily influence runoff and recharge, vary distinctly from one region to another. Topography varies from the hills in the south and southeast to alluvial plains in the north. The Saurashtra peninsula to the southwest is a plateau with a central, higher area, which causes a pattern of radial drainage and high runoff rates. Numerous small ephemeral streams drain into the sea from the Saurashtra peninsula. The district of Kachchh in west Gujarat constitutes the driest parts of the state, with scanty rainfall and a mean annual rainfall of 377mm. In this peninsula large continental wetlands and vast saline flats of the Great and Little Ranns are complemented by extended areas of sandy and rocky desert.

The Great Rann is a vast saline plain, stretching eastwards from the eastern edge of the Indus delta, covering an area of over 700,000 ha at an altitude of 0–10m. Temperatures range from about 7C in winter to well over 40C in the summer. To its east and north lies the Great Thar desert. The Great Rann is linked by a narrow channel to the Little Rann, which in turn is linked to the Gulf of Khambhat through a marshy depression in which Nal Sarovar survives as a freshwater remnant of the seas that once covered all three sites. The Great Rann is flooded between May and October, when sea water driven by high winds and tides from the Arabian sea and monsoon run off from the hills of Gujarat and Rajasthan mix together over the plains. For the remainder of the year the Rann is a vast expanse of sun–baked mud and sand with halite and gypsum. Vegetation is confined to the "bets" areas of relatively salt free land, a few meters higher than the surrounding plains and several hundred hectares in extant. These become true islands during the monsoons and support grasses and a few scrubby bushes.

The Little Rann is an extensive area (495,300ha) of salt impregnated flatlands. It lies to the southeast of the Great Rann, along the line of marine recession between the Indus delta and the Gulf of Khambhat. During the southwest monsoon (July–September) large areas are inundated by up to two meters of water, much of which is pushed up from the Gulf of Kachchh by strong westerly winds. The salt flats dry out almost completely during the long dry season. Similar to the Great Rann, "bets" of various sizes and slightly more elevation have salt free soil and host some scrubby vegetation. During good rainfall years the floodwaters overflow into the Nal Sarovar depression to the southeast. Soil conditions are saline to hyper saline. Average rainfall is less than 300mm.

3.1 Humans in the PAs of Gujarat

Apart from its urbanized and settled rural populations, Gujarat has 29 independent tribal populations (declared as scheduled tribes by the government for aid and advancement purposes) of which at least five live predominantly in and around PAs. Most of these people still live a life entirely dependent upon forest produce and have a vast knowledge about the distribution and use of species, particularly plant species in their area. The forest areas along the eastern boundaries of the state are mainly inhabited by a tribal population spread over eight districts. Though management plans have been prepared for many of the parks and sanctuaries in Gujarat, implementation has not taken place. These plans were meant to be guiding documents for the maintenance and upkeep of these PAs, and as such most of them constitute plans for wildlife conservation. However, few take into account the use of PA resources by local populations or the populations' interests (CEE 1997, GEC 2000).

The National Forest Policy of 1988 categorically emphasizes the involvement of people in the management of forest resources through the concept of Joint Forest Management (JFM). Although efforts are being made through various projects and organizations to implement the JFM, most efforts have not succeeded because local populations have been treated as passive beneficiaries (often as temporary employees) rather than involving them in real decision making throughout the project cycle.

3.2 Management Issues

Although Gujarat has over 25 PAs, with two of them more than 4900 sq. km. in area, the status of conservation and management in most remains unclear. Most PAs do not have legally settled boundaries and approved management plans. Except for two PAs, all others have yet to be divided into different zones for management and management priorities are not set. About 590 positions in the section of the forest department of Gujarat dealing with the protected areas are vacant; most vacancies are for guards and foresters. With inadequate protection measures and a lack of community development programmes, protected areas face serious threats from illegal cutting, grazing and poaching. Baseline ecological information for many PAs is missing entirely, long–term monitoring programmes are absent, and species lists are superficial. Pressures from an increasing population, their cattle, tourism and industry are however rapidly growing. A few of the protected areas in the arid districts of Gujarat are described below, including the major issues and threats they face.

3.2.1 Nal Sarovar Wildlife Sanctuary

The Saurashtra peninsula is divided from mainland Gujarat by a narrow depression linking the Gulf of Khambhat in the south with the Little Rann of Kachchh to the north west. Tectonic uplift, increased sedimentation and aeolian infill reduced this former estuary to a series of saline flats and a large freshwater lake, Nal Sarovar. The wetland is fed entirely by the ephemeral rivers Bhogavo, Bharmani and catchment's run off, depending upon the rainfall. The terrain is uniformly flat and the lake very shallow. However, it is unique in the Saurashtra peninsula by virtue of the extensive marshes that surround it. It has a catchment's of about 2300 sq km, and water spread area of 115 sq km. Water depth varies from 0.5 to 2m. The lake has about 360 islands, most exposed only when the water level drops. As the lake's hydrology depends entirely upon incoming rainfall and runoff, it has a remarkably strong seasonal character. It dries out completely in most years with the water turning saline, while during and after good rains, it is freshwater, over 2m deep in most areas and replete with wildlife. Even in years of good rainfall, however, the wetland begins to dry out by early summer.

An area of 115 sq km of the Nal Sarovar's wetland was declared a bird sanctuary in 1969. In 1982, 5.82 sq km were added. A core zone of about 100 ha is kept free from disturbance. The sanctuary area comprises a shallow lake and adjoining reed beds. In general, however, vegetation on the lake shore is sparse and scattered, while species like *Salvadora* and *Prosopis* are dominant. Dominant aquatic flora include, algae, bryophytes and angiosperms, including *Typha.* In the shallows, reeds and sedges are common. As the waters recede, grasses become abundant.

Nal Sarovar supports over 200 species of birds, 13 mammals, 71 angiosperms, 20 species of fish and numerous other smaller species. Aquatic invertebrates include nematodes, snails, bivalves, crustacea and insects. The waterfowl census of 1991–92 recorded over 200,000 birds. In years of adequate rainfall Nal Sarovar is crowded with a variety of avifaunal species, notably pelicans, flamingos, ducks, cranes, and shorebirds. The numbers of ducks have however been decreasing steadily over the past century and the wetland is now considered most important for its wintering cranes.

The park is surrounded by 12 villages, with populations ranging from 2000 to 8000 households per village. A rough estimate of the cattle population of these villages is about 10–15 thousand, of which the buffalos graze in the wetland area and the cows in the surrounding, temporarily exposed uplands. Other activities include fishing, cutting and removal of aquatic vegetation for fodder, and water sports. The lake is a popular boating and picnic area for people from the nearby cities. In the late 1980s, parts of the dry lakebed were under cultivation.

Major threats include over exploitation of biomass via overgrazing and removal, illegal shooting and trapping of birds and small mammals, and excessive tourists during weekends that causes considerable disturbance. Unregulated fishing at Nal Sarovar, along with the increasing and unchecked use of boats. Although the management has made numerous plans for streamlining tourism and integrating local people in management of the park, none of this has yet been put into practice (Ramani et al. 1995).

3.2.2 Gir National Park and Sanctuary

Established in 1965, the sanctuary was created to help restore the declining population of Asiatic lions in the area. Hunted by sportsmen and poisoned by the Maldhari tribes whose herds they preyed upon, only a few family groups of lions remained. However, with the creation of the sanctuary the population has recovered and plans are being developed to transfer part of the population to different parks in similar ecological regions. Despite the recovery, threats to the lions and their habitat remain; these include increasing human settlements, loss of habitat, decline in prey populations, poaching, encroaching industrial activity and mining and vehicular traffic.

There have been numerous management schemes to reduce conflicts between humans and the park's resources. These have included afforestation, economically viable plantations and other such eco–development schemes. The relocation of the Maldhari's, a nomad tribe, has been a controversial issue with the forest department. The Maldhari population has moved with its herds of cattle and camels throughout the park's area for centuries. Now, their presence within numerous protected areas is a matter of contention. A number of families were relocated from Gir, but promises of infrastructure facilities, housing and land failed to materialize and this has made the remaining families extremely wary of any suggestions or recommendations of the forest department, even to the point of passive and active resistance. In return, the forest department curbed all rights the Maldhari's had had in the park area (Abdi 1993). A management plan was drawn up in 1996 that quite effectively covers most of the problems in the areas and suggests possible alternatives. The plan however is yet to be implemented.

3.2.3 Dhangadhra Wildlife (Wild Ass) Sanctuary

The Dhangadhra Wildlife Sanctuary was created in 1973 to protect the Indian wild ass within the area of the Little Rann of Kachchh. The sanctuary covers an area of about 4840 sq km. The Rann itself is a flat, saline plain with little vegetation and most parts of the sanctuary are flooded during the rainy season, forcing the herds to move to patches of higher ground within the Rann itself, known as "bets." The nearest town is Dhangadhra, a distance of about 22km from the sanctuary. Along with the Wild Ass, the sanctuary provides habitat for other species of mammals and many species of both aquatic and terrestrial birds.

The local population of Kachchh is primarily dependent upon grazing and milk sales for a livelihood. In Kachchh alone about 4000 families survive on sheep and goat rearing (Rahamani et al. 1997). A rapid increase in sheep and goat populations have contributed to the degradation of the habitat. The sanctuary is also facing serious threats from air and noise pollution caused by numerous trucks that pass through the sanctuary carrying salt produced in salt works surrounding the sanctuary. Approximately 28 lakh tonnes of salt is produced annually, which requires about 280 thousand trips for its transport. The salt works have hundreds of salt pans (ponds), which use the sun's energy to concentrate saline waters in small ponds. When the salt pans of one area are exhausted they are abandoned and new pans are created in different locations. Consequently, the industry is now encroaching further into the sanctuary. An additional disturbance to the animals of the sanctuary is the use of part of the area as a firing range.

As a result of these disturbances, the population of Wild Asses, which has gone up from 720 in 1976 to around 3000 at present, is found mostly at the periphery of the sanctuary away from the disturbances. In the 1960s the number of Wild Asses had declined significantly due to outbreaks of the Surra disease (insect vector) in 1958 and 1960, and then of the African Horse sickness in 1961. This was followed by a severe drought in 1968–1969, which also took a heavy toll (Gee 1963, Singh 1999). Further loss of habitat is occurring due to the spread of *Prosopis juliflora* that is rapidly replacing the open grasslands the animals require. The combined situation of rapid habitat degradation and disturbances is pushing the animals out of the sanctuary boundaries, where they often stray into adjoining agricultural fields, causing conflicts with the villagers (Singh et al. 1999). Litigations contesting the government's decision to grant leases to the salt pans have been filed with the Gujarat High Court. Apparently, though most of the newer salt pans are unlicensed units, mounting pressure from the influential salt manufacturers lobby delayed matters considerably. On

August 1, 1999, the Gujarat government denotified a 589 sq. km. area of the sanctuary spread over five neighbouring districts for the development of the salt industry.

3.2.4 Jessore Wildlife Sanctuary

Located in a part of the Aravalli ranges where it links with the Jessore Hills on the northern edge of the Thar desert, the sanctuary covers an area of 180.66 sq. km. Established in 1978 predominantly to protect a wild population of sloth bear, the sanctuary is also inhabited by numerous other rare and endangered species including the leopard, jungle cat, civet, caracal, wild boar, blue bull, wolf, and hyena. Reptiles include the cobra, krait, different species of viper, and monitors. The vegetation is primarily dry deciduous. The sanctuary's forests were originally well connected with forests further up in the mountains, which facilitated the migration of the sloth bear and other animals. Now however, the forest area survives in patches, as do the wildlife populations.

There are two temples inside the sanctuary that attract large crowds, especially during religious festivals during October and November. There is also a well–protected sacred grove within the sanctuary near the temples.

Threats stem predominantly from the spread of *Prosopis juliflora,* which is suppressing the growth of native species. Increasing sheep and goat populations are encouraging this spread. Grazing of local and migratory livestock within the sanctuary area is another problem. It has been a tradition for villages in drier areas to send their cattle to greener pastures during the dry season. Traditionally, cattle were turned over to nomadic tribes that moved the collected herds from place to place for a fee. Now, however, cattle populations have increased tremendously, and green areas have shrunk. The pressure of fodder and fuel wood collection by local populations is also seen as an encroachment by the forest department. Forest fires are a common occurrence during the dry season. A management plan was developed by Singh for this sanctuary area, which however was neither assessed nor implemented (Singh undated).

4. COMMON ISSUES FOR PAS IN THE HOT ARID REGIONS OF INDIA

The major concerns of PAs in the hot arid areas of India stem from conflicting, multiple demands on their resources, changing hydrology and industry. Unsustainable agricultural practices and water being brought from outside the catchments promotes the creation of water logged areas and salinity. Mining and salt works create chemical risks by dumping untreated and toxic wastes into sensitive ecosystems. The situation is complicated by greater demands for water and wetland resources and dwindling supplies of accessible water, competing claims, over–exploited and mismanaged systems, and lack of knowledge. For example, rapid growth in goat and buffalo populations in the past ten years stems from increasing human populations and their needs, and results in rampant, unsustainable over grazing in every protected area with any greenery. In protected areas in the Kachchh, the bets are overgrazed by herds of camels, goats and sheep, as are the grasslands in PAs of the Ranns. The Nal Sarovar Wildlife Sanctuary's wetland's rich vegetation is collected for fodder and thatch, grazing by domestic livestock occurs, poaching is common and fishing is a large–scale activity. Fishing activities in the open water areas of Nal Sarovar cause considerable disturbance to birds. In the wetlands and grasslands in PAs of South and Central Gujarat, grazing pressure is very high, particularly in years of below average rainfall. Large scale hunting and trapping of waterfowl is common. In many areas mist nets are used to trap ducks at night. Fishing, though technically banned in most PAs as per the WLPA, continues. These issues are further complicated by conflicting governance policies, change in animal varieties, change in migration patterns and the changing economy of the regions.

The increasing conversion of land area within PAs for agriculture due to the increasing availability of water is a direct loss of land. Indirectly, the increase in the number of reservoirs and the reduced water flow in rivers, is causing increases in soil salinity in vast areas. Reduced water availability also occurs due to water logging in areas adjoining canals and reservoirs, and also in coastal areas due to the excessive use of water from wells and wetlands via saline ingress.

Inequities in resource utilization and distinctions due to economic, gender, age and caste within communities inevitably develop. In Gujarat, as in much of India, caste distinctions play an extremely significant role in resource use patterns. Inevitably, the lowest castes utilize areas not normally used by others and this often leads to habitat degradation, which under more equitable conditions would not occur.

Lack of well defined governmental policies for PA's and the leasing of land areas within PAs boundaries to industries highlights a remarkable lack of awareness of biodiversity issues. About 6382 ha of mangrove area and mudflats have been leased out for industrial and saltpan use in the Gulf of Kachchh between 1947–85. Since 1988 the renewal of leases is permitted with mandatory compensatory afforestation of the mangroves. Since the afforestation procedure is not monitored, results are questionable. The Kachchh is surrounded by a large number of salt works producing hundreds of thousands of tones of salt every year. In coastal areas these activities take in fresh seawater at 35 ppt and release liquid concentrate at 250 ppt back into the sea. The implications of this release on coastal mangroves and coral reefs have not been studied. In inland areas like in the Rann and in the Wild Ass Sanctuary the highly saline waters are collected in shallow ponds and solar energy is used for crystal formation.

Invasive species are also a problem. The extension of irrigation canals in parts of Gujarat and Rajasthan have carried with them weeds (predominantly *Eichhornia crassipes*) that readily establish along the banks and in waterlogged areas. They have also covered the open water areas of nearby water bodies. These weeds reduce the water storage capacity of reservoirs, increase siltation and evapo–transpiration rates, create stagnant and eutrophic conditions and hinder direct use.

One of the greatest threat to PAs is from commercial demands, development projects, especially mining, hydrology changes, illegal logging, poaching and encroachment. Forest department officials express their limitations, as government functionaries in countering strong commercial and vested interests, as do local communities. Furthermore, local communities are alienated as their traditional access to grazing areas/forests is been curtailed by the imposition of the WLPA and by the creation of the PAs. In many cases local communities express their inability or unwillingness to help in conservation efforts, as there is little recognition of their traditional access to these resources. In others, however, communities are willing to participate in conservation and natural resource management, usually in PAs that attract larger numbers of tourists and hence offer the local populace an alternative employment (CEE 1997).

The conservation of wildlife and natural habitats can only be achieved if there is strong and sustained cooperation between local communities, forest officials, conservation and social action groups as well as scientists. Clear national and state policies on protected areas is urgently required, as is an official recognition of the legitimate rights and needs of local traditional communities and measures to meet these needs. Furthermore, these communities must play a central role in the planning, protection, management and monitoring of these PAs. Protected area management must be based upon both ecological studies of the areas and traditional practices that have helped conserve and sustainably use these areas for centuries. Finally, there must be a more equitable sharing of the benefits of protected areas, including biomass rights, tourism income, employment in wildlife and forest related works, and alternative livelihood opportunities.

5. CONCLUSIONS

The main lesson learnt from this analysis of PA management in hot arid regions of India is that the survival of these ecosystems cannot necessarily be insured by simply leaving them alone. An understanding of the science behind the resilience of many of these protected areas in the face of occasional disturbance, and the fact that they have thrived for so long in continuous interaction with human and cattle populations is required. One conclusion from the Keoladeo National Park case, is that continuation of ecological linkages of the protected area with the surrounding area and its biotic communities may be necessary for ecosystem health. Also that the maintenance of biodiversity in these protected areas depends upon numerous smaller management issues linked with human and cattle communities from outside the area, and on a holistic approach integrating the economics, industry, health and resources of the region.

Protected areas should not simply be islands of biodiversity in an otherwise degraded landscape. Under a bioregional approach to conservation PA's are planned and managed as part of a mosaic of land uses that together seek to satisfy environmental and social needs. The design of a PA system therefore has to take into account a range of inter–related social and biophysical factors, both inside and outside the PA's themselves. This marks a step beyond the fulfilment of targets for PA's and requires consideration of conservation values, social issues and the biogeographical characteristics of the area (Dudley and Stolton 1998).

It is not reasonable to expect underprivileged communities of arid environments, living in and around resource rich protected areas, to live austerely for the sake of PAs. Resource degradation in developing countries, normally attributed to the common property system, actually originated in the loss of local–level institutional arrangements originally developed to give rise to sustainable resource use patterns (Cernea 1987). To secure effective and active participation of communities, programmes must restore the local institutions concerned. In addition, institutional linkages with mainstream development programmes need to be formalized. Further, it should not be assumed that poverty leads to overexploitation of the natural resource base; in many areas of northern, eastern and northeastern India the poor have managed their environment sustainably. Conversely, reducing poverty through development activities may not automatically reduce environmental degradation. Different dimensions of poverty such as social inferiority, vulnerability, seasonal deprivation and powerlessness, are related to environmental change in specific ways and require different measures to bring about successful change (Badola 1999). People–oriented conservation projects in arid regions need to address the different resource priorities and requirements among the various sectors of a community and must establish equitable partnerships so that all stakeholders participate in decision making about biodiversity use and benefits.

6. REFERENCES

Abdi, R.D. 1993. Maldharis of Saurashtra–Glimpses into their past and present. Suchitra offset, Bhavnagar, India. 57 pages.

Ali, S. and Vijayan, V. S. 1986. Keoladeo National Park ecology study. Summary report 1980–85. Bombay Natural History Society, Bombay. 186 pages.

Auffenberg, W. 1980. Autecological notes on *Xenochrophis piscator* (Reptilia: Serpents) from Keoladeo Ghana National Park. International Journal of Ecology and Environmental Sciences 6: 77–82.

Badola, R. 1999. People and protected areas in India. Wildlife Institute of India, Dehradun, Uttar Pradesh, India. Unasylva Issue.

Baviskar, A. 1998. Social concerns in eco–development. Paper presented at the National Workshop on eco–documentation, Dehradun, India, 24–26 November.

CEE, 1997. An evaluation study for the centrally sponsored scheme of eco–development around protected areas. Centre for Environmental Education Report submitted to the Ministry of Environment and Forests, Government of India, Ahmedabad, India.

Cernea, M. 1987. Farmer organisation and institution building for sustainable development. Regional Development Dialogue 8: 1–24.

Chauhan, M. and Gopal, B. 2001. Biodiversity and management of Keoladeo National Park (India)–A wetland of international importance. Pages 217–256 In: Gopal, B., Junk, W. J. and Davis, J. A. (Editors) Biodiversity In Wetlands: Assessment, Function and Conservation, Volume 2. Backhuys Publishers, Leiden, The Netherlands.

Dudley, N. and Stolton, S. 1998. Protected Areas for the New Millennium. The Implications of IUCN's Protected Area Categories for Forest Conservation. A joint IUCN and WWF Discussion Paper. WWF International and IUCN, Gland.

GEC. 2000. State environmental action plan, subcomponent–biodiversity conservation: Overview and problems. Report prepared by the Gujarat Ecological Society for the Gujarat Ecological Commission. 299 pages.

Gee, E. P. 1963. The Indian wild ass: A survey. Journal of the Bombay Natural History Society 60: 517–529.

Gopal, B. 1991. Wetland (mis) management by keeping people out: Two examples from India. Landscape and Urban Planning, Special Issue –Wetlands 20 (1–3): 53–60.

http://sdnp.delhi.nic.in/headlines/nov01/06-11-2001-news12.html. 2001. Desert's gains spell desert's losses. The Hindu, New Delhi, 6th November, 2001

(IBWL) Indian Board for Wildlife. 1970. Wildlife conservation in India. Report of the expert committee, Government of India. Forest Research Institute, Dehradun.

IUCN. 1994. Guidelines for protected area management categories. IUCN, The World Conservation Union, Gland, Switzerland.

John, G. M., and Nanjappa, C. 1988. *Paraponyx diminutalis* Snellen (Pyralidae: Lepidoptera) as a serious pest of *Nymphoides cristatum* in Keoladeo National Park, Bharatpur, Rajasthan. Journal of the Bombay Natural History Society 85:12.

Kothari, A., Pandey, P., Singh, S., and Variava, D. 1989. Management of national parks and sanctuaries in India. Status report. New Delhi, India, Indian Institute of Public Administration.

Rahamani, A. R., Samanth, J., Rao, Y. N., Manakadan, R., Ajitkumar, C. R., Natrajan, V., Tiwari, J. K., Javed, S., and Kumar, S. 1997. Pages 341–433 In: A Study on Ecology of Grasslands of the Indian Plains with Particular Reference to Their Endangered Fauna. Final report submitted to the Bombay Natural History Society and Center of Wildlife and Ornithology, Bombay.

Ramani, K. R., Suthar, D. C. and Patel, G.A. 1995. Management plan for Nal Sarovar Bird Sanctuary. Report commissioned by and submitted to the Forest Department of Gujarat. 200 pages.

Singh, H. S. 1999. Ecological study of Wild Ass Sanctuary–Little Rann of Kachchh (A comprehensive study of management issues), GEER Foundation, Gandhinagar. 123 pages.

Singh, H. S., Patel, B. H., Parvez, R., Soni, V.C., Shah, N., Tatu, K. and Patel, D. 1999. Ecological of the Wild Ass Sanctuary, Little Rann of Kachchh–A Comprehensive Study on Biodiversity and Management Issues. RR–Ecos/19/99 Gujarat Ecological Education and Research Foundation Gandhinagar. 300 pages.

Singh, J. (Undated). Management plan for Jessore Wildlife Sanctuary. Banaskantha Forest Division, Palanpur, Gujarat State. 60 pages.

UNESCO. 1995. The vision from Seville for the 21st century: The Seville strategy for biosphere reserves. United Nations Educational, Scientific and Cultural Organization, Paris.

Vijayan, V. S. 1991. Keoladeo National Park ecology study. Final report 1980–90. Bombay Natural History Society, Bombay.

Wildlife Institute of India (WII), 1998. National wildlife database. Dehradun, India.

Chapter 18

CO–MANAGEMENT PROCESSES TO MAINTAIN LIVESTOCK MOBILITY AND BIODIVERSITY IN ALPINE RANGELANDS OF THE TIBETAN PLATEAU

Camille E. Richard
International Centre for Integrated Mountain Development, G.P.O. Box 3226, Kathmandu, Nepal; email crichard@icimod.org.np

1. INTRODUCTION

The Tibetan Plateau is an uplifted geographical region that is encircled by the western Karakorum massif, the northern Kunlun mountains, the southern Himalayas, and a multitude of deeply incised mountain ranges that drain the Plateau to the east (Gurung 1999). The plateau is primarily located in western China (Tibetan Autonomous Region, Qinghai, Gansu, Sichuan and Yunnan Provinces), but also extends into parts of northern Pakistan, northwest and northeast India, northern Nepal and northern Bhutan. Encompassing about 1.65 million sq. km. and with an average elevation exceeding 4000m, it is the world's highest, and certainly one of the most important rangeland landscapes, possessing distinct cultural and biological resources (Miller in press). Vegetation types range from cold deserts to semiarid steppe and shrublands to alpine steppe and moist alpine meadows and forests, which support a rich array of unique floral and faunal assemblages. Given its conservation significance as one of the most outstanding and diverse alpine ecosystems, WWF has identified the Tibetan Plateau as a site of "global significance" and nominated it to be one of its Global 200 Ecoregions (Olson and Dinerstein 1998, Sherpa 2000).

Despite the dry and cold climate, these rangelands are also quite productive and provide forage for an estimated 12 million yaks and 30 million sheep and goats, which in turn support the livelihoods for about five million Tibetan pastoralists and agropastoralists (Miller in press). The people of this region have sustained themselves in one of the harshest landscapes on earth, relying on extensive and opportunistic mobility of livestock to procure forage from native grasslands and shrublands and through cultivation of a rich knowledge base, including elaborate mechanisms to collectively manage resources.

These "Common Property Regimes" (CPR's), where a select group of people possess either formal or informal rights to access and manage a particular set of resources, have been documented as an effective institutional means to manage common pool resources such as forests, pastures and water (Agrawal 2001). These indigenous systems are especially apparent across marginal dryland regions of the world as an adaptive strategy to survive a harsh and uncertain environment. Despite rhetoric to the contrary, extensive livestock grazing and the diverse array of CPR's to manage human and livestock movement have been shown to help maintain floral and faunal diversity in rangeland environments, especially if pastoralists can maintain a degree of mobility that fosters optimal use of pasture resources (Steinfeld et al. 1997, Naimer–Fuller and Turner 1999). Fortunately, many local institutions are still in place that regulate spatial and seasonal access to pasture resources in the remote rangeland regions of the Himalaya and Tibetan Plateau (Goldstein and Beall 1990, Rai and Thapa 1993, Ura 1993, Richard 1994, McVeigh 1994, Saberwal 1996, Wu 1997).

Despite their skills, pastoralists of the Tibetan Plateau, like those throughout the world, face a number of growing challenges that constrain them from exercising their full traditional rights and practices (Miller 1995, Wu and Richard 1999, Naimer–Fuller and Turner 1999). These include natural factors such as desiccation of pastures due to changing climate, and significant loss of livestock during severe drought or excessive snowfall events. In addition, livestock mobility is increasingly becoming restricted due to a number of socioeconomic factors, such as regional population increase, generally poor infrastructure, social services, market access for mobile communities, increasing education and employment opportunities outside the pastoral sector, and a shift to a more monetary economy.

Government policies and development programs also impact the way local pastoral communities access and manage rangeland resources. Although often well–meaning, these programs are driven by a general disdain for "backward" and "irrational" nomads and a poor understanding of the efficacy of pastoral production systems amongst policy makers who hail from agricultural areas (Zhang 1989, Wu and Richard 1999). Such actions, depending on the geopolitical environment, include (Naimer–Fuller and Turner 1999): (1)appropriation of the more productive pastures by the state (perceived as "vacant" land) for crop cultivation, where investments and political authority are vested in sedentary agricultural populations; (2)closure of land for "protection" resulting in loss of access to high quality pastures; (3)separation of the legal ownership of natural resources from their users, leading to breakdown in indigenous institutions for managing natural resources; and (4)centralisation of social and technical services through the settlement of nomadic communities, causing concentration of livestock and overgrazing near settlements.

In the name of "sustainability," these policies have often resulted in outcomes opposite of what was originally intended, leading to increased environmental degradation and the marginalisation of pastoral communities. Consequently, such actions have globally increased risk for pastoral households and reduced livestock mobility, threatening the very social capital necessary for effective community–based conservation and development initiatives (Naimer–Fuller and Turner 1999).

Given these alarming trends among pastoral communities, the question arises whether livestock mobility can be maintained as a sustainable livelihood option. However, the integrity of rangeland ecosystems, from both an ecological and economic perspective, depends on it. There is a critical need to explore innovative institutional arrangements that can favor effective mobility, which will require a high degree of collaboration among a variety of stakeholders, from the local herder to the policy maker. This will require policies and programs that are flexible and responsive to local needs and aspirations, in an environment of mutual respect and effective organisational cooperation, a situation that is far from the current reality. To facilitate effective collaborative policies first requires a theoretical foundation that legitimises pastoral mobility and collective action. A far greater challenge is developing working strategies to promote collaborative management models for pastoral regions such as the Tibetan Plateau. This paper summarises the theoretical literature and outlines the approach being taken by the International Centre for Integrated Mountain Development (ICIMOD), to foster community–based rangeland management in the Hindu Kush–Himalayan (HKH) region.

2. ENVIRONMENT, RESOURCE TENURE, AND GOVERNMENT POLICY

The theoretical discussion presented here revolves around the following three key themes which form the basis of the strategies presented in this paper:

1. Pastoral institutions that regulate human and livestock movements have an inherent rationale based on the socioeconomic and ecological environment within which a particular system evolves. These flexible and often transient relationships become increasingly complex as the environment becomes more harsh and unpredictable.

2. Policies focused on private land tenure and intensification become less appropriate with increasing ecological stress and variability, expressed by lower rates of adoption. Resource tenure models are used to illustrate these trends along an ecological gradient.

3. The perceptions and discourse of various levels of governance are inherently not compatible when implementing land use policies, thus leading to misconceptions and misinterpretations of policy. This trend also becomes more pronounced in more heterogeneous environments where policies are either more flexibly interpreted or ignored. If these attitudinal barriers are not bridged, collaborative efforts are severely compromised.

2.1 The Ecological and Economic Rationale for Mobility

In areas of erratic climatic conditions and highly variable vegetation dynamics, such as the desert steppes of Central Asia, pastoralists must range over large areas to procure sufficient forage for their herds. They do so through institutional arrangements regarding access and collective herding, thereby minimising the costs associated with protection and mobility (Sandford 1983, Naimer–Fuller and Turner 1999, Banks 2001, Fernandez–Gimenez 2002). Although seemingly counter intuitive, a very large piece of communal rangeland will support more animals per square kilometre than a smaller piece of private land (Gilles and Jamtgaard 1982). This is because a larger area typically has greater vegetation patch dynamics (thus more varied forage production and availability) in response to erratic precipitation. A group of pastoralists with a high degree of mobility can better exploit this resource than if the area was subdivided into private ranches, without the added cost of fences and enforcement or access negotiations.

In addition, pastoralists tend to adopt an opportunistic (or reactive) strategy (Sandford 1983, Blench and Marriage 1999). An "opportunistic" pastoral strategy is one in which herders vary the number of livestock in accordance with the current availability of forage, as opposed to a "conservative strategy" which involves maintaining a population of livestock at a relatively constant level, during both good and bad years. Opportunism allows the forage available in good years to be converted into an economic output such as meat or milk or into productive capital (larger breeding herd). In the case of severe snowfall or drought, heavy livestock losses often occur, potentially resulting in severe economic loss. Because of the risk associated with livestock rearing in such extreme environments, it makes sense for pastoralists to maintain large diverse livestock herds. This way they can accumulate capital to use as a reserve during lean years and ensure against catastrophic loss.

A conservative strategy is typically promoted by land management agencies throughout the world. This strategy does not allow much fluctuation in livestock numbers to exploit excess forage during good years. While this strategy is proposed as the best for the health of the range, an exclusively conservative approach is usually not economically profitable on native range. Exclusive adoption of this strategy is only pertinent for fenced, privately owned or leased pastures that are well connected to markets, and within ecological zones that benefit from relatively stable climatic regimes (Narjisse 1996). Most pastoralists pursue a combination of strategies based on perceived need, and will likely reject imposed systems that do not reflect these perceptions (Sandford 1983).

From the hot and humid lowlands of Sichuan and Yunnan provinces, China, up in elevation and across to the northwestern cold deserts of the Tibetan Plateau, in Tibet and Xinjiang, environments range from "resource rich" (those that produce high vegetative biomass in warm and humid climates) to those that are "resource poor" (cold and dry with low productivity and high variability). The areas' farming and pastoral systems, including their forms of customary tenure, evolve with their environment and often prove to be the most suitable for a given locale. In this light, extensive grazing and the communal institutions regulating livestock movement are seen as rational responses to high uncertainty and risk, with highly dynamic and sophisticated systems of cooperation and risk mitigation.

2.2 Resource Tenure and Management Options in China

Despite the effective rationale evident among pastoral cultures, much of the theoretical literature has argued for privatisation or state control as a means to alleviate degradation of common lands. This is based on the now much refuted theory of Garrett Hardin (1968), "The Tragedy of the Commons", which in practice has contributed to rangeland degradation in much of Africa and Asia, rather than prevented it. China, more than any other country in the Asia region, has embraced Hardin's paradigm by promoting settlement policies and the individualisation of rangeland stewardship as a means to mitigate the effects of perceived grassland degradation. Consequently, the government, citing the success of Deng Xiao Peng's reforms of the early 1980's, specifically the "Individual Household

Responsibility System" in cropping areas, formulated the Grassland Law in the mid–eighties, and has been implementing it throughout western China (Thwaites et al. 1998, Williams 1996, Wu 1997, Banks 2001, Richard 2002). This approach is now being promoted in grassland regions with the goal to transform these marginal zones into a "prosperous modern livestock base" to meet increasing demand for wool, meat, and dairy in expanding urban areas (Longsworth and Williamson 1993). However, implementation is proving to be difficult in nonarable lands (Li in press), especially remote, socially and environmentally marginal landscapes such as Inner Mongolia and the Tibetan Plateau.

The Chinese government offers sound rationale to justify these policies, such as difficulty in providing nomads with social services such as education and health care, and heavy snowfalls that have historically led to livestock loss (Wu and Richard 1999). They felt that fencing could help provide reserve pastures during these critical periods. But underlying the theory is the belief that pastoral migratory practices are destructive, despite overwhelming evidence to the contrary. In the process of implementing such standardised policies, those locally mandated to implement are often at the mercy of high level decision makers and thus there is poor local representation in the grassland allocation process. Other issues are those related to spatial and temporal heterogeneity of pasture resources and the unsuitability of allocation in regards to fair distribution of resources to individual households. These factors become significantly more pronounced as the environment becomes more marginal, rendering such policies ineffective at the local level.

Numerous studies have shown that degradation of pastures tends to accompany situations where communal management systems break down, usually precipitated by outside induced changes in tenure, leading to concentrations of livestock and overgrazing. Globally, extensive areas of both private and state–managed rangeland have been found to be highly degraded, often more than communal lands (Gilles and Jaamtgard 1982, Sandford 1983, National Research Council 1994, Sneath 1996, Williams 1996, Naimer–Fuller and Turner 1999). A number of researchers working across Central Asia and western China have documented increasing environmental degradation with changes in land tenure. Sneath (1986) compared land use change in Inner Mongolia of China, where grasslands were converted to cropland over the past five decades, to outer Mongolia, where land use is still primarily pastoral and communally used. He found that land degradation was extremely high where land had been converted, forcing herds onto more marginal range. Thwaites (1996), Williams (1996) and Wu (1997) have observed increasing erosion and weed invasion surrounding areas where individual allotments were fenced off, forcing those without fenced allotments to concentrate livestock on surrounding commons.

Although some positive outcomes have arisen from implementation of the Grassland Law in the eastern Tibetan Plateau, such as reduced overall labour demand for households and increased survivability of herds in the winter, researchers have noted several instances where individual allotments have restricted access of many households to water sources, forcing them to travel long distances to riparian areas (Du and Zhang 2000, Yan and Luo 2000). This has lead to increased bank erosion along water courses (author's observations). The studies have also noted significant social impacts, including increasing conflicts due to poor allocation of pastures, widening gender gaps, and reduced access to schooling for children. Table 1 summarises these findings. Given these realities, arrangements for collective action will continue to be the norm until the socioeconomic environment is such that the majority of pastoralists can engage in alternative forms of livelihood, while those remaining can access the capital and pasture that is sufficient to maintain economically viable herds on an individual household or cooperative level, similar to situations in more developed nations.

Table 1. Impacts of the Grassland Law through individual household tenure and management. Sources of information: Field visits by author to Maqu County, Gansu, Dari County, Qinghai, and Hongyuan County, Sichuan, in 1999 and 2000; field study conducted by participants of a Participatory Rangeland Management training in Hongyuan, Sichuan, July 2000; Regional Rangeland Programme Annual reports by Yan and Luo (2000), Ma et al. (2000), and Du and Zhang (2000).

	Positive	**Negative**
Allocation process	On paper the process is perceived to be fair and equitable	In reality, poor allocation of pastures in many areas; some receive good quality land and others poor land
Size of pastures	Has forced herders to fix number of livestock	Individual pastures often too small; herders liquidate herds/rent pasture from those with excess land
Water availability	None documented	Lack of water on individual pastures and lack of access to neighbour's water sources; high cost of water development
Risk management	Where fencing is used for reserve pastures, livestock mortality is reduced	Costs per household too high for improvements; high subsidies required by the government
Social services	Better access to veterinary care and government services where holding pens constructed	Greater isolation of individual households

Table 1 (continued)

	Positive	Negative
Household labour distribution	Reduced labour for overall household; gaps between men's and women's labour has increased as men spend less time long distance herding	Increased labour for children; parents find time for employment in towns, children required to maintain herds and have less opportunity for school
Social conflicts	Boundaries are demarcated which has reduced conflicts	Increased conflicts over water and pasture access
Market access	Increased access to markets where fencing used for holding pens and feedlots	None documented
Ecosystem protection	Improved productivity within the fence due to protection during the growing season	Degradation of "commons;" no communal responsibility for landscape amenities, such as riparian areas, which are heavily grazed "outside the fence"

2.3 Resource Tenure Models along the Ecological Gradient

Recent revisions of legislation that affect pasture allocation in China do inherently allow flexibility in interpretation, although this is not apparent when one speaks to those mandated to implement, reflecting a lack of understanding of policy (Yan and Luo 2000). Although vague, China's newly revised Rural Land Contracting Law (adopted 2002) does allow for some degree of collective tenure and management, specifically... "contracting parties may voluntarily join their land contracting and operation rights as stock shares for the purpose of engaging in cooperative agricultural production." This gives households the security to access resources but also allows them to engage in collective management arrangements. What is unclear from the policy is whether a "contracting party" is strictly at the household level or whether household groups or villages can also contract land. It is also unclear if "contracting parties" can legally break from the collective in the case of irreconcilable conflict. Table 2 highlights a variety of tenure–management arrangements that are potentially possible, given that the above issues are addressed in future policy implementation guidelines.

Table 2. A typology of tenure and management for pastures and rangeland landscapes. The left column represents tenure or legal contracts; the columns on the right represent management arrangements (de jure or de facto)

	Household (HH)	**Household group (HG)**	**Collective contract holders (CH)**
Household	1. Grassland contract with individual HH 2. Management by individual HH; benefits at HH level	1. Grassland contract with individual HH 2. Management by HH; group benefits at HH level	1. Grassland contract with individual HH 2. Cooperative of individual CHs for management 3.Benefits at HH level
Household group		1. Grassland contract with HG 2. Management by group 3. Benefits based on HH and livestock	1. Grassland contract with HG 2.Cooperative of HGs for management 2. Benefits based on HH and livestock numbers
Collective (village or larger)			1. Grassland contract with collective 2. Management by collectives 3. Benefits based on HH and livestock numbers

Such arrangements range from individual household contracts where land is individually managed to large–scale collective arrangements for the protection and management of landscape amenities such as riparian zones or watersheds. In the latter case, this would involve the formation of cooperatives of contract holders (individual, household group or village level) that enter into management agreements for protection of ecosystem–level amenities, which to date has not been promoted in policy. Actual case study sites are given as examples in the ensuing discussion to illustrate some of these arrangements, along with hypothetical scenarios for comparison.

Table 3 describes three simplified models of land allocation and management to illustrate how the Grassland Law could conceivably be implemented and the current and potential impacts given three scenarios: local autonomous control; strict enforcement of Individual Household Responsibility; and a co–management model that brings together indigenous and scientific strategies and allows for more flexible policy interpretations. The three models are then compared in three regions of the Tibetan Plateau that follow the gradient going from "resource rich" to "resource poor." The comparisons are simplified for the sake of this discussion.

Table 3. A comparison of tenure and management systems in the different research counties.

Counties:	Maqu	Naqu	Rutok
Pasture area (ha)*	1,600	1,850	25,000
Individual Household Tenure–Management			
Tenure system	Individual household (50 year lease); State owned land		
Management system	Individual household		
Fencing costs/HH**	$3,155	$3,400	$11,900
Actual situation (in Maqu and Hongyuan counties)	winter pasture allocated; summer pasture held; high costs/ conflicts at local level when arrangements enforced	few areas allocated to HH level; most to administrative village	implementation at administrative village level
Flexible Interpretation–Individual Tenure–Household Group Management			
Tenure system	Individual household (50 year lease); households allowed to recombine pasture land into management units of 10 families; State owned land		
Management system	Communal (10 families share labor with equal access to pasture)		
Fencing costs/HH**	$1,220	$1,345	$4,180
Actual situation	winter pasture to to HH level; effective management for families recombining	informal tenure at HH group, legally at village level; soon to allocate to HH level with group management	informal tenure arrangements at HH group level but legally at administrative village level
Local Autonomous Control–Indigenous CPR			
Tenure system	State owned land; access rights legally allocated to villages		
Management system	communal (informal sharing of labor among kinship groups but share costs among all village households)		
Fencing costs/HH**	$0	$0	$0
Actual situation	some villages ignore policy; potential problems with encroachment due to high population	some villages ignore policy; potential problems with encroachment due to high population	traditional routes followed. Population dispersed and conflicts minimal in most areas; potential conflict with wildlife in protected areas

* for ten households (HH) each with 300 SEU; **with 30% subsidy

At the more "resource rich" end of the scale (Table 3) is Maqu County in southern Gansu Province, an extremely productive rangeland type dominated by alpine meadow communities. Average elevation is 3,700m with average precipitation of 616mm. Average stocking rates are 1.88 SEU/ha. This is not to imply even distribution across the county; many areas are sparsely populated while areas near settlements or townships are heavily grazed. In addition, the average calculation does not reflect stocking rates for different rangeland types within the county. Also the stocking rate does not necessarily reflect rangeland carrying capacity so it is not an indication of pasture quality or intensity of grazing but merely the number of animals currently stocked in each county, which can allude to carrying capacity. Alpine meadows also dominate in Naqu County, Northern Tibetan Autonomous Region (TAR), but the elevation of this region exceeds 4,500m with lower average precipitation of 400–500mm. Rangeland productivity in Naqu is much lower than Maqu due to the higher elevation, colder temperatures, and more variable climatic conditions. This is reflected by the average stocking rate, which is 1.62 SEU/ha, which is considered by many to be too high (Duobujie 2000, Richard and Yonten 2002). At the extreme end of the "resource poor" gradient lies Rutok County in western Tibet, in the cold desert steppe of the Changtang Plateau. Average elevation exceeds 4,800m, and precipitation is scant and variable, ranging from 30–180 mm per year. Stocking rates are a mere 0.12 SEU/ha. Tables 4 and 5 summarise the statistics for each county.

Table 4. Ecological and demographic information for Maqu, Naqu and Rutok counties. Information sources: Maqu (Du and Zhang 2000, Miller in press); Naqu (Duobujie 2000, Naqu Poverty Alleviation Office 1999); Rutok (Dorje and Tsering 2001).
* Population of those engaged in some form of pastoral livelihood. ** Households are on average 5 persons per household.

	Maqu, Gansu	Naqu, TAR	Rutok, TAR
Average elevation (m)	3,700	4,500	4,800
Average precipitation (mm)	616	400–500	30–180
Mean annual temp (C)	1.1	–0.4–0.3	0.1
Total area (ha)	1,019,100	1,606,700	7,520,000
Usable pasture area (ha)	1,000,000	1,387,100	3,560,000
Main vegetation type	alpine meadow	alpine meadow	desert steppe
Pastoral population*	28,940	58,541	6,495
Number households**	5,788	11,000	1,299

The types and degree of intensification that can be adopted in the different environments have an inverse relationship with the degree of environmental stress. High risk and capital–intensive management options such as cropping, stall feeding and large–scale pasture and hay development are simply not economically or ecologically feasible in extremely limiting environments due to slow growth rates and high variability. For example, costs associated with rangeland improvements (mainly fencing) drastically increase as the environment becomes more limiting and carrying capacity decreases. In the eastern Tibetan Plateau, higher and more reliable moisture means benefits from small–scale fencing and other improvements are more quickly realized and more readily adopted, either at an individual or kinship group level. However, in the dry and cold desert regions of the western plateau, the cost to contain pastures large enough to support an economically viable herd size are simply too high. In addition, the labour required to guard such a large piece of land at the individual household level would be unmanageable.

Table 5. Livestock statistics for Maqu, Naqu, and Rutok counties. Information sources: Maqu (Du and Zhang 2000, Miller in press); Naqu (Duobujie 2000, Naqu Poverty Alleviation Office 1999); Rutok (Dorje and Tsering 2001).

Livestock type	Livestock numbers	% total	Number per H	SEU's***	% SEU	SEU's per HH	Stocking rate (SEU/ha)
Maqu, Gansu	600,000	100	104	1,884,000	100	326	1.88
Yak	276,000	46	48	1,380,000	73	238	
Sheep	288,000	48	50	288,000	15	50	
Goat	0	0	0	0	0	0	
Horse	36,000	6	6	216,000	11	37	
Naqu, TAR	1,066,000	100	97	2,249,000	100	204	1.62
Yak	280,000	26	25	1,400,000	62	127	
Sheep	600,000	56	55	600,000	27	55	
Goat	170,000	16	15	153,000	7	14	
Horse	16,000	2	1	96,000	4	9	
Rutok, TAR	388,334	100	299	432,604	100	333	0.12
Yak	11,650	3	9	58,250	13	45	
Sheep	155,334	40	120	155,334	36	120	
Goat	217,467	56	167	195,720	45	151	
Horse	3,883	1	3	23,300	5	18	

***Sheep Equivalent Unit (SEU) is calculated as follows: 1 sheep = 1 SEU; 1 yak = 5 SEU; 1 goat = 0.9 SEU; 1 horse = 6 SEU.

Obviously, these three areas vary significantly in their production potential for livestock and in the adaptive strategies employed by pastoral communities residing there. They have also differed significantly in how the

Grassland Law has been implemented (see "actual situation" in Table 3). With a "one size fits all" grassland policy implemented under very different environmental conditions (along the ecological gradient), adoption of fencing and settlement would more likely happen where the environment is more mesic and climatically stable, as in Maqu. Here, many families have been legally allocated individual winter pastures and manage at an individual level, although they express dissatisfaction with the allocation process (Du and Zhang 2000). However, this county has allowed groups of up to ten households in size to pool their pastures together and fence the outer boundary (Tables 2 and 3). The benefits, as perceived by the pastoralists themselves, include the lowering of fencing costs and the continued realization of economies of size with respect to herd supervision, as households take turns at supplying labour for supervision of the joint herd. With the Maqu County approach, the total number of stock units that can be grazed on the joint pasture and each household's share of this are calculated. Households that graze fewer livestock than the hypothetical carrying capacities of their portions of the joint pasture are compensated by those households that graze more. Poor households are thus guaranteed access to forage equivalent to that produced by their pasture if they need it, and can earn supplementary income in the form of land rents to the extent that they don't.

In the nearby county of Hongyuan, Sichuan Province, a pilot program has been established by the government as a demonstration site for livestock and pasture development programs. Here families have been forced to settle on individual allotments, for year round use and individual management. What differentiates this area from Maqu is that Hongyuan county has been designated a milk production zone, while Maqu is a butter and meat production zone. This has had a dramatic impact on herd distribution, as most families wish to keep their lactating herds near the road and milk collection points, renting tent sites and pasture from those who were allocated roadside allotments. Here, impacts of overgrazing have become quite severe along the roadside (Yan and Luo 2000). In Maqu, herds are more evenly distributed across the landscape, regardless of the tenure arrangements, because commercial livestock products in the form of butter and meat are more durable.

Pastoral communities in Naqu County are currently still managing pastures communally with legal rights given to administrative "villages," although official contracting under the Rural Land Contracting Law will start in 2003 (Richard and Yonten 2002). In some areas the Naqu County government of the TAR has established a number of fattening pastures that have been formally contracted to the village (either administrative or natural) as a whole. The location of feedlots were selected through consultation with communities at the natural and administrative village levels, and fenced, particularly marshlands, where the fencing serves to protect wetland functions and facilitate rapid growth response. Rules for the use of the collective pastures, including stocking rates and timing of grazing, have been set by the village government and vary from site to site, with criteria including household labour contribution and the number of livestock per household. Once the formal grassland contracting process begins, households may choose to take individual winter allotments or to combine land access rights at the group or natural village levels, provided that they decide to do this prior to the land division process. The use right per family would be calculated depending upon household population (70%) and livestock numbers (30%). For these collective agreements, the county has established a use tax of 0.05 RMB/day for each SEU (Sheep Equivalency Unit). The grazing fee will be collected by the village or group leader and redistributed among the member households within the village or group, based on the access rights formula above.

In comparison, nomad families in Rutok are still highly mobile and few families have settled, compared to Maqu and even Naqu counties. No pastures have been fenced at the individual household level, and the only areas where fencing is seen are a few wetland areas along rivers that are communally accessed and managed. No dry upland areas have been allocated to units beneath the administrative village level and households or small kinship groups still, more or less, follow traditional migratory routes.

These actual and hypothetical examples show that when communities are given the choice, they will often choose collective arrangements because they are in keeping with customary practice, and they are more affordable. To substantiate this, very few grasslands of the Tibetan pastoral area have actually been allocated to individual household levels, despite the government rhetoric (Banks et al. in press). And in cases where allotments have legally been granted, actual use and management practice follows a more customary pattern of group tenure and management, functioning in a de facto manner. At the heart of these indigenous regimes is the presence of a well functioning system of cooperation, with cooperation generally being strongest when: (1)the environment is more stressful and unpredictable; (2)potential for livestock profitability is greater; (3)there is low use by non–community members, thus reducing potential for conflict; (4)numbers of households within the community are low; (5)wealth distribution within the community is relatively even; (6)stock densities are relatively low; and (7)land has not been privatised to a great extent (Williams 1996, Ngaido et al. 2002). Often the argument is that herders will exploit the commons when potential for economic gain is greater, but actually, these and other studies have illustrated that when herders form cooperative groups, they gain more individually, especially in harsh environments.

2.4 A Clash of Cultures: The Discourse of the "Expert"

Despite the growing body of evidence regarding the negative impacts of pasture privatisation, the knowledge of the "expert," influenced by nationalistic ideologies and political interests, still holds supreme in the making of policy, promoting resource tenure models more appropriate for more "resource rich" environments (Williams 2000). Although scientists insist that their investigations are based on objective reasoning, many cultural, political and economic factors influence the direction that research takes. In the case of China, scientific inquiry in grassland management is strongly influenced by centrist socialist ideologies and a rigid belief that the Individual Household Responsibility System is the only path toward achieving sustainable management of China's grasslands. In addition, scientific representations are always culturally embedded, e.g., Han Chinese dominate the scientific and state sphere of influence while ethnic minorities such as the Mongols and Tibetans are the ones mandated to implement top–down policy directives. Chinese history has been charted by conflicts between pastoral Mongols and Han agriculturalists, most dramatically represented by the Great Wall. However, this cultural divide is also expressed in scientific interpretation, particularly the use of grassland science as a form of social control over pastoral communities in the new "socialist market economy. In many instances, it has been noted that officials from local areas are reluctant to allocate pastures to individual households, but are forced to do so by higher level decision makers (Yeh unpublished data, Yan et al. in press) who are swayed by scientific discourse that is rooted in heuristic learning and static scientific models inappropriate to semiarid grassland regions (Williams 2000, Naimer–Fuller 1999). In addition, narrowly focused mandates and rigid administrative structures found in research institutions tend to meet government needs rather than those of local pastoral decision making bodies.

This cultural disharmony in perception and discourse ultimately leads to misconceptions and misinterpretation of the Grassland Law at various levels. The discourse at high levels remains based on heuristic models of lowland agriculture and livestock development, but closer to the local situation the perspectives become grounded more in local knowledge, especially as the environment becomes more harsh. In these environments, local perceptions and knowledge have stronger influence on the way that policies are interpreted and implemented (or even ignored). So while these communities may be out of the mainstream in terms of technological inputs, they maintain more autonomy in decision making in the local and county or district polity. This is clearly illustrated in places such as Naqu or Rutok where communities still use and manage pastures as a collective, despite government desires to settle and divide.

3. CONCEPTUAL FRAMEWORK AND STRATEGY FOR CO–MANAGEMENT

Co–management involves: (1)the process for stakeholders to make decisions and exercise control over resource use; and (2)the actual management arrangements that are based on a set of negotiated rights and privileges (tenure) that are locally acceptable and legitimized by the government (Ingles et al. 1999). Given this definition, it is important to understand that participation is not a means to achieve some end, but rather, entry points such as forage development or livestock improvements are merely a means to achieve participation of key stakeholder groups, principally the local communities. In other words, any co–management support program should be striving to create a self sustaining partnership of stakeholders able to identify and address development issues as they arise.

Primary constraints to such an approach, as discussed, are the harsh environment and relatively slow response time with resource improvements, the incompatibility of policies such as the Grassland Law with a diverse, heterogeneous landscape, and a bureaucracy untrained and unable to cope with site–specific and diverse concerns at the local level, all leading to poor implementation. This is only accentuated by the often rigid mindset of government bureaucracies and the lack of coordination among relevant but discordant organisations that are mandated to work in pastoral areas. Strategies for overcoming such constraints require a multitiered approach, one in which varied government actors become more engaged in a process of critical reflection on process, and immerse themselves in development implementation. This invariably will bring about changes in attitudes, allowing strategic "change agents" within organizations to slowly transform the way governments work.

To address these issues, the International Centre for Integrated Mountain Development (ICIMOD) initiated the Regional Rangeland Program (RRP) in 1999, currently being implemented in the high altitude semiarid rangelands of the Hindu Kush–Himalaya and Tibetan Plateau (HKH). The primary objective of the RRP in these past four years has been to build the capacity of partner institutions working in pastoral development to conduct participatory action research with relevant stakeholders, through formal training, research grants, and networking opportunities, as a grounding for policy advocacy. This approach has, in the process, facilitated a better understanding of the socioeconomic and environmental realities of rangeland areas and has generated a greater awareness of integrated and participatory approaches among a variety of actors, from local communities to researchers and politicians. Of growing importance in our work is the role of community–based institutions in the management of common pool resources and the means to strengthen them, which will continue to be the focus of our future program.

Any co–management scheme requires a dynamic inquiry and planning process that stresses mutual respect, transparency, joint ownership of information generated, and continuous learning through group reflection and decision making. Participatory action research is simply a systematic and collaborative process involving iterative cycles of planning, implementation, observation on the outcomes of actions, and group reflection on the assumptions and attitudes applied to analyze resulting situations (Greenwood and Levin 1998). This process leads to the next phase of fine–tuned planning, action, observation and reflection. The methods used in participatory action research can be the same as those used in conventional research. The difference lies in how the methods are utilized, a participatory paradigm versus an extractive one. Thus, participation action research is not a methodology; rather, it is an approach to research that aims to democratize knowledge generation (Greenwood and Levin 1998). It is appropriate to use when the issues are complex and people have diverse opinions but do not know where to begin to address them (Fisher 2001) such as in pastoral areas where common property arrangements are still the norm.

The first step requires an understanding of the factors that influence how local pastoral communities use and manage their resources, ranging from biophysical limitations to geopolitics diagnostic studies. Objectives of a project must be based on perceived needs and environmental realities. Communities need secure access to resources, information, and marketable options for improving productivity and livelihoods. They need strategies to minimize risks due to uncertain socioeconomic and climatic environments. The next step involves insight into the characteristics of key stakeholder groups in a co–management process: local communities, government line agencies, NGO's and private businesses, research institutions, and policy makers at various levels (stakeholder and organizational analyses). What is the nature of a community's organizations, social norms and rules, individual managerial and technical skills, systems of decision making and collective action, and their access to financial resources and information? What skills and attitudes do facilitators possess in terms of appropriate participatory action research? Do extension staff work in a supportive organizational environment that allows flexibility in delivery of services to rural communities with long–term planning and funding mechanisms in place? Are they also supported by relevant and timely inputs by research personnel who also possess good participatory skills and methodologies? These organisations must be properly networked so that if they cannot respond to a particular local need, they know who can, or they know where to get the necessary information. To support this type of multi–level learning and networking requires policy makers who are aware of and respect the needs of local people and can adjust policies to accommodate equitable innovations. And what role does a supporting organization such as ICIMOD play to develop necessary skills and support collaborative processes? If the proper conditions are lacking, what capacity building measures are needed to build such skills among the various actors?

Table 6 describes the strategy that ICIMOD is employing to implement its rangeland program, initiated in late 1999. The building blocks of the program are: (1)to build the capacity of partner institutions who are mandated to work in pastoral regions of the HKH in co–management processes in pastoral regions; (2)to help partners develop and promote institutional and technical innovations in pilot areas; and (3)to enable policy dialogue and commitment to support co–management initiatives at a larger scale, building on the outcomes of the capacity building and innovation phases. The project began with formal training programs in participatory assessment and planning, which led to diagnostic and stakeholder analyses in pilot areas. ICIMOD is currently in a stage of implementing a more rigorous pilot site program to build on the diagnostic studies and is initiating participatory research projects in several sites (Maqu and Naqu counties included). Entry points for innovation are based on the priorities set by the local stakeholder groups and include pasture development, livestock marketing, ecotourism planning, and water development, varying from site to site. Our partners have also conducted several policy level meetings to present findings and garner commitments. To date, the primary achievement has been of a more intangible nature, a change in attitude and management style among the staff of participating institutions (research, extension and administrative), forming the cadre of "change agents" required to foster a transformation in the way government agencies think and work. A co–management process cannot proceed until government institutions' staff and decision makers have the proper mindset for sharing power, and this can be a long and tedious change to bring about. But now that this is happening in some sites, the subsequent steps in our strategy are beginning to rapidly fall into place.

It should be mentioned here the pitfalls that many supporting organizations encounter when implementing co–management schemes. Often not enough budget is allocated for meetings, planning, and monitoring.

Management agreements can take many rounds of negotiation, especially when many stakeholders are involved and communities are highly heterogeneous in economic and social characteristics. Another limiting factor is that rangeland ecosystems, especially on the Tibetan Plateau, are remote and travel can be time–consuming and costly. Also, no program will be successful without a committed institutional partner who takes the lead in facilitation at both local and state levels. A supporting organization such as ICIMOD simply cannot play this role effectively. In our experience, the sites where we have had the greatest success are where we had a good local organization that followed up on outcomes of diagnostic studies, who were supported by key local decision makers, and who had the trust of the local pilot area communities. However, this is not to say that trials in a site without these characteristics will inevitably fail. These areas require more follow up and opportunities to learn through exchange visits with stakeholders from other more successful trial areas. ICIMOD as a regional organization is perfectly suited to provide these networking opportunities and has contributed to the spread of technologies and institutional innovations in the region.

Overall, the expected outcomes of our co–management strategy are the following: (1)new and improved technical and institutional innovations identified and tested to address complex rangeland management and pastoral development issues, incorporating biodiversity conservation values; (2)increased skills among partner institutional staff working with pastoral people regarding multidisciplinary, participatory approaches to rangeland management and pastoral development; (3)increased capacity among various decision making bodies to engage in collective management processes; (4)increased participation in decision making by vulnerable groups in pilot communities; (5)improved organizational performance among partner institutions in delivering services and information to nomadic communities; (6)improved networking among stakeholder groups at national and regional levels; (7)improved capacity of partner institutions to develop and support local community networks and advocacy groups; and (8)a strategic policy framework developed that accommodates and supports the complexity and flexibility needed for development and conservation in rangeland ecosystems, including effective tenure arrangements that favor collaborative management schemes.

Table 6. Long–term articipatory action research strategy for the Regional Rangeland Programme, 1999–2005.

Phase 1 Inception →	**Phase 2 Diagnostic →**	**Phase 3 Feedback →**	**Phase 4 Participatory Project Design →**
Steps: 1. Identify initial organizational partners 2. Obtain support from relevant government authorities for general program 3. Training in PAR methodology, methods and tools ↓ Action plan for diagnostic phase (why, what, when, where, who, and how)	Steps: 1. Improve knowledge on pastoral systems 2. Identify local priorities and issues 3. Identify key local actors and change agents 4. Identify key NGO's and GO's able to address issues 5. Conduct local level training to enhance partner and local knowledge and skills in PAR ↓ Action plan for Feedback phase	Steps: 1. Conduct meetings with strategic service organizations 2. Present local perspectives and issues 3. Identify organizational perspectives/priorities 4. Identify strategic change agents 5. Form interdisciplinary and multiorganizational team ↓ Action plan for participatory project design	Steps: 1. Present GO/NGO perspectives to communities 2. Prepare Action Matrices (Identify main beneficiaries and potential impacts 3. Present to community focal groups for feedback and confirmation 4. Resolve conflicts if necessary 5. Prepare draft plan for government approval ↓ Action Plan for Feedback Phase

Phase 5 Feedback →	**Phase 6 Implementation →**	**Phase 7 Organizational Reflection**
Steps: 1. Present plans to local government authorities 2. Adjust plan given policy, human resource, or funding constraints 3. Confirm government commitment ↓ Action plan for implementation phase	Steps: 1. Present revised plan back to community for feedback 2. Revise original plan with new tasks and responsibilities 3. Work out final budgeting details and commitments 4. Design an initial participatory monitoring plan ↓ Action Plan for monitoring and reflection activities	Steps: 1. Form PAR teams within and among organizations 2. Form PAR teams among community groups 3. Plan reflection exercises among teams about outcomes and performance 4. Present back to other groups 5. Refine monitoring system to reflect learning ↓ Action Plan for next cycle of steps

4. CONCLUSIONS AND RECOMMENDATIONS

Across the globe, CPR's and collective management arrangements have been shown to be effective and sustainable, especially with the right supportive environment (Fisher 2002). Although not perfect, CPR's tend to be more biodiversity friendly because resource demands are more diverse. Because they reflect indigenous systems of management they are more acceptable and adaptable. Collective agreements help to foster coalitions for management of larger landscape values (biodiversity and watershed amenities). These agreements tend to be more equitable than private tenure arrangements and therefore contribute toward poverty alleviation. And, in the long–term, they reduce the cost (in terms of both money and labor) for individual households, thus reducing socioeconomic risk. Given that we are working in stressful environments where collective action has been the cultural norm for centuries as an adaptive response, we should build on these local strengths and enhance them through legitimate tenure and management arrangements that reflect local aspirations, skills and environment. The fast disappearing knowledge base of pastoral communities can potentially be a valuable asset for developing new technological and institutional innovations in pastoral development and rangeland conservation, provided that those who work with pastoral people know how to tap into this rich source of wisdom and promote it.

A co–management approach is needed to work in these diverse cultural and ecological landscapes, an approach that embraces diversity and melding of knowledge systems to create sustainable and appropriate management models, and that promotes flexibility in programming. A long–term strategy is needed to improve the capabilities of relevant organisations to engage in a true process of collaboration, creating a supportive external environment that protects the rights of users, facilitates conflict resolution, and promotes timely financial and technical inputs in the face of uncertainty.

Critical steps in this process on the part of those responsible for decision making in semiarid and arid regions are the following: (1)to create a cadre of professionals who understand and respect the fundamental nature and value of opportunism evident among pastoral people in response to variable dryland environments; (2)to grant greater decision making and legal authority to customary (or newly formed) pastoral institutions so that their adaptive mechanisms form the core of development and conservation initiatives; (3)to promote technical approaches that match the environment in which they are tried through the process of action research; (4)to develop effective communication and network mechanisms to feed back results of action research to key decision makers; (5)for collaborating organisations to commit themselves to a process of internal reflection and change if they are to effectively implement flexible and appropriate strategies and policies; and (6)to identify and sustain the role of the "change agents" and build their capacity to facilitate processes of change among various governance levels and within their own organisations.

Clearly, promoting democratic mechanisms for research and development planning and implementation may seem like a lofty goal. However, based on its experience ICIMOD has found that engaging policy makers and researchers in local level action research initiatives has begun to bring about fundamental changes in attitudes and organizational performance, which is the foundation to developing supportive policies for decentralised decision making. This is certainly a slow and painstaking process, but one, which in the long–term, will prove to be more cost effective and sustainable than what has been promoted in the last five decades of failed pastoral development and rangeland conservation around the globe.

5. REFERENCES

Agrawal, A. 2001. Common property institutions and sustainable governance of resources. World Development 29: 1649–1672.

Banks, T. 2001. Property rights and the environment in pastoral China: Evidence from the field. Development and Change. 32: 717–740.

Banks, T., Richard, C.E., Li, P., and Yan, Z. In press. Community–based grassland management in western China: Rationale, pilot project experience and policy implications. Mountain Research and Development.

Blench, R., and Marriage, Z. 1999. Drought and Livestock in Semi–Arid Africa and the Far East. Working Paper 117. Overseas Development Institute, London. 138 pages.

Borrini–Feyerabend, G., Farvar, M. T., Nguinguiri, J. C., and Ndangang, V.A. 2000. Co–management of Natural Resources: Organizing, Negotiating and Learning by Doing. GTZ and IUCN, Kasparek Verlag, Heidelberg. 96 pages.

Dorje, Y., and Tsering. 2001. Study on Cashmere Goat Production System on Changtang, Tibet. Final Report of ICIMOD Regional Rangeland Programme (Phase I). Submitted to International Centre for Integrated Mountain Development. Unpublished.

Du, G., and Degang, Z. 2000. Final Report of ICIMOD Regional Rangeland Programme–Maqu County, Gansu (Phase I). Submitted to International Centre for Integrated Mountain Development. Unpublished.

Duobujie, C. 2000. The status and harnessing of the grassland ecological environment in Naqu, Tibet Autonomous Region. Pages 106–112 In: Zhi, L. and Springer, J. (Editors) Tibet's Biodiversity: Conservation and Management. China Forestry Publishing House, Beijing.

Fernandez–Gimenez, M.E. 2002. Spatial and social boundaries and the paradox of pastoral land tenure: A case study from post–socialist Mongolia. Human Ecology 30: 49–78.

Fisher, R.J. 2001. Experiences, challenges, and prospects for collaborative management of protected areas: An international perspective. Pages 81–96 In: Buck, L.E., Geisler, C.C., Schelhas. J., Wollenberg, E. (Editors) Biological Diversity: Balancing Interests through Adaptive Collaborative Management. CRC Press, Boca Raton.

Fisher, R.J. 2002. Common property regimes in mountain regions: Risks and the governance of biodiversity. Paper presented at The Asia High Summit, Kathmandu, Nepal May 6–10, 2002. CD–Rom. International Centre for Integrated Mountain Development, Kathmandu.

Gilles, J.L., and K. Jamtgaard. 1982. Overgrazing in pastoral areas: The commons reconsidered. Nomadic Peoples 10: 1–10.

Gilmour, D.A., and Fisher, R.J. 1991. Villagers, Forests and Foresters: The Philosophy, Process and Practice of Community Forestry in Nepal. Sahayogi Press, Kathmandu. 212 pages.

Goldstein, M.C., and Beall, C.M. 1990. Nomads of Western Tibet: Survival of a Way of Life. University of California Press, Berkeley. 200 pages.

Greenwood, D.J., and Levin, M. 1998. Introduction to Action Research: Social Research for Social Change. Thousand Oaks, CA: Sage Publications. 264 pages.

Gurung, H. 1999. Mountains of Asia. International Centre for Integrated Mountain Development, Kathmandu. 86 pages.

Hardin, G. 1968. The tragedy of the commons. Science 162: 1243–1248.

Ingles, A.W., Musch, A., Qwist–Hoffmann, H. 1999. The Participatory Process for Supporting Collaborative Management of Natural Resources: An Overview. Food and Agriculture Organization of the United Nations, Rome. 84 pages.

Li, P. In press. Tenure and management arrangements for China's forestland and grassland resources: Fieldwork findings and legal and policy recommendations. In: Richard, C.E., and Hoffman, K. (Editors) The Changing Face of Pastoralism in the Hindu–Kush Himalaya Tibetan Plateau Highlands: Forging a Sustainable Path for the Future. Proceedings of a Strategy Workshop Conducted in Lhasa, TAR, PRC, May 2002. International Centre for Integrated Mountain Development, Kathmandu.

Longsworth, J.W., and Williamson, G.J. 1993. China's Pastoral Region: Sheep and Wool, Minority Nationalities, Rangeland Degradation and Sustainable Development. CAB International, Canberra. 368 pages.

Ma, Y., Li, Q., and Li, Y. 2000. Final Report of ICIMOD Regional Rangeland Programme–Dari County, Qinghai (Phase I). Submitted to the International Centre for Integrated Mountain Development. Unpublished.

McVeigh, C. 1994. Indigenous Resource Management Systems Among Tibetan Speaking Herders in Western Nepal. Report submitted to USAID, Nepal Mission. Unpublished.

Miller, D.J. 1995. Herds on the Move: Winds of Change Among Pastoralists in the Himalayas and on the Tibetan Plateau. Discussion paper series No. MNR 95/2. ICIMOD, Kathmandu. 27 pages.

Miller, D. In press. The Tibetan Steppe. In: Reynolds, S. (Editor) Grasslands of the World. FAO, Rome.

Naimer–Fuller, M. 1999. Toward a synthesis of guidelines for legitimizing transhumance. Pages 266–290 In: Naimer–Fuller, M. (Editor) Managing Mobility in African Rangelands: the Legitimisation of Transhumance. Intermediate Technology Publications Ltd., London.

Naimer–Fuller, M., and Turner, M.T. 1999. A review of recent literature on pastoralism and transhumance in Africa. Pages 8–46 In: Naimer–Fuller, M. (Editor) Managing Mobility in African Rangelands: The Legitimisation of Transhumance. Intermediate Technology Publications Ltd., London.

Narjisse, H. 1996. The range livestock industry in developing countries: Current assessment and prospects. Pages 14–21 In: West, N.E. (Editor) Proceedings of the Fifth International Rangeland Congress. Logan, UT. Vol. II.

National Research Council. 1994. Rangeland Health: New Methods to Classify, Inventory, and Monitor Rangelands. National Academy Press. Washington D.C. 180 pages.

Ngaido, T, McCarthy, N., and Di Gregorio, M. 2002. International Conference on Policy and Institutional Options for the Management of Rangelands in Dry Areas: Workshop Summary Paper. CAPRI Working Paper No. 23. International Food Policy Research Center, Washington. 70 pages.

Olson, D., and Dinerstein, E., 1998. The Global 2000: A Representation Approach to Conserving the Earth's Distinctive Ecoregions (Draft manuscript). WWF–US Conservation Science Program, Washington DC, USA.

Rai, N.K., and M.B. Thapa. 1993. Indigenous Pasture Management Systems in High Altitude Nepal: A Review. HMG Ministry of Agriculture/Winrock International. Research Report Series No. 22. 79 pages.

Richard, C.E. 1994. Natural Resource Use in Protected Areas of the Himalaya: Case Studies from Nepal. IOF Technical Paper. Institute of Forestry, Pokhara, Nepal. 120 pages.

Richard, C.E. 2000. An agro–ecosystem perspective on grassland management in the Terai and mountain protected areas. Pages 15–21 In: Richard, C.E., Sah, J.P, Basnet, K., Kharki, J. Subba, B., and Raut, Y. (Editors) Grassland Ecology and Management in Protected Areas of Nepal: Workshop Action Summary (Volume I). International Centre for Integrated Mountain Development, Kathmandu.

Richard, C.E. 2002. The potential for rangeland management in yak rearing areas of the Tibetan Plateau. Pages 11–20 in: Jianlin, H., Richard, C., Hanotte, O., McVeigh, C. and Rege, J.E.O. (editors) Yak Production in Central Asian Highlands. Proceedings of the Third International Congress on Yak Held in lhasa P.R. China, 4–9 September 2000. Ilri (International Livestock Research Institute), Nairobi.

Richard, C.E., and Yonten, K. 2002. Sustainable Rangeland Development in Two Counties of Naqu Prefecture, Tar–Preliminary Consultant Report. Submitted to Tibet Poverty Alleviation Fund, USA. Unpublished.

Saberwal, V.K. 1996. Pastoral politics: Gaddi grazing, degradation and biodiversity conservation in Himalchal Pradesh, India. Conservation Biology 10: 741–749.

Sandford, S. 1983. Management of Pastoral Development in the Third World. John Wiley & Sons, Chichester. 316 pages.

Sherpa, M.N. 2000. Ecoregion–based conservation: a global perspective. Pages 16–19 In: Zhi, L., and Springer, J. (Editors) Tibet's Biodiversity: Conservation and Management. China Forestry Publishing House. Beijing.

Sneath, D. 1996. State policy and pasture degradation in Inner Asia. Science 281: 1147–1148.

Steinfeld, H., de Haan, C., and Blackburn, H. 1997. Livestock–Environment Interactions: Issues and Options. FAO, Rome. 56 pages.

Thwaites R., De Lacy, T., Li, Y.H., and Liu, X.H. 1998. Property rights, social change, and grassland degradation in Xilingol Biosphere Reserve, Inner Mongolia, China. Society and Natural Resources 11: 319–338.

Ura, K. 1993. Nomad's Gamble: Pastoralists of Northern Bhutan. South Asia Research. 13: 81–100.

Williams, D.M. 1996. Grassland enclosures: Catalyst of land degradation in Inner Mongolia. Human Organization. 55: 307–313.

Williams, D.M. 2000. Representations of nature on the Mongolian Steppe: An investigation of scientific knowledge construction. American Anthropologist 102: 503–519.

Wu, N., and Richard, C.E. 1999. The privatisation process of rangeland and its impacts on the pastoral dynamics in the Hindu Kush Himalaya: The case of western Sichuan, China. Pages 14–21 In: Eldridge, D., and Freudenberger, D. (Editors) People and Rangelands. Proceedings of VI International Rangelands Congress, Townsville, Australia.

Wu, N. 1997. Ecological Situation of High–Frigid Rangeland and Its Sustainability–A Case Study on the Constraints and Approaches in Pastoral Western Sichuan. Dietrich Reimer Verlag, Berlin. 281 pages.

Yan Z., Wu, N., and Richard, C.E. In press. Nomad people should be the major concern in grassland policy: Case study from the eastern Tibetan Plateau, China. In: Richard, C.E., and Hoffman, K. (Editors) The Changing Face of Pastoralism in the Hindu–Kush Himalaya Tibetan Plateau Highlands: Forging a Sustainable Path for the Future. Proceedings of a Strategy Workshop Conducted in Lhasa, TAR, PRC, May 2002. International Centre for Integrated Mountain Development, Kathmandu.

Yan Z., and Luo, G. 2000. The Pastoral Production System in Hongyuan, Eastern Qinghai–Tibet Plateau. Final Research Report submitted to the International Centre for Integrated Mountain Development. Unpublished.

Zhang, R. 1989. Case Study on Mountain Environmental Management: Nyemo County (Tibet). ICIMOD Occasional Paper No.13. International Centre for Integrated Mountain Development, Kathmandu. 66 pages.

Chapter 19

THEORETICAL MODELS FOR REGENERATION OF MEDICINAL PLANTS AND THEIR APPLICATION IN SUSTAINABLE WILD–HARVESTING

Gloria Montenegro[1], Miguel Gómez[1], Ana Maria Mujica[1], and Barbara N. Timmermann[2]
[1]Department de Ciencias Vegetales, Facultad de Agronomia e Ingeniería Forestal, Pontificia Universidad Católica a de Chile, Santiago, Chile, email gmonten@puc.cl; [2]College of Pharmacy, University of Arizona, Tucson, AZ, USA

1. INTRODUCTION

Mediterranean–type ecosystems, such as Central Chile, the Mediterranean Basin, California, South Africa and Southwestern Australia are some of the 25 biodiversity hotspots on earth, selected because of their high plant diversity and endemism. These ecosystems represent 20 percent of the world's flora in only 5 percent of the earth's surface (Cowling et al. 1996). The risk to their flora from an increase in population density is another reason for their selection as hotspots (Cincotta et al. 2000).

Around half of all plants species recorded in continental Chile have been found in the Mediterranean area of Central Chile, in a surface area of 155,000 km^2 accounting for 20 percent of the total country. This flora has a high degree of endemism at 46 percent. This zone has also supported a long history of human occupation and a relatively high density of population. In fact, 80 percent of the country's population is distributed in this area thus implying a high demand for land to develop cities and agriculture that causes a significant reduction in area occupied by the wild vegetation (Arroyo and Cavieres 1997). Furthermore, the survival of some species, particularly the endemic species and those of restricted distribution is threatened by indiscriminate extraction for medicinal use.

Approximately 30 percent of the Chilean population, especially those living in rural areas, uses the wild plants for medicinal purposes following traditional or indigenous prescriptions handed down by "word of mouth." The demand for medicinal plants has also been growing, as much in the internal market as in the external market with the increase in exports of certain native or naturalized species in Chile. Thus, market trading in medicinal plants is currently of special concern to public policy makers and local people, mainly because of the lack of knowledge regarding in situ regeneration and the conservation status of native and endemic plants intensively harvested for commerce in public markets.

A study on the conservation and sustainable use of medicinal plants in Central Chile was developed at the Catholic University of Chile, as a part of the research program of the International Cooperative Biodiversity Group (ICBG). The objective was to strike a balance between exploitation and conservation of natural resources by collaborating with people who collect and sell medicinal plants, explaining the impact of local practices on the conservation and sustainable use of species intensively harvested in natural communities for commercial purposes. During this process, the species threatened by over exploitation is evaluated and data are collected in the field about areas of exploitation, the amount of biomass gathered, the frequency of collections and methods of plant harvesting. The biology of these plants is studied in order to develop morphological models of regeneration based on the position and location of renewal buds. In order to calculate their rate of regeneration after harvesting, the dynamics of growth of these species is followed in the field using experimental plots and by checking the growth of tagged branches after clipping. By monitoring the dynamics of growth of the species under study, it is possible to evaluate the rate of recovery and the plant's ability to cope with the rate of exploitation. Cultivation and propagation for reforestation purposes are developed along with the scientific studies. Databases are established with the information gathered for each species.

The matorral is the natural shrubby sclerophyllous vegetation growing in the semiarid Mediterranean region of Central Chile, which covers the slopes of the coastal range and those of the Andean foothills between 32° and 36° S (Di Castri 1968, Arroyo et al. 1995). This vegetation is adapted to a harsh environment that includes extended drought, unstable landforms, desiccating winds and low nutrient availabilities in the soil (Aljaro and Montenegro 1981, Miller 1981, Montenegro et al. 1989, Montenegro 2001). Matorral vegetation is diverse in growth forms with ancient woody groups (e.g., phanerophytes) of tropical origin coexisting today with short–lived herbaceous species.

According to the position of renewal buds, models of regeneration based on the growth form of the species were developed. By analyzing the regeneration model of dominant species of two plant communities along an altitudinal transect in Central Chile from the coast up to 1,200 meters above sea level, the potentiality of regeneration of the vegetation after severe human impact, including harvesting for medicinal purposes, fire and logging was predicted.

2. CASE STUDY

2.1 The Plant

The case study presented here focuses on *Chorizanthe vaginata* Benth (Montenegro et al. 2001). *Chorizanthe vaginata* is an endemic, commercially harvested medicinal plant. It is used traditionally to treat diarrhea, rheumatism and headache and to help prevent formation of gall stones. Moreover, it is believed to have analgesic and anti–inflammatory properties. Usually, its leaves and stems are sold at the markets, but it is quite common to also find whole roots. Morphologically, this species corresponds to a hemicryptophyte, a plant that maintains its renewal buds in a root crown at the base of the stems located at the level of the soil surface. Renewal buds are meristematic areas that allow the plant to resprout and regenerate the aerial biomass after being harvested. The morphological study of *Ch. vaginata* roots showed an anomalous secondary growth that develops an interaxillary cork between xylem rings, which allows the splitting of the main root into separate longitudinal sections without experiencing dehydration. This type of morphological feature appears in plants growing under extreme drought conditions, and allows them to survive and reproduce vegetatively (Montenegro et al. 1979).

2.2 Growth Characteristics

Monitoring of the phenology of this species for one year indicates that *Ch. vaginata* is dormant from the middle of March (late autumn) to the middle of July (early winter) when buds begin to sprout, developing small branches with short internodes or brachyblasts (Figure 1). Some of the elongated developing dolichoblasts, in turn, develop short shoots from their axillary buds. The long shoots, and some short shoots flower at apical positions in early September and fruit in early December. Since all flowering takes place in apical positions, renewal buds are located on the axillary buds that remain dormant and on those buds located at the upper side of the root forming a regenerative structure known as the root crown.

Experimental harvesting was conducted on ten plants growing in the field in El Canelillo, a coastal scrub vegetation area. Plants were harvested to the ground level, over the root crown, at the beginning of the growing period (middle July). Plants were capable of resprouting a few days after harvesting. For a mean biomass 104 ± 49 g (n = 10) extracted in the beginning of the growing season (middle July), the mean regenerated biomass after 140 days at the end of the growing period (middle December) was 35 ± 9 g (n =10). The biomass regeneration per plant ranged from 15–47 g. The recovery of biomass allows the prediction that in favorable climatic conditions plants would take 1.2 years to recover the initial biomass, or under unfavorable conditions would take about 5.8 years. This period was calculated using the values of the regeneration rates in mg/day obtained from growth curve dynamics (Montenegro et al. 2001).

2.3 Impact of Harvesting In Situ

Fieldwork was conducted with the Alarcón family who sell this medicinal plant at the Mercado Central Oriente in Santiago in order to follow up and document their harvesting procedures. Harvesting activities at two sites, El Canelillo and El Mirasol, were recorded, including the number of collection trips made, the weight of the collected harvest, and the average number of plants in each collection. In general, when the average plant canopy diameter was over 120 cm, 20 percent of the aerial biomass was left intact. An interview with the family revealed that they believed their harvesting techniques allowed the plants to regenerate. However, when plants collected by this family was examined in the laboratory, some unsustainable harvesting methods were detected.

Bundles or plant parts indicated four methods of harvesting: (1)plants cut from the aerial canopy to the root along the longitudinal splitting; (2)plants cut above ground; (3)plants harvested with the root crown included, but not the root itself; and (4)plants harvested with the entire roots. The method of harvesting showed that less than half the plants (44 percent) were harvested in a sustainable way (i.e., longitudinal splitting and/or plants cut above ground). The study family harvested a total of 120 plants accounting for 31 kg of biomass in one year. Of this, 53 plants with a biomass of 14 kg (44 percent) were taken as sustainable harvest, while 67 plants with a biomass of 17 kg (56 percent) were removed as unsustainable harvest.

Figure 1. Phenomorphological cycle of Chorizanthe vaginata Benth.

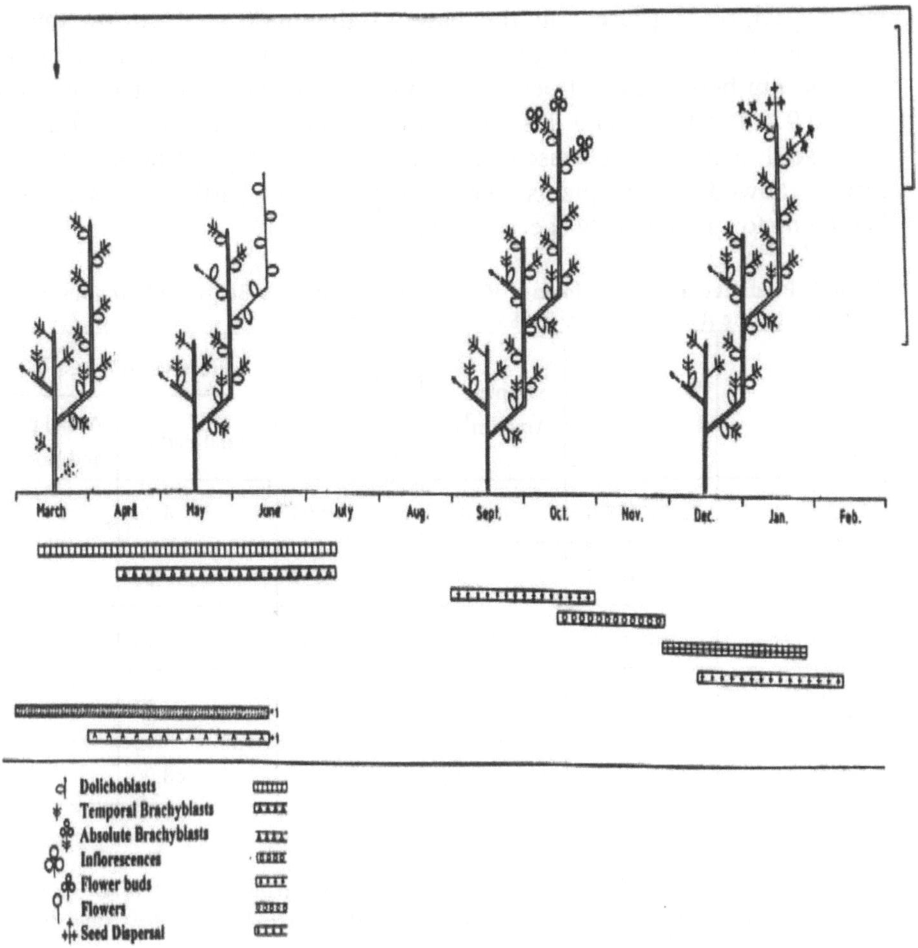

Field studies estimated the relative cover and biomass still available in situ at four coastal sites where populations of *Ch. vaginata* are still growing. At three of these sites, people were allowed to collect plants, while the fourth site was left undisturbed as a control. Table 1 shows the predicted biomass available for harvesters, separated into aerial plant parts and into aerial plant parts plus roots at different sites.

Roots found as a result of unsustainable harvesting throughout one year at El Canelillo indicated that harvesting by extraction of the root crown mainly occurs in individuals between the ages of 8 and 12 years old (Table 2). The frequency of dead roots found in this age group was significantly higher than those of other age groups. The age groups determinations were based on the morphological analysis of the roots. A good correlation ($r^2 = 0.63$, $p = 0.002$, 95 percent confidence level) between the root diameter and the annual rings present at the base of the main root allows the age determination of plants (logarithmic regression, $y = a + b \ln x/x$). The relationship between plant perimeter and age showed a stronger correlation ($r^2 = 0.978$ $P < 0.01$), and this allowed the calculation of the age of a given plant sold at the market based on measurements of plant perimeters. Our results showed that the plants were very old in this particular case, despite belonging to a non–woody species group.

Table 1: Predicted available biomass of Chorizanthe vaginata Benth. In four sites of collection by the Alarcon family 1998.

Site	Location Latitude S	Sampled Area (m^2)	Relative Cover (percent)	Predicted Biomass (Kg)	
				Aerial Parts	Aerial parts and Roots
El Canelillo	33°21’51” S	15,000	18.93	4.70	8.90
Salinas	33°23’40” S	300	5.23	17.20	32.95
Mirasol	33°18’30” S	20,000	24.68	22.93	43.64
Tunquen (Control)	33°17’30” S	20,000	24.73	23.23	44.53
Total	—	70,300	—	68.06	130.02

Table 2: Frequency of dead roots from different age groups found in El Canelillo: Evidence of unsustainable extraction of Chorizanthe vaginata Benth. (From January 1998 to November 1998).

Root diameter (cm)	0.0–0.5	0.55–1.0	1.10–1.5	1.55–2.0	2.05–3.0
Range age (year)	< 7	8–12	13–17	18–22	> 23
Frequency	8	31	15	6	1
Percentage	13.1	50.8	24.6	9.8	1.6

The dynamic and seasonal observations of the harvesting methods employed for *Ch. vaginata* by the local collectors indicate that it is possible to work with them to teach sustainable harvesting methods. They are also able to learn about the morphological aspects of particular plants in order to practice sustainable harvesting that leaves the meristematic tissues responsible for the resprouting process intact. In order to reach these goals of sustainability at the local level, several workshops were conducted on conservation and the sustainable use of plant resources at the Catholic University of Chile.

2.4 Developing Models of Regeneration Based on Plant Morphology in Communities of Central Chile

Evergreen sclerophyllous shrubs and trees, succulents and drought–evading herbs predominate in Central Chile at an altitudinal gradient from the coast to about 1,200 masl (Mooney et al. 1970, Mooney 1977, Montenegro et al. 1979). Evergreen shrubs predominate on polar–facing slopes, while drought–deciduous shrubs and succulents are mostly found on equatorial–facing slopes (Rundel 1975, Parsons 1976, Mooney 1977, Armesto and Martínez 1978). The coastal scrub and middle elevation sclerophyllous matorral found on the slopes of the coastal range and on the foothills of the Cordillera de Los Andes are replaced at about 1,850 m by a montane evergreen scrub community (Mooney et al. 1970, Rundel and Weisser 1975, Hoffmann and Hoffmann 1978, 1982; Montenegro et al. 1979). Changes in plant growth forms with altitude also are evident when the positions of renewal buds are observed (Montenegro et al. 2001). Camaephytes predominate at coastal scrub and at higher elevations. Phanerophytes and geophytes predominate at the foothill matorral on the coastal range and on the Andes Mountain. Cushion plants are dominant above 2,000 masl.

We have compiled lists of dominant species in the central region of Chile (Montenegro et al. 2002), with their growth forms (e.g., phanerophyte, chamaephyte, hemicryptophyte, geophyte, terophyte) and organs of regeneration (e.g., lignotuber, bulb, root crown, seed, rhizome). The analysis of this information allows the prediction of regeneration at the community level, calculating the percentage of plants that are able to regenerate from renewal buds and the percentage of plants that are only able to recover by seed germination (Figures 2–3).

Hemicryptophytes are well represented in the Chilean matorral (Figure 2–3). This plant growth form is also well adapted to seasonal climates such as the Mediterranean–type climate. The herbaceous growth form presents a root crown with dormant buds at the ground level that are usually protected by a dense rosette of dead leaves. About half of the shrubs, or plants that show slight lignification at the base of the main stems, also develop a root crown, so that they can regenerate from root crown buds or from the axillary buds that remained dormant under the periderm or suberous cortex of the main stems. In this type of shrub, the leaves usually die every season and form and develop again from the renewal buds on the root crown or on the base of the old stems.

Phanerophyte adaptive regeneration traits after harvesting showed that they might regenerate from buds located in underground, lignified tubers, the so-called lignotuber, or from epicormic buds located on lignified stems or trunks where the renewal buds are protected by the cortex (Figure 3). Along these lines, it has been shown that most matorral species are able to resprout after fire (Cody and Mooney 1978, Araya and Ávila 1981, Montenegro et al. 1983, Montenegro and Ginocchio 1995) due to the presence of lignotubers. Lignotubers have been defined as a source of dormant epicormic buds buried in a modified stem (Montenegro et al. 1983). Such buds are capable of resprouting after the crown is killed, consumed by fire or removed by mechanical means, thereby regenerating the aerial part of the plant. The presence of adventitious buds in root crowns of woody plants is nearly a ubiquitous trait in dicotyledonous plants (Wells 1969). Storage of carbohydrates plays an important role in adventitious buds and there is strong circumstantial evidence from comparison of burning and cutting experiments that seasonal depletion of carbohydrates may strongly affect regenerative capacity (Rundel et al. 1987).

The Chilean matorral of the Mediterranean climate is characterized by the dominance of species that are able to develop lignotubers. These morphological structures are ontogenetic features adapted to initiate their development early at the seedling stage (Wells 1969, Montenegro et al. 1983). Buds of lignotubers develop from the vascular cambium and not from the cortical parenchyma. Thus, bark thickness is of great importance in protecting these buds, as well as the cambium during human impact or fire. Therefore, harvesters of bark in medicinal plants must leave certain parts of the bark intact in order to allow regeneration. Experimental bark cutting in *Quillaja saponaria*, an evergreen shrub from the matorral vegetation of central Chile, showed that it takes approximately 36 months to recover 10 cm^2 of periderm in 30 year–old shrubs.

The resprouting capability differs among species, both in the proportion of individuals that exhibited the response and in the amount of foliage produced (Ginocchio et al. 1994). Resprouting can immediately occur after a harvest, independent of the time of year, since the formation of new branches takes place at the expense of the carbohydrate reserves located in underground organs. Species that normally begin growth later in the season are able to resprout in autumn when climatic conditions are not optimal (Montenegro et al. 1981, 1989). The resprouting capability from lignotubers is related to plant age; older plants with larger lignotubers had increased resprouting capabilities (Montenegro et al. 2002). This may be due to higher starch reserves in parenchymatic cells of larger underground woody structures (Montenegro et al. 1983). Carbohydrate levels as high as 4.5 to 10.2 percent dry weight have been detected in lignotubers of *Erica australis* in Mediterranean–type ecosystems of Spain (Cruz and Moreno 2001) and as high as 7.5 to 12.4 percent dry weight in lignotubers of *Cryptocarya alba* in Chilean matorral.

The only way to conserve and use plant resources in a sustainable manner is to study the biology of every single species threatened differently. It is urgent to do so with rare and endangered plant species, and even more so with endemic species.

Figure 2. Plant species on the Coastal matorral, middle elevation matorral and Montane matorral and their regeneration properties

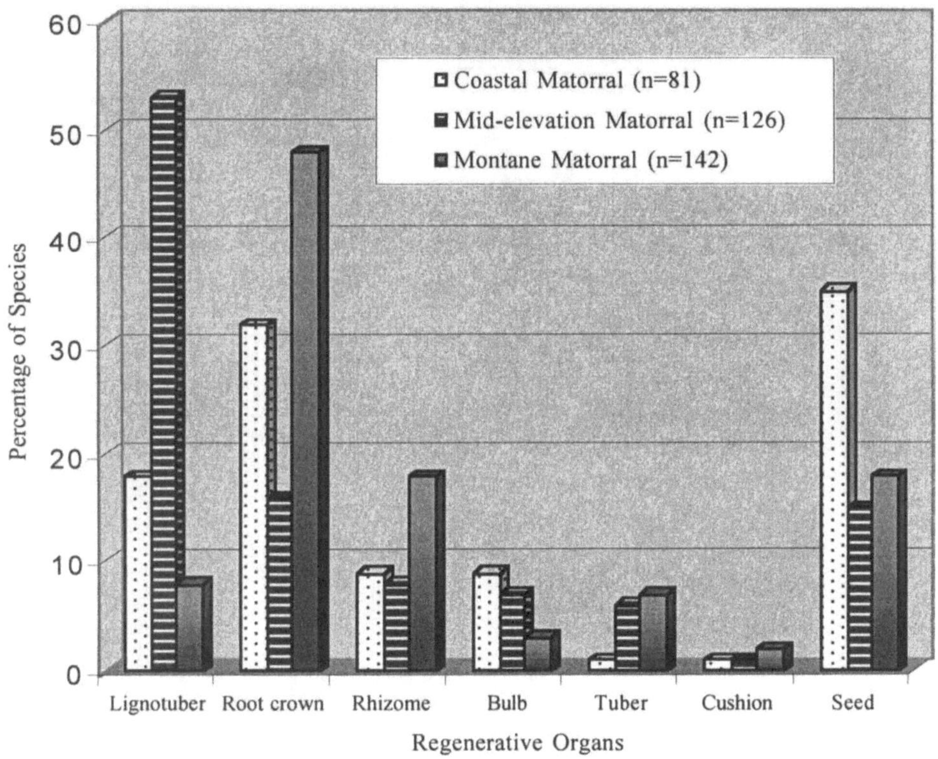

Figure 3. Theoretical Models of Regeneration

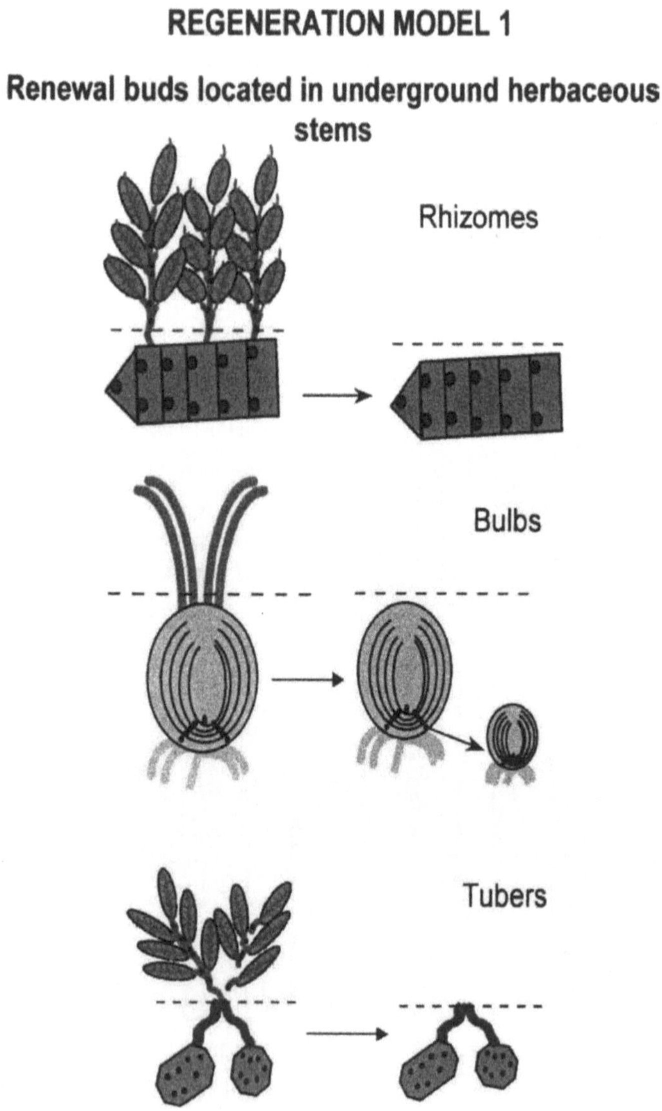

Figure 3. Theoretical Models of Regeneration (continued)

REGENERATION MODEL 2

Renewal buds located at the soil surface in root crowns

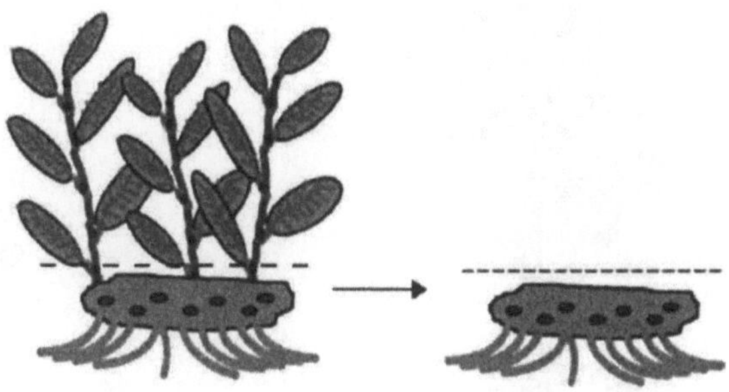

REGENERATION MODEL 3

Woody plants with a life form in a cushion structure at the soil surface. Renewal buds in cushion structure

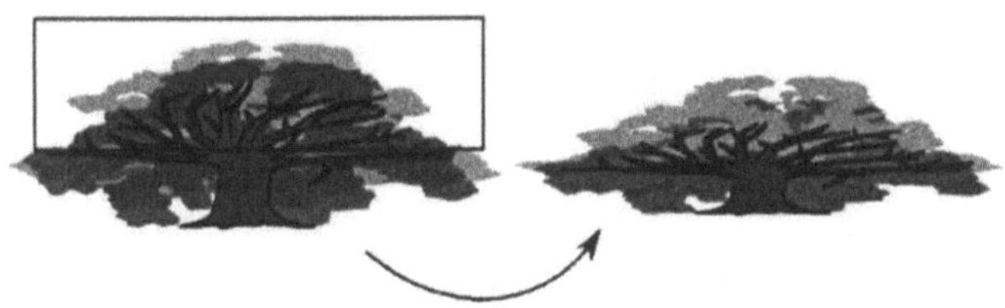

Figure 3. Theoretical Models of Regeneration (continued)

REGENERATION MODEL 4

Woody plants, lignotubers as underground stems.
Renewal buds in Lignotubers

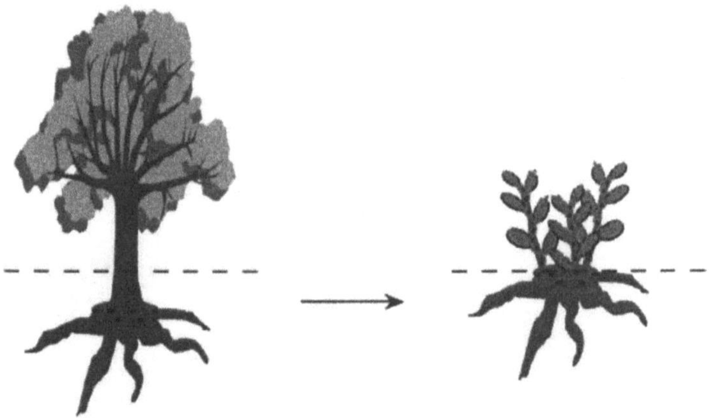

REGENERATION MODEL 5

Renewal buds located only in the seeds

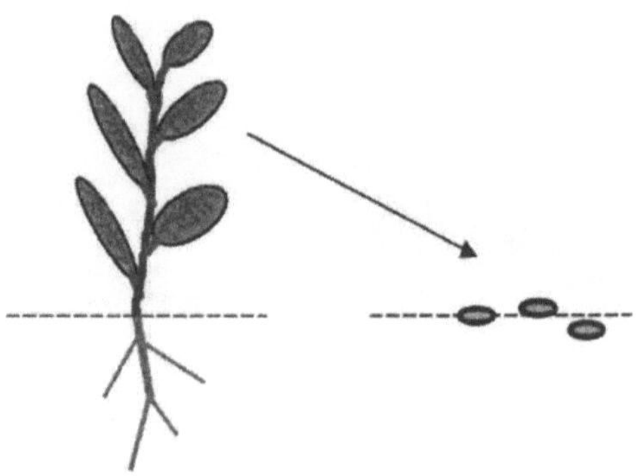

3. LESSONS LEARNED

Science and technology are fundamental tools to achieve a smooth transition toward the conservation and sustainable use of biodiversity. Local and rural population participation is essential to change patterns of harvesting, consumption and production. It is vital to improve communication and cooperation among scientists and other stakeholders of biodiversity, as well as to promote codes of practices and guidelines. Capacity building and education at all levels are essential to promote conservation and sustainability. Scientists must play a key role in developing, planning and decision making.

4. ACKNOWLEDGEMENTS

This work was supported by grant NIH-NSF-USDA 2UO1 TW00316-09 to B.N. Timmermann and Proyecto FIA C01-1-G-002 to G. Montenegro.

5. REFERENCES

Aljaro, M. E. and Montenegro, G. 1981. Growth of dominant Chilean shrubs in the Andean Cordillera. Mountain Research and Development 1: 287–291

Araya, S. and Ávila, G. 1981. Rebrote de arbustos afectados por fuego en el matorral chileno. Anales del Museo de Historia Natural de Valparaíso 14: 107–13

Armesto, J. and Martínez, J. A. 1978. Relations between vegetation structure and slope aspect in the Mediterranean region of Chile. Journal of Ecology 66: 881–889.

Arroyo, M.K., Cavieres, L., Marticorena, C. and Muñoz, M. 1995. Convergence in the Mediterranean floras of Central Chile and California: Insights from comparative biogeography. Pages 43–88 In: Arroyo, M. K., Fox, M. and Zedler, P. (Eds.) Ecology and Biogeography of Mediterranean Ecosystems in Chile, California and Australia. Springer–Verlag, NY.

Arroyo, M.K. and Cavieres, L. 1997. The Mediterranean–type climate flora of Central Chile. What do we know and how can we assure its protection? Pages 48–56 In: Montenegro, G. and Timmermann, B.T. (Eds.). Taller Internacional Sobre Aspectos Ambientales, Ideológicos, Éticos y Políticos en el Debate Sobre Bioprospección y Uso de Recursos Genéticos en Chile. Noticiero de Biología, Vol. 5, No.2. Sociedad de Biología de Chile, Santiago, Chile.

Cincotta, R. P., Wisnewski, J. and Engelman, R. 2000. Human population in the biodiversity hotspots. Nature 404: 990–992.

Cody, M.L. and Mooney, H.A. 1978. Convergence versus nonconvergence in Mediterranean–type ecosystems. Annals Review of Ecology and Systematics 9: 265–321.

Cowling, R.M., Rundel, P.W., Lamont, B.B., Arroyo, M.K. and Arianoutsou, M. 1996. Plant diversity in Mediterranean–climate Regions. Trends in Ecology and Evolution 11: 362–368.

Cruz, A. and Moreno, J. 2001. Seasonal course of total non–structural carbohydrates in the lignotuberous Mediterranean–type shrub *Erica australis*. Oecologia 128: 343–350

Di Castri, F. 1968. Esquisse Ecologique du Chili. Biologique de l'Amerique Australe. Centre National de la Recherche Scientifique, Paris 4: 1–52.

Ginocchio, R., Holmgren, M. and Montenegro, G. 1994. Effect of fire on plant architecture in Chilean shrubs. Revista Chilena de Historia Natural 67: 177–182.

Hoffmann, A.J. and Hoffmann, A.E. 1978. Comportamiento fenológico de plantas de la Cordillera de los Andes. Archivos de Biología Experimental 11: 188.

Hoffmann, A.J. and Hoffmann, A.E. 1982. Altitudinal ranges of phanerophytes and chamaephytes in Central Chile. Vegetatio 48: 151–163.

Miller, P.C. (ed). 1981. Resource Use by Chaparral and Matorral. A Comparison of Vegetation Function in Two Mediterranean–Type Ecosystems. Springer–Verlag, Berlin.

Montenegro, G., Aljaro, M. E. and Kummerow, J. 1979. Growth dynamics of Chilean matorral shrubs. Botanical Gazette 14: 114–119.

Montenegro, G., Hoffmann, A.J., Aljaro, M.E. and Hoffmann, A.E. 1979. *Satureja gilliesii*, a poikilohydric shrub from the Chilean Mediterranean vegetation. Canadian Journal of Botany 57: 1206–1231.

Montenegro, G., Aljaro, M. E. and Arrieta, A. 1979. Dinámica de crecimiento y productividad de especies dominantes en un transecto altitudinal de la Cordillera de los Andes. Pages 96–108, In: Seminario sobre el Programa de Investigación Integrada "Impacto del hombre en los ecosistemas de montaña" MAB 6. UNESCO.

Montenegro, G. Aljaro, M.E., Walkowiak, A. and Saenger R. 1981. Seasonality, growth and net productivity of herbs and shrubs of the Chilean Matorral. Pages 135–141 In: Conrad, C. and Oechel, W. (Eds.). Dynamics and Management of Mediterranean–Type Ecosystems. Department of Agriculture and Forest Service, General Technical Report PSW 58. USA.

Montenegro, G., Avila, G., and Schatte, P. 1983. Presence and development of lignotubers in shrubs of the Chilean matorral. Canadian Journal of Botany 61: 1804–1808.

Montenegro, G., Avila, G., Osorio, R. and Gómez, M. 1989. Chile. Pages 347–387 In: Orshan, G. (Ed.). Plant Phenomorphological Studies in Mediterranean–Type Ecosystems. Kluwer Academic Publishers, Dordrecht, The Netherlands.

Montenegro, G. and Ginocchio, R. 1995. Ecomorphological characters as a resource for illustrating growth–form convergence in matorral, chaparral and mallee. Pages 160–176 In: Arroyo, M.K., Zedler, P.H. and Fox, M.D. (Eds.) Ecology and Biogeography of Mediterranean Ecosystems in Chile, California and Australia. Springer–Verlag, NY.

Montenegro, G. 2001. Biodiversity's Importance in Dryland Regions. TWAS Newsletter 13 (2): 5–7.

Montenegro, G., Patrick, G., Echenique, P., Gómez, M. and Timmermann, B. 2001. Mechanisms toward a sustainable use of *Chorizanthe vaginata* Benth, var. *maritima* Remy: A medicinal plant from Chile. Phyton International Journal of Experimental Botany 2001: 91–106

Montenegro, G., Díaz, F., Gómez, M. and Ginocchio, R. 2002. Regeneration potential of Chilean matorral after fire: An updated view. Pages 375–403 In: Veblen, T., Baker, T., Montenegro, G. and Swetnam, T. (Eds.) Fire and Climate Change in Temperate Ecosystems of the Western Americas. Springer–Verlag, NY.

Mooney, H.A., Dunn, E.L., Shropshire, F., and Song, L. 1970. Vegetation comparisons between the Mediterranean climatic areas of California and Chile. Flora 159: 480–496

Mooney, H. A. (ed.). 1977. Convergent evolution of Chile and California Mediterranean Climate Ecosystems. Dowden, Hutchinson and Ross, Pennsylvania. 244 pages.

Parsons, D.J. 1976. Vegetation structure in the Mediterranean climate scrub communities of California and Chile. Journal of Ecology 64: 435–447.

Rundel, P.W. 1975. *Trichocereus* in the Mediterranean zone of Central Chile. Cactus and Succulents Journal 46: 86–88.

Rundel, P.W. and Weisser, P. J. 1975. La Campana, a new national park in central Chile. Biological Conservation 8: 35–46.

Rundel, P.W., Baker, G.A., Parsons, D.J. and Stohlgren, T.J. 1987. Postfire demography of resprouting and seedling establishment by *Adenostoma fasciculatum* in the California Chaparral. Pages 575–596 In: Tenhunen, J.D, Catarino, F.M., Lange, O.L. and Oechel, W.C. (Eds.) Plant Response to Stress. Functional Analysis in Mediterranean Ecosystems: Springer–Verlag, Berlin.

Wells, P.V. 1969. The relation between mode of reproduction and extent of speciation in woody genera of the California Chaparral. Evolution 23: 264–267.

Chapter 20

PLANTS OF NORTHEASTERN BRAZIL: A PROGRAMME IN SUSTAINABLE USE OF PLANT RESOURCES

Ghillean T. Prance
School of Plant Sciences, University of Reading, Whiteknights, Reading RG8 3AS, U.K.; email Gtolmiep@aol.com

1. INTRODUCTION

The northeast of Brazil is a typical semiarid region with chronic social and environmental problems resulting from decades of overexploitation, inappropriate land use and extended periods of drought. This area, two–thirds the size of western Europe is home to 30 percent of the Brazilian population. Millions of people have migrated from the region to the large cities of southern Brazil or the Amazon region.

There is a considerable diversity of ecosystems in the northeast of Brazil varying from coastal rainforest to semi–desert. Much of the region is covered by semi–arid scrubland locally named "caatinga." This vegetation type contains many useful plant species, but they are still poorly studied.

Prior to 1992 the Royal Botanic Gardens, Kew, had successful collaboration with Brazilian botanical institutions, especially in Bahia and São Paulo. In 1992, after discussion with many Brazilian collaborators, it was decided to establish a larger and more relevant programme of botanical work in the region and an organisation was established under the name "Projeto Nordeste" with the central theme of "local plants for local people." This quickly developed into a registered association called "Plantas do Nordeste" (PNE). The governing board of the "Associação Plantas do Nordeste" consists of botanists from a number of regional organisations and representatives of the Royal Botanic Gardens, Kew and the Brazilian Research Council (CNPq).

The project began with the establishment of three subprogrammes: biodiversity, economic botany and information and some of the activities of the subprogrammes are described below. Recently, the Association has focused on integrated projects that combine elements of each subprogramme in order to encourage greater cooperation. The goal of PNE is to promote the preservation of the biodiversity of northeastern Brazil and the sustainable use of native plant resources for the benefit of local people.

2. BIODIVERSITY

A serious problem facing development and conservation programmes involving the use of plants was the poor state of knowledge of the flora and the vegetation dynamics of northeastern Brazil. Building upon previous experience of collaboration projects of the Royal Botanic Gardens, Kew in the region, a component on biodiversity was an obvious selection for PNE. Work was already underway on the flora of Pico das Almas in Bahia (Stannard 1995) and on a survey of various high altitude savannahs on rocky ground (Campo rupestre). The aim of the biodiversity subprogramme is to conduct studies in each of the different vegetation types found in northeast Brazil.

2.1 Campo Rupestre

This type of vegetation occurs over wide expanses of rocky areas and elevated grasslands at high altitudes up to 2000 m. The patches of this vegetation are isolated from each other and so a large number of endemic species occur. The vegetation is especially rich in orchids, bromeliads, vellozias and Eriocaulaceae. The latter called "sempre vivas" (everlasting), are much sought after for dried flower arrangements. Though many of these species are unique to localised areas they are being threatened by overexploitation.

One of the early biodiversity projects of PNE was on "Floristic Diversity and Plant Distribution in the Chapada Diamantina" (Guedes et al. 1999). This mountain has become a tourist attraction due to its natural beauty and rich natural diversity. The PNE personnel have carried out comparative studies between two peaks, Morro do Pai Inácio and Chapadinha. This project was made possible by a grant from the UK based Rufford Foundation which enabled a much greater participation of botanists from Brazilian institutions.

2.2 Brejos

Rising up from the lowlands are areas of hills on slightly higher ground on which thrive islands of forest, locally called brejos (Lima 1977). The hills are high enough to attract water from the clouds and so are islands of forests within the surrounding caatinga vegetation. The British government "Darwin Initiative" provided funding to PNE for a study of the 11 remaining brejos of Pernambuco State. This survey project was carried out in collaboration with the Botany Department of the Federal Rural University of Pernambuco and Kew (Rodal et al. 1998, Sales et al. 1999 a,b).

2.3 Caatinga

More than half of northeast Brazil is covered by caatinga. The caatinga vegetation project is based at the Federal University of Ceará in Fortaleza, but has had several other participating institutions. The first phase, financed by the company Souza Cruz, studied the caatinga of Aiuaba in Ceará state and Chapada de Apodi on the border between Ceará and Rio Grande do Norte. Monthly trips collected 6,400 specimens of over 360 species including two new to science. This project is notable for its training and education aspects. Classes were given for local teachers and schoolchildren and workshops on conservation and sustainable use of the caatinga were held. Maps of the semiarid vegetation of Ceará were distributed at Aiuaba and teaching material such as posters were regularly displayed at the Aiuaba Ecological Station (Figueiredo and Lima–Verde 1999).

The biodiversity subprogramme has included several other projects on caatinga since it is the dominant vegetation of the region. A good example is the study of the caatinga in Pernambuco state by Rodal and Melo (1999).

The Biodiversity subprogramme is also trying to build up the capacity within the regions to identify their plants. The herbaria in several of the cities involved in the programme are growing in size and, more importantly, in the number of correctly identified specimens.

An important aspect of the UK side of PNE is the repatriation of data to Brazil. Much information needed for taxonomic work lies in European institutions, where many original type specimens and publications are housed, often without duplication in Brazil. The initiation by Brazil of a flora of the biodiverse state of Bahia, emphasised this difficulty and led to the application for funding from the Darwin Initiative to repatriate data on Bahian plants to Brazil. The funding was granted in August 1988 and a Brazilian, Teo Sena was appointed as the Bahia Darwin Initiative Officer. The project was later extended to all states of NE Brazil and a database was prepared of Kew's holdings, from NE Brazil, of specimens of eight important plant families including the grasses and legumes. During this process a preliminary checklist of the plant families studied was produced and photographs of all type material was added to the database. The data gathered is also incorporated into PNE's Information Centre (CNIP) database so that it is readily available in Brazil. This project is an excellent model of what can be done to improve the resources available for research in a developing area.

3. ECONOMIC BOTANY

The economic botany subprogramme focuses on the useful or potentially useful plants of the region and their sustainable use.

3.1 Forage Plants

From early on in PNE's work an emphasis has been placed on forage plants for livestock, especially goats and cattle. One of these projects concentrated on fodder plants for goats at Brazil's National Goat Research Centre in Sobral on Ceará State. These studies showed that using native nitrogen fixing legumes and rotating pasture for goats through a series of corrals the native woody species could satisfy the entire nutritional needs of goats throughout the year. The dominant plant group in the caatinga is Leguminosae, many of which are good fodder. One project in Bahia State collected and recorded 280 species of legumes from which 26 were selected as potentially useful forage (Queiroz 1999). The study of native forage plants of the Rio Parnaíba Basin in Piauí State included chemical analysis of the protein and fibre content of potential forage plants (Nascimento 1966, Nascimento et al. 1999). These studies all found a better potential for forage if the caatinga is properly managed and they indicated that it is important to maintain areas of caatinga as a food resource for cattle during the dry season.

3.2 Medicinal Plants

Medicinal plants to provide cures for local people that cannot afford to buy pharmaceuticals have been a strong focus of PNE. Perhaps the flagship project has been the Farmácia Viva or "Living Pharmacy" programme. Chemist Professor Francisco de Abreu Matos in collaboration with medical doctor Adalberto Barreto developed this. The first project was located in the low income community of Quatro Varas in Fortaleza, Ceará. There, street children who have been rescued by the project, in return for education and food, tend the crops of medicinal plants and herbs and also promote the use of their herbal medicines through street theatre (Matos 1999).

PNE funds have been used to improve the medicinal plant nursery at the Federal University of Ceará, to build an extension to Professor Matos's chemical laboratory and to improve the water system at Quatro Varas. The programme has gradually expanded to other disadvantaged urban communities. The nurseries and gardens provide herbal medicines that have been proven to be effective and nontoxic. It is offering an effective, low cost alternative to commercial medicine for the treatment of many common ailments.

Brazil's first lady, Ruth Cardoso, coordinated the printing of 30,000 copies of the PNE brochure "Make Your Own Medicinal Plant Garden" and eight folders on medicinal plant species used in the "Living Pharmacies." These have been widely distributed in the region.

The "Living Pharmacy" programme is one of the several projects on medicinal plants within PNE. For example, an ethnobotanical study of medicinal plants was carried out in Paraíba State. This project screened 14 species, trained several people and produced a series of publications on their work (Agra 1996, 1999). Another medicinal plant project has been set up in Alagoas State in collaboration with the Chemistry Department of the Federal University of Alagoas. This has carried out an ethnobotanical survey with local healers, indigenous herbalists and local users of medicinal plants. A medicinal plant garden was also established to support a living pharmacy project.

3.3 Other Projects

Several other projects in economic botany are planned or in early stages of implementation. One that is just starting is on the production of fuel wood and charcoal from the region. This programme will draw on the experience of Kew with a similar programme in East Africa. Another project, which has been discussed but not yet implemented, is the cultivation of everlasting flowers to take pressure off the collection of some of the rare endemic species. This too, obviously has future potential. It is vital to develop as many sustainable uses of arid region plants if any of the native vegetation is to survive.

4. INFORMATION

The third and very important aspect of PNE is the processing of all the information gathered. From the outset of the programme this has been emphasised and Dr. Bob Allkin, an information specialist at Kew, has been involved. The spreading of information has taken many forms, from books, leaflets, training courses and workshops to computer databases. Early in the life of PNE, the ALICE database system was set up in many of the collaborating institutions in Brazil. For example, it was used for a database on 537 medicinal plant species used in NE Brazil (Pereira et al. 1999). After a long period of solicitation PNE was able to obtain funds from the British Department for International Development (DfID) to set up a Plant Information Centre (CNIP) at the Botany Department, Federal University of Pernambuco. Information is available from: cnip@elogica.com.br or on the website http:/www.cnip.org.br. Dr. Allkin has been funded by the DfID project to be seconded to Brazil for a four–year period. He has carried out many training programmes and workshops on information throughout the region of NE Brazil.

5. FUNDING

Some of the varied sources of funding for the activities have been mentioned in the text. Kew's participation in the programme began with a small grant from Shell Corporation. The international cooperation really got underway with a grant of £1 million from the Weston Family Foundation. This helped to stimulate funding from Brazil and the Brazilian Research Council (CNPq) has made regular grants for fellowships to students and researchers from the NE region. This began with a sizeable grant of $500,000. The largest single grant so far has been one of £1.91 million from the Department for International Development (DfID) to support the Information subprogramme. That subprogramme has also received support from Reuters Foundation. The success of PNE is that it has obtained support from both the UK and Brazil making it truly binational in all aspects. The support in Brazil has been from both government and private sources. A programme of this size and nature is costly and so constant work on fund raising has been necessary to ensure its continuation. Since it is a programme that is doing much to help an arid region where there is still much poverty and much environmental destruction, it has been possible to obtain support for this important programme.

6. FURTHER INFORMATION

In addition to the references cited above, further information about the work of PNE can be found in the publications of Sampaio et al. (1995) and

Araujo et al. (1999). PNE puts out a regular Newsletter available through Amélia Baracat (a.baracat@rbgkew.org.uk). The PNE email address is pne@netpe.com.br. Information is also available at CNIP's website at www.cnip.org.br.

7. ACKNOWLEDGEMENTS

I thank Dr. Simon Mayo for reading an earlier draft of this paper and Brenda Carey for typing it.

8. REFERENCES

Agra, M.F., Rocha, E.A. Lucatelli, E., Baracho, G.S. and Formiga, S.C. 1996. Plantas da Medicina Popular dos Cariris Velhos, Paraíba, Brasil: Espécies Mais Comuns. Editora União, João Pessoa, Paraíba. 112 pages.

Agra, M.F. 1999. Plantas medicinais dos Cariris Velhos, Paraíba: Um projeto interdisciplinar. Pages 16–21 In: de Araujo, F.D., Prendergast, H.D.V., and Mayo, S.J. (Editors.) Plantas do Nordeste. Anais do I Workshop Geral. Royal Botanic Gardens, Kew. 150 pages.

Araujo, F.D. de, Prendergast, H.D.V., and Mayo, S.J. (Editors). 1999. Plantas do Nordeste: Anais do I Workshop Geral. Royal Botanic Gardens, Kew. 150 pages.

Figueiredo, M.A., and Lima–Verde, L.W. 1999. Caatingas e carrasco, comunidades xerófilas no nordeste do Brasil. Pages 104–112 In: de Araujo, F.D., Prendergast, H.D.V., and Mayo, S. J. (Editors.) Plantas do Nordeste. Anais do I Workshop Geral. Royal Botanic Gardens, Kew. 150 pages.

Guedes, M.L.S., Harley, R.M., Giulietti, A.M., Carvalho, A.M.V. de, Batista, H.P., Melo, E. de, and Funch, R. 1999. Diversidade florística e distribuição das Plantas da Chapada Diamantina–BA. Pages 76–82 In: Araujo, F.D. de, Prendergast, H.D.V., and Mayo, S.J. (Editors) Plantas do Nordeste. Anais do I Workshop Geral. Royal Botanic Gardens, Kew. 150 pages.

Matos, F.J. de A. 1999 Farmácias vivas–um projeto para preservação de plantas medicinais do nordeste e seu aproveitamento no sistema de saúde pública local. Pages 83–88 In: Araujo, F.D. de, Prendergast, H.D.V., and Mayo, S.J. (Editors.). Plantas do Nordeste. Anais do I Workshop Geral. Royal Botanic Gardens, Kew. 150 pages.

Nascimento, M. do P.S.B. 1996. Forrageiras da Bacia do Parnaíba. Halley S/A. 85 pages.

Nascimento, M. do P.S.C.B., Nascimento, H.T.S. do, Oliveira, M.E.A., Carvalho, J.H. de, Alcoforado Filho, F.G., and Santana, C.M.M. de. 1999. Levantamento preliminar, identificação botânica e valor nutritivo de plantas forrageiras nativas da Bacia do Parnaíba. Pages 22–32 In: de Araujo, F.D. de, Prendergast, H.D.V., and Mayo, S.J. (Editors.) Plantas do Nordeste. Anais do I Workshop Geral. Royal Botanic Gardens, Kew. 150 pages.

Pereira, M.F.C., Matos, F.J. de A., Lodes, P.C., and Allkin, R. 1999. Cadastro informatizado de plantas medicinas do nordeste. Pages 89–94 In: Araujo, F.D. de, Prendergast, H.D.V., and Mayo, (Editors) Plantas do Nordeste. Anais do I Workshop Geral. Royal Botanic Gardens, Kew. 150 pages.

Queiroz, L.P. de. 1999. Leguminosas da Caatinga, espécies com potencial forrageiro. Pages 63–75 In: de Araujo, F.D., Prendergast, H.D.V., and Mayo, S.J. (Editors) Plantas do Nordeste. Anais do I Workshop Geral. Royal Botanic Gardens, Kew. 150 pages.

Rodal, M.J.N., Sales, M.F.S., and Mayo, S.J. 1998. Florestas Serranas de Pernambuco: Localização e Conservação dos Remnascentes dos Brejos de Altitudes, Pernambuco, Brasil. Impressa Universitária da UFRPE, Recife. 30 pages.

Rodal, M.J.N., and Melo, A.L. de. 1999. Levantamento preliminar das espécies lenhosas da caatinga de Pernambuco. Pages 53–62 In: de Araujo, F.D, Prendergast, H.D.V., and Mayo, S.J. (Editors) Plantas do Nordeste. Anais do I Workshop Geral. Royal Botanic Gardens, Kew. 150 pages.

Sales, M.F., Mayo, S.J., and Rodal, M.J.N. 1999a. Plantas Vasculares das Florestas Serranas de Pernambuco. UmChecklist da Flora Ameaçada dos Brejos de Altitude, Pernambuco, Brasil. Impressa Universitária da UFRPE, Recife. 130 pages.

Sales, M.F., Rodal, M.J.N., Mayo, S.J., Zickel, C.S., and Sales de Melo, M.R.C. 1999b. Composição florística e diversidade dos "brejos" de Pernambuco. Pages 42–52 In: de Araujo, F.D., Prendergast, H.D.V., and Mayo (Editors.) Plantas do Nordeste. Anais do I Workshop Geral. Royal Botanic Gardens, Kew. 150 pages.

Sampaio, E.V.S.B., Mayo, S.J., and Barbosa M.R.V. (Editors). 1995. Pesquisa Botânica Nordestina: Progresso e perspectiva. Sociedade Botânica do Brasil, Recife, Pernambuco. 415 pages.

Stannard, B.L. (Editor). 1995. Flora of Pico das Almas. Royal Botanic Gardens, Kew. 853 pages.

Chapter 21

CHOICE OF SPECIES FOR RECOVERING A DEGRADED MINING AREA IN THE SEMIARID ZONE OF BRASIL

Paulo César Fernandes Lima, Adriano Quadros Lima, Marcos Antonio Drumond

Embrapa, Caixa Postal 23, CEP 56300–000 Petrolina, Brasil; email pcflima@cpatsa.embrapa.br

1. INTRODUCTION

Human activities are one of the main causes of ecological changes and plant species disappearance in Brazil's semiarid ("caatinga") zone. The primary activities are: the formation of pastures, implementing irrigation projects, the production of energy, overgrazing by livestock, and burning (Queiroz et al. 1992). According to Sá et al. (1994), about two–thirds of the driest area of the semiarid zone of Northeast Brazil has experienced significant degradation and loss of biodiversity.

This case study describes a joint project of the Mineração Caraíba S.A. and the Semi–Arid Agricultural Research Center, Embrapa Semi–Árido, by analyzing the actual and possible losses of diversity of native species of the flora affected by copper mining exploration in the semiarid zone of Northeast Brazil; it also discusses some of the processes used to choose species to be used in the recovery of the area degraded by the mining. Some nonendemic species have demonstrated good potential for use in the recovery of degraded areas because they grow fast and have high survival rates, while others, including many native species, do not have high survival and regeneration rates in soils affected by mine residue.

2. THE AREA OF STUDY

The area of study is located at Pilar, municipality of Jaguarari, State of Bahia, at 9° 87' latitude S and 39° 86' longitude W, 400 m of altitude, in the semiarid zone of Brazil. The climate of the region is the BSW type according to Köppen climatic classification, characterized by distinct wet and dry seasons, with irregular rainfall concentrated in 2 to 4 months of the year. The annual average rainfall is 439 mm with a wet season beginning in November/December and finishing in March. The average evaporation is 2000 mm/year and the annual average of relative humidity is approximately 50 percent. The annual average temperature varies from 23° C to 25° C, with a difference of 8 ° C to 10° C between the average maximum and minimum temperatures. Landscape is gently undulating, broken by a few rocky hills. In general the soils are shallow with a low water–holding capacity and low organic content.

The agroecological zoning of Northeast Brazil, based on various criteria such as water potential, farming and livestock areas, the most viable crops to be planted, preservation and reforestation and the socioeconomic profiles of farmers, shows 172 geoenvironmental units where 110 units are semiarid but that with implementation of suitable farming development and environmental protection sustainable agriculture can be carried out (Silva et al. 1993). The municipality of Jaguarari is located in the "Depressão Sertaneja" environmental unit that corresponds to 22 percent of the northeast region.

The socioeconomic base of the Northeast region depends on rural activities such as livestock and mining. Limiting factors for livestock production are the low quantity and poor quality of the forage in the dry season. Deforestation in the region has increased both because of wood extraction for firewood and because of the demand to clear land for pasture areas. The wood productivity in the area is low and the exploration is extensive due, in part, to its commercial use as an energy source (charcoal). On the farms, the wood is used for small constructions, fences and firewood. The area's land is further degraded because farmers and ranchers attempt to graze too many animals in most areas thereby exceeding the fodder capacity of pastures.

In 1969, the company Caraiba Metais S.A. was established to operate mining in Pilar, a municipality of Jaguarari; at the same time a metallurgic company in Dias D'Avila municipality was established to produce electrolytic copper. Since that time, there have been various changes in private and state ownership of the companies, and the mining company now is known as Mineração Caraiba.

The mining operations are performed at depths as much as 800 m below the surface. From the underground mine approximately 181,500 tons of metallic copper have been extracted. After being mined, the ore is crushed, milled and concentrated in an industrial plant. This concentration plant has been in operation since 1980, processing approximately 1.2 million tons of ore per year, with an average grade of 3.05 percent of Cu. Since the beginning of its operation it has produced 1.79 million tons of concentrated ore with 615 thousand tons of copper. The predominant minerals in the mining are chalcopyrite and bornite, drawn from the underground mine at a depth of 800 m.

3. MINING IMPACTS AND THE LOSS OF BIODIVERSITY

Despite attempts by the Mineração Caraiba to mitigate environmental consequences of its activities, nevertheless significant impact of the activities occur. During normal operations, the original vegetation of caatinga and the superficial layers of soil are removed to other areas. The impacts on topography, vegetation, and water resources are evident even upon casual visual examination. The flora and fauna of the area and the diversity and number of the microorganisms in the soil are greatly affected during mining operations. The effects of the mining process extends for several kilometers from the mine, and vary from small modifications to large ecosystem degradations. The effects include topographical modifications, excavations of craters, and accumulation of thick and fine mine residues resulting from the crushed ore. Increased erosion by water and wind also has resulted in the transport of fine dust debris and pollutants over long distances and, in fact, pollutants have accumulated in lakes and flats along watercourses. Environmental impacts in these areas are exacerbated by increasing demands of people to use the areas (e.g., cattle raising) and in part this demand is brought on by the economical growth in the area caused by the mine itself.

Overall, ecological alterations affect largely the natural vegetation which is removed and/or buried by debris or modified in its original floristic composition. This process also alters the population of arbuscular mycorhizal fungi, obligatory symbionts of vegetative roots that multiply exclusively when associated to a host plant. There are also effects on other symbionts of vegetation such as the nitrogen–fixing bacteria that form nodules in the roots of some species (leguminosae), known as rhizobia.

4. PARTNERSHIP BUILDING CAPACITY

To promote recovery of the areas degraded by mining, the Mineração Caraíba invited the Brazilian Agriculture Research Corporation Embrapa to participate in the project, strengthening and increasing the relationship with this government enterprise both for scientific research and environmental restoration.

Embrapa, a public company linked to the Ministry of Agriculture, Animal Science and Food Supply, created in 1973, has generated and adapted technologies, products and services for the Brazilian agribusiness complex in order to permit rational farming and forest exploitation while reducing adverse environmental consequences. Embrapa has contributed to partnerships and capacity building with public institutions as well as nonpublic organizations (companies, NGOs, associations of farmers) in the country and abroad.

Embrapa is present throughout Brazil with thirty nine research centers and specialized service units. The unit of Embrapa in the semiarid zone is the Semi–arid Agricultural Research Center, Embrapa Semi–Árido. It was founded in 1975, in Petrolina, State of Pernambuco, with the mission to develop and adapt technologies for the Brazilian semiarid region in order to offer alternative expertise in identifying new cropping systems for strengthening the small farmers sector.

The administrative policy of Embrapa Semi–Árido is divided into two principal sectors: "Policies of Research and Development" (PRD) and "Policies of Technological Business and Communication" (PTBC), respectively. The PRD's aim is to ascertain the adequacy of information and knowledge in strategic areas and development of new technologies and adaptation of available ones. The PTBC identifies clients and establishes agreements and conditions to apply the emerging technologies.

When conserving and managing biodiversity of the caatinga biome, Embrapa Semi–Árido focuses on surveys, biological control, reforestation of degraded areas and agricultural biotechnology. The scientific research has a fundamental role in attempting to obtain high production compatible with conservation of the environment through the generation of technologies, products and knowledge to guarantee the sustainability of agricultural activities for the benefit of present and future generations. Due to the ecological diversity of the Brazilian ecosystems their exploration can result in diverse and understudied impacts.

The Brazilian federal government is the main financial source of Embrapa Semi–Árido for payment of salaries, funding of the research costs, and for modernization of the physical base. Nontraditional financial sources for research activities receive high priority in the field of sustainable agriculture. To increase and obtain financial support and to diversify studies, Embrapa Semi–Árido looks for partners and collaboration among national and international institutions interested in making projects feasible for the sustainable development of the region. Presently, Embrapa Semi–Árido has about 60 researchers, with master and doctoral degrees obtained at Brazilian and foreign universities working directly or indirectly in areas of environment or agricultural sustainability in the semiarid region.

For the technical cooperation project on recovery of an area degraded by mining, a formal agreement was signed by Mineração Caraíba and Embrapa Semi–Árido, where the Mineração Caraíba provides all the material, physical and local human infrastructure necessary for carrying out the work in the area affected by the mine. Embrapa Semi–Árido is in charge of managing the technical information and supervising researchers and identifying new technologies to solve the problems of rehabilitation of a degraded mining area.

Embrapa Semi–Árido has held meetings with its research staff to discuss the problem involving several disciplines that study environmental disturbances. Consequently, a research staff was formed to conduct necessary impact and recovery studies; the staff was comprised of researchers in forestry, botany, biology, microbiology, soil and plant improvement, and mining engineering. Embrapa Semi–Árido also contacted Universidade Federal de Pernambuco–UFPE in Recife, and Embrapa Agrobiologia in Rio de Janeiro for assistance with subprojects dealing with microorganisms in soils under degraded conditions. A contract was established with UFPE and graduate students presented plans of surveys of fungi in the soil degraded by mining and in areas not influenced by the mining activities.

In order to obtain financial support for the students of UFPE, their projects were submitted to the Research National Council (Conselho Nacional de Pesquisa–CNPQ), a financial agency of scientific research in Brazil, for paying for the identification and fungal multiplication in controlled laboratory settings. The costs of soil and seed collecting, nursery, site preparation, tree planting and maintenance was paid by Mineração Caraíba.

5. PUBLIC POLICIES AND MANAGEMENT

The Brazilian government has developed at local and national levels several environmental policies that focus on the short and long–term social and economic sustainable development of the country. One of the most recent environmental policies is on desertification control, which has the objective of recovering and expanding the protected areas and providing facilities for scientific research on conservation of biodiversity.

The Forestry Code (Law 4.771 of November 15th, 1965) has regulated the vegetation areas of permanent conservation, use of planted and permanent forest, commerce of plants, and other determinations related to conservation and management of the Brazilian forest. The control and inspection of the forest situation is under responsibility of the Brazilian Institute of Environment and Natural Resource (IBAMA).

The Bahia State Constitution obligates the state and municipalities to define areas and components representing the ecosystems to be protected and species of the fauna and flora in danger of extinction. The resolution 1.009 CEPRAM of December 6th, 1994, prohibits the cutting, storage and commercialization of *Miracrodruon unrudeuva* (Engl.) Fr. All., *Schinopsis brasiliensis* Engl. and *Andenanthera macrocarpa* (Benth.) Brenan. In the State of Bahia, the IBAMA, through PORTARIA 37–N of April 3, 1992, prepared a list of species in danger of extinction. Among the species that occur in the semiarid zone *Miracrodruon urundeuva, Pilocarpus trachylophus* Holmes and *Schinopsis brasiliensis* are listed. For preserving plant genetic resource, especially those at risk, the research institutes preserve them in ex situ areas.

For the mining sector, Laws 6.938 and 9.605 apply policy for the enterprises that operate with activities that cause environmental impacts, promulgating actions that aim to eliminate or to minimize the impacts caused by its operations. Law 6.938 of January 17th, 1981 (Federal Law of the Environment), disposes on the national policies of the environment, its objectives, formulation mechanisms and application, and the Law 9.605, of February 12th, 1998 (Law of Environmental Crimes), disposes the penal and administrative sanctions applied to enterprises that harm the environment.

Law 6.938 is one of Brazil's most important environmental laws. It stipulates that the polluter is required to compensate the environmental damages that it caused, independent of blame. The public prosecution service can propose actions of civil responsibility for damages to the environment, imposing to the polluter the obligation of recovering and/or compensating the damages caused.

Law 9.605 strengthened Brazilian environmental legislation that refers to the infractions and punishments. Those responsible for environmental infractions can be punished with the liquidation of the company if it has knowingly created or hidden an environmental crime.

Despite the laws, IBAMA does not have a rigorous control and inspection service concerning tree exploration and hunting in the region. This institute does not have sufficient staff and infrastructure to operate in all of the Bahia State. Small groups are located at different regions of the state and attend several municipalities. The Mineração Caraíba does not permit tree exploitation and hunting in their property and it has policies for protection of vegetation. For operations that eliminate or minimize the impacts caused by mining there are regular inspections by the Government.

6. THE STRUCTURE OF THE PROJECT

The fundamental purpose of this case study project is to be able to determine the name, number, and structure of species that occur in caatinga vegetation surrounding the mine of Mineração Caraíba, species in risk of disappearance because of the mining operation, and potential species suitable for relatively rapid rehabilitation of the degraded area.

The work was divided in three phases. In the first phase, an analysis of the soil, climate and vegetation characteristics, and local agricultural systems of production and commercialization was conducted. In the second phase, an experimental planting was done using technologies developed by Embrapa Semi–Árido concerning selection of species for reforestation. In the third phase, technological alternatives for restoration of the area were analyzed.

6.1 Assessment of the Area

The agricultural systems of production and the main commercial activities used in the municipalities were analyzed by an economist of Embrapa Semi–Árido. It was observed that local farmers rely on livestock as their main economic activity, and they raise sheep and goats in an extensive system using the caatinga vegetation. Embrapa Semi–Árido specialists in soil classification studied the area according to Brazilian soils classification. They mapped seven types of soils (Podzólico Eutrofico, Vertissolo, Planossolo Solódico Vértico, Bruno não Cálcio Vértico, Regassolo, Litossolo and Aluvial) present in the property, where the type Bruno não Cálcio and Planossolo Solódico Vértico correspond to 41 percent and 40 percent of the area, respectively. The area was also classified according to annual rainfall, altitude, depth of the soil, capacity of retention of water, natural fertility and relief. The rainfall is uniform and the area presents light glides with small elevations in its southern part.

Scientific studies on wildlife were not conducted. Instead, knowledge about the animals that occur in the region was gathered from people living in the region. Hunting in the mining area may occur but most likely on a small scale due to the efficacy of the mining company's restrictions on hunting. However, it is known that hunting reduces the population densities of several species. The most vulnerable species are the carnivores and birds. There is a general lack of scientific information about the fauna of the region, which makes it difficult to develop effective management plans for animal preservation in the caatinga biome.

6.1.1 Flora

For the flora survey, a group of two specialists in botany and forestry inventory, and three other local people that helped in the measurements and identification of species by local name was formed. The scientific identification was made in the section of botany of Embrapa Semi–Árido where there is a collection with around 3000 botanical specimens of caatinga vegetation, corresponding to 120 families, 400 genera and 1000 species.

The methodology used in the floral survey focused on identification of specimens found in the mining area (Embrapa 1979, Rodal et al. 1992). The intensity of mixture of species found in the flora was obtained by the Quotient of Mixture (QM), that is, the relationship between the number of species and the total individuals found per hectare. The horizontal structure analysis showed the sociological importance of each species in the community, calculated through the values of "abundance," "frequency," "dominance" and "Index of Importance Value" of the species. The abundance refers to the number of individuals of each species in the vegetation association per unit of surface. The frequency indicates the regularity of the horizontal distribution of each species in the area (medium dispersion). Through the frequency it is possible to calculate the degree of homogeneity of the forest, expressing the vegetation association. The dominance allows the measurement of the productive potentiality of the forest and it constitutes a useful parameter for the determination of the quality of the site.

For the analysis of the vegetation it is important to utilize a value that allows an assessment of the importance of each species in the total vegetation. The Index of Value Importance that results from the combination of the three parameters previously mentioned generally is taken by ecologists as a quantitative indicator of the ecological importance of the species in a particular environment. This index is obtained by adding the values of abundance, frequency and dominance of each species.

Based on floristic analysis, the structure of the present vegetation showed that the composition and the distribution of the tree and bush species are heterogeneous, containing species generally distributed in the caatinga biome. In the total area of study, 35 families, 60 genera and 85 species were found. The main species found were analyzed in relation to their time of regeneration after human interference. In the vegetation surrounding the mine (caatinga vegetation), the survey showed a flora with about 902 plants per hectare, corresponding to 67 tree/bush species, 31 genera and 30 families. The area directly affected by mine activities yielded a flora with about 18 plants per hectare, corresponding to two species, two genera and two families. Both species, *Prosopis juliflora* (SW) DC and *Nicotiana glabra* Graham, were found in the area although they are exotic to the caatinga biome. The structural analysis of this area showed the high rate of the frequency, abundance and dominance of *Prosopis juliflora,* indicating the first symptoms of invasion of this species in the community.

6.1.2 The Most Important Species in the Community

The importance of a species (tree) in the community is characterized by the number of trees and its dimensions (abundance and dominance), which determine its space among the vegetation, independently of the type of occurrence of the species (isolated or in groups). In the analysis of the caatinga vegetation data using the Index of Value Importance, *Caesalpinia pyramidalis* Tul., *Commiphora leptophloeos* (Mart.) J.B.Gillet, *Schinopsis brasiliensis*, *Mimosa arenosa* (Willd.) Poiret. and *Jatropha* sp were the most important tree and shrub species of the community. Although *Prosopis juliflora* is found in the region it is not endemic to the caatinga, and in fact is classified as an invader in areas directly affected by mining operations.

Caesalpinia pyramidalis is a typical caatinga species, widely distributed throughout most of the semiarid Northeast, growing in various soil types. It presents an average height of 4–6 m and a twisted trunk. The young shoots are eaten readily by all livestock and can be an important source of green forage at critical times of the year. The leaves emit a pungent odor, but after dropping from the tree, their odor disappears. The uses of this species are incipient. According to local people, the leaves, flowers and bark are used for treatment of common colds, diarrhea, hepatitis and anemia.

The wood of *Commiphora leptophloeos* is very important for the commercial manufacturing of artifacts, furniture and other household utilities in the region. Despite being an important commercial species for the region, programs of reforestation using this species are not common.

Schinopsis brasiliensis is a large tree species and is valued highly as a source of wood in the semiarid region. Consequently, it is among the most exploited species and its density has decreased to low levels in comparison to the past when it was not exploited so extensively.

There are several *Mimosa* species occurring in the caatinga vegetation making field identification very difficult. These species are important for the production of charcoal, firewood and the construction of fences in the region. The forage is high in caloric content and therefore is a preferred source for grazing by livestock, especially in dry seasons.

Another important species present in the semiarid region studied was *Anadenanthera macrocarpa* Benth., a fast growing tree that occurs naturally in all states of Northeast Brazil. It is adaptable to a wide range of soil types but grows best in fertile and deep soils. The wood is used for fine furniture, tool handles, stakes, etc. The leaves can be toxic to cattle when green due to its high tannin content but can be eaten by livestock when dry. The bark contains up to 32 percent tannin and therefore it is used for tanning leather in the region.

Myracroduon urundeuva is becoming rare due to its high wood and forage values. The wood is highly valued for building construction and fence purposes. According to Kirmse (1983), local people believe the bark has hemostatic qualities. A tea made from the bark is used for menstrual relief, and the bark is chewed to alleviate stomach ache and to stop diarrhea.

6.1.3 Soil Microorganisms

For soil microorganisms studies, a group of one specialist in fungi and some students of biology of the Federal University of Pernambuco–UFPE was formed. In areas degraded by mining the absorption of nutrients by the plants is limited by the toxicity of the mineral elements. The symbiotic microorganisms have a fundamental role in the balance of nutrition of the plants and possibly they are indispensable for the regeneration of vegetation in these areas.

For the fungi survey, the area was divided in six different sub areas according to use and condition and the soil sampling was made in two periods during the year, one in the dry and other in the wet season. In general, the density of spores and number of infective propagates were low (<2 per g of soil) and there was no significant difference between number of spores in the dry or in the rainy season (Silva et al. 2001). Species of *Acaulospora*, *Entrophospora*, *Glomus* and *Scutellospora* genera were found in the area, and their diversity is directly relate to the plants found in the region.

6.2 Trials of Species for Reforestation

Generally, the first step in the restoration of degraded areas is the incorporation of large quantities of organic compounds or the transfer of fertile soil to the affected areas. The use of leguminous plants associated with bacteria (*rhizobia*) with capacity of utilizing atmospheric nitrogen and fungi (*mycorrhiza*) which are more efficient in transferring soil nutrients to plants, allows revegetation of degraded soils where organic material has been lost. This technology makes possible the low cost recuperation of areas which have been excavated by mining activities or where landslides have occurred due to mining activities.

In order to assess species for reforestation, the studies focused on the characteristics of some native species and on exotic species planted in the areas affected by mine debris. The analyzed species were: *Leucaena leucocephala* (Lam) de Wit., *Ceiba pentandra* Gaertn., *Cnidosculus phyllacanthus* Pax.S.K. Hoffm., *Spondias tuberosa* Arruda, *Tabebuia caraiba* (Mart.) Burret, *Mimosa caesalpiniifolia* Benth., *Eucalyptus camaldulensis* Dehnh., *Tamarix* sp, *Parkinsonia aculeata* L., and *Prosopis juliflora*.

The preliminary evaluation showed satisfactory results for *Leucaena leucocephala*, *Tamarix* sp and *Prosopis juliflora* species, i.e., the species had fast rates of growth and high rates of survival. All of these species are exotic in the region. Among the native species the best results were observed with *Ceiba pentandra* and *Mimosa caesalpiniifolia*, but neither belongs to this local ecosystem. It is necessary to test more native species, especially the ones identified in local surveys as having a high Index of Value of Importance in the community.

The analysis of the techniques employed in the reforestation showed best performance of the species when the seedlings were planted in pit with substratum composite by manure or other mixture. The seedlings, when planted directly in the mine debris, presented low height and showed symptoms of intoxication by the ore. Some species like *Spondias tuberosa* and *Eucalyptus camaldulensis* presented high quantity of leaves with toxic symptoms due to absorption of heavy mineral elements such as copper.

The preliminary results confirmed the potentiality of *Leucaena*, *Prosopis* and *Mimosa* for use in recovery of degraded areas. These multipurpose trees also are used by local communities for wood for fence, firewood and animal feed by providing arboreal fodder in the dry season, especially if the period of drought is extended. Under this situation, they serve as an emergency fodder reserve for livestock.

6.2.1 Prosopis juliflora

Tolerant to drought and highly saline and alkaline soils, species of *Prosopis* provide high quality fuel wood and charcoal and, from pods, a nutritious animal fodder. Another important aspect of *Prosopis* is the possibility for pod use as human food in arid and semiarid environments. The fruits do not contain toxic substances and can be used for preparing cake and bread by rural people. Protein and sugar content vary among different species. Pods' protein content ranges from 10 to 17 percent and sugar content from 10 to 44 percent.

Prosopis juliflora also is a good candidate species for rehabilitation of areas affected by salt; it can grow in a soil salinity regime equivalent to sea water. It also can be raised with saline water with an electrical conductivity as high as 20 dS/m. According to Singh and Singh (1993), applications of phosphorus and zinc help in accelerating the growth of *P. juliflora* in saline soils. Plants of this species also help reclaim salt–affected soils more effectively than other species of the same age and stocking rate, and they also are useful in reclamation of alkali soils.

In the Brazilian semiarid zone, the wood production of *Prosopis juliflora* is estimated at 27 t/ha of woody biomass, harvesting with 8 years of age (Lima 1994). For fodder, the average production in Petrolina, Pernambuco State, with 15 years of age is 78 kg by tree, spacing 10 x 10 m (Lima 1987). After the 20th year of age, pod production begins to decrease, so the trees should be harvested for furniture around this time. In the Northeast of Brazil, the pod production of *P. juliflora* is estimated to be between 2 to 8 t/ha/year, being most frequently 2 to 3 t/ha/year for the semiarid zone (Nobre 1982).

6.2.2 Leucaena leucocephala

Despite some literature reports that *Leucaena* only grows in the humid and subtropics, it also grows well in the semiarid regions. In the semiarid zone of Brazil *Leucaena* has a good survival rate and performance in sites with 300 mm of rainfall and 8 months of dry season. Forage from *Leucaena* is highly palatable, digestible and nutritious. Young or mature, green, dry or ensiled, the foliage is relished by livestock. The plant can be harvested and given fresh to the animals, dried into a leaf meal, or fermented to silage. The animals can be allowed to browse the standing bushes (National Academy of Science 1977). The studies with *Leucaena leucocephala* in the Northeast of Brazil, for wood and fodder production, showed that the species is suitable for reforestation and recovery of degraded areas. For firewood production, the volume found was 32 m^3/ha corresponding to 14.5 t of firewood per hectare. For forage, the mean production was 7 t/ha/year of dry matter, with three cuttings per year (Lima et al. 1986).

6.2.3 Mimosa caesalpiniifolia

Among the native species of the Northeast of Brazil, *Mimosa caesalpiniifolia* is the most recommended and used tree for the recovery of degraded areas because of it high rates of survival and growth. This species occurs naturally in the vegetation of Ceará and Piauí States and also in other regions of the Northeast. *Mimosa* is adaptable to a wide spectrum of conditions; however, it grows best on shallow soils.

Mimosa caesalpiniifolia reaches up to 8 m of height, normally with aculeus (thorns), and roots with many nodules that fix nitrogen in the soil. The dry or green leaves are probably one of the most preferred browse to all livestock. The leaves are soft and contain about 17 per cent of crude protein when green. The species is an important source of fence posts in the northeast region, being highly resistant to decomposition in the soil. The wood is also used as firewood and charcoal. The plant has the ability of regrowing readily from the stump when cut (Kirmse et al. 1983).

7. VALIDATION OF BEST PRACTICES

Based on data analysis of the survey, trial species, environmental laws and local economy, it was decided for the short–term to plant species with fast growth and high economic value to animal production. The planting of *Leucaena* or other fast growing fodder tree solves the problem of rehabilitation of the area in the short–term and also benefits animal production.

The planting data showed the importance of the fast–growth species in reducing the action of the wind in the area covered with mine debris. The plants, disposed in two or three rows perpendicular to the predominant wind direction act like a windbreak and reduce the fine soil particles carried in the wind when velocity is strong. In this case, the exotic fast–growth species protect the native species from impacts from pollution–laden soil and dust.

Because of the potential problems which can be caused by the use of nonnative species, concerns about the use of such species in the project were discussed extensively. However, given that the area affected by mining is relatively small in comparison to the total area of native vegetation combined with the fact that management of other exotic species for fodder production also occurs with apparent minimal risk to the area's native biodiversity, the decision was made to use *Leucaena* in mine recovery programs.

8. SCIENTIFIC LESSONS LEARNED

Because of lack of development, most local communities in the semiarid region of Brazil support natural resource management plans based on short–term economic gains and benefits and secondarily on conservation and sustainability. Public policy and decision makers should try to create systems of management that while due to economic necessity might entail some environmental risks while nevertheless at the same time presenting the least amount of damage to the local biodiversity as feasible. Thus, it is necessary to have detailed information about the ecological diversity of a site and to keep growing representative species in the population and maintaining ex situ plantations in places where local species are in danger of disappearance.

The survey presented in the project has an important specific value to the local and regional environment, particularly regarding flora diversity and possibilities of mitigating some environmental impacts associated with mining. Mitigation of impacts might require the use of native or exotic species. Generally, exotic species are relatively fast growing and highly productive. The use of exotic species in recovery of degraded areas creates the controversy of favoring the economical development in spite of environmental risks to native species. The concept of sustainable development embraces economical, social and environmental aspects. Yet, the role of biodiversity in sustainable development is not clear. For example, although Brazil has laws to minimize the impacts caused by mining the laws do not clarify or specify if species used in recovery or revegetation must be native or exotic.

9. CONCLUSION

Much information on areas subject to desertification in Brazil's Northeast are obtained as a result of the Government's strategic planning for integration of "Conservation and Use of the Biodiversity Program." This planning is intended to facilitate integrated models for Country, State and Municipality Plans of Development and formulation of proposals for the National Plan of Combat to Desertification (PNCD).

For the case of managing the biodiversity and recovering the degraded area of Mineração Caraíba, it is expected that reduction of up to 40 percent of the negative environmental impacts on the areas covered by mine debris through the recovery processes using *Leucaena*, *Prosopis* and/or other fast growth species in the degraded areas will be achieved.

Meanwhile, Dias (1992) reminds us of six factors which contribute to destruction of the biodiversity; (1)the expansion of the agricultural borders that results in the destruction of the vegetative covering and the fragmentation of that covering; (2)the introduction of exotic species which cause significant destruction to native vegetation but is an understudied problem; (3)the increasing exploration of species of plants and animals for subsistence and commercial enterprises; (4)the contamination of soil, water and atmosphere by pollutants; (5)the modification of the global climate; and (6)agricultural and forest exploration that increasingly restricts genetic biodiversity.

In our project some native species of caatinga vegetation have been studied and showed some promising attributes for recovery of the degraded mining areas, but only exotic species like *Leucaena* and *Prosopis* were recommended for recovery of mining degraded areas in the short–term due to economical considerations. These species are fast growing and are established easily, even in adverse soil conditions. Recommendations on the use of *Leucaena* and *Prosopis* were also based on the size of the affected area (relatively small in relation to the total area where native biodiversity is found in the semiarid region), ecological conditions of all species in the community, as well as the importance for livestock feeding and for local economic products from farms of the region, e.g., *Leucaena* is considered an excellent fodder. On the other hand, it is possible that new plantings with the best species identified in the local vegetation survey will be undertaken in the future– species like *Caesalpinia pyramidalis, Commiphora leptophloeos* and others due to their high importance value for the vegetation structure and the fact that being native species they might avoid some of the potential risks of using introduced species.

10. REFERENCES

Dias, B.F. de S. 1992. Estratégia Mundial Para a Biodiversidade. Revista do Instituto Florestal, São Paulo, Brazil. v.1. Pages 62–76.

(EMBRAPA) 1979. Centro de Pesquisa Agropecuário do Trópico Semi–Árido, Petrolina–PE. Relatório Técnico Anual do Centro de Pesquisa do Trópico Semi–Árido. EMBRAPA–DID, Brasília, Brazil.

Kirmse, R.D., Pfister, J.A., Vale, L.V. and Queiroz, J.S. 1983. Woody Plants of the Northern Ceará Caatinga. Embrapa Caprinos: Utah State University, Technical Report Series, n.14. Logan, UT.

Lima, P.C.F. 1987. Produção de Vagens de Algaroba. Revista da Associação Brasileira de Algaroba, Mossoró, Brazil. v.1 (2). Pages 151–170.

Lima, P.F.L. 1994. Comportamento Silvicultural de Espécies de *Prosopis*, em Petrolina–PE, Região Semi–Arida Brasileira. Tese de Doutorado. Setor de Ciências Agrárias–Universidade Federal do Paraná, Curitiba, Brazil. 110 pages.

Lima, P.C.F., Drummond, M.A., and Albuquerque, S.G. 1986. Freqüência de Corte em *Leucaena Leucocephala* (Lam.) De Wit Visando a Produção de Forragem, na Região de Petrolina, PE. Petrolina, EMBRAPA–CPATSA. 2 pages.

National Academy of Sciences. 1977. *Leucaena:* Promising Forage and Tree Crop for The Tropics. Washington, D.C. 115 pages.

Nobre, F.V.A. 1982. Algarobeira no Nordeste Brasileiro, Especialmente no Rio Grande do Norte. Pages 257–282 In: Simpósio Brasileiro Sobre Algaroba, v. 1 Documentos 7, EMPARN, Natal, Brazil.

Queiroz, M.A. de, Nascimento, C.E. de S., and Silva, C.M.M. 1992. Fruteiras Nativas do Semi–Árido do Nordeste Brasileiro: Algumas Reflexões Sobre os Recursos Genéticos. In: Simpósio Nacional de Recursos Genéticos de Fruteiras Nativas. Cruz das Almas, Brazil.

Rodal, M.J.N., Sampaio, E.V., and. S.B. Figueiredo. 1992. Manual Sobre Métodos de Estudos Florísticos e Ftossociológicos–Ecossistema Caatinga. Brasília: Sociedade Botânica do Brasil. 24 pages.

Sá, I.B., Fotius, G.A., and Riché, G.R. 1994. Degradação Ambiental e Reabilitação Natural no Trópico Semi–árido Brasileiro. Conferência Nacional e Seminário Latino–Americano da Desertificação, Fortaleza, Brasília. Pages 260–275.

Silva, F.B.R. e., Riché, G.R., Tonneau, J.P., Souza Neto, N.C., Brito, L. T. de L., Correia, R.C., Cavalcanti, A.C., Silva, F.H.B. da, Silva, A.B. da, Silva, J.C. de A. 1993. Zoneamento Agroecológico do Nordeste: Diagnóstico do Quadro Natural e Agrossocioeconômico. Petrolina, PE. EMBRAPA–CPATSA/Recife: EMBRAPA–CNPS. Coordenação Regional Nordeste, Brazil.

Silva, G.A. da, Maia, L.C., Silva, F.S.B. da, Lima, P.C.F. 2001. Potencial de Infectividade de Fungos Micorrízicos Arbusculares Oriundos de Area de Caatinga Nativa e Degradada por Mineração. Revista Brasileira de Botânica no Estado da Bahia, Brasil 24: 135–143.

Singh, G., Singh, N.T. 1993. Mesquite for the Revegetation of Salt Lands. Central Soil Salinity Research Institute, Karnal, Haryana, India. Bulletin 18.

Chapter 22

USE OF CREEPING FOG WATER AS A NON–TRADITIONAL WATER RESOURCE IN CHILE

Jorge Araya–Valenzuela and Roberto Espejo–Guasp
Departamento de Física, Universidad Católica del Norte, Antofagasta, Chile; email jaraya@ucn.cl

1. INTRODUCTION

The Atacama Desert located between latitude 8° and 29° S and longitude 68° and 72° W is one of the most arid regions of the world. This northern and arid region of Chile runs 1600 km from north to south, with a width varying from 160 km to 350 km from east to west. The transverse topography from east to west, from the Pacific Ocean to the Andes Cordillera, shows marked altitude differences varying in a narrow strip of 250 km from sea level to 5000 m, with some Andean mountain peaks of almost 6000 m.

The coastal zone, consisting of a very narrow seashore strip bordered by a high mountain range that rise over 1200 m (Coastal Cordillera) is frequently covered with a layer of low clouds (stratocumulus). This layer is about 300 m high and lies between 600 m–1200 m. The Coastal Cordillera mountains intercept the stratocumulus forming creeping fog known as "Camanchacas."

Most of northern Chile coastal cities, particularly those near Antofagasta, are supplied with water piped at an altitude of 3400 m and transported for a distance of 300 km. In a zone where annual rainfall is null between sea level and approximately 2500 m, and reaches a maximum of 300 millimeters per year between 2500 m and 5500 m, water is a scarce resource; this situation is aggravated by the growing competition for water among mining, industrial and agricultural sectors and the needs of the local population residing in the area. Agricultural activity, traditionally scarce and non–intensive, has been decreasing due to the expansion of mining companies and the economic pressure they exert through the purchase of water rights.

In the coastal zone, fog water capture may become an invaluable complementary resource to be used for reforestry, ecosystem recovery, small–scale agriculture, human consumption in small local communities, etc. The National Forestry Corporation (CONAF), an institution at the Ministry of Agriculture of Chile that is in charge of the protection and administration of the country's flora and fauna ecosystem, has published the "Libro rojo de los Sitios Prioritarios para la Conservación de la Diversidad Biológica en Chile" (CONAF 1996). This book reports the conclusions of a 1988 symposium organized together with the Chilean Botany Society for the VII Botany National Meeting. Sites where biodiversity is in danger are identified in the book.

Within this framework to protect biodiversity, the Catholic University of the North (UCN) has conducted long–term research to capture water from creeping fog by using static structures called fog collectors. For example, various vertical structures have been mounted as obstacles to capture fog passing along the coastal mountain peaks and to serve as barriers to the clouds coming from the Pacific Ocean which pass over the mountains on their way to the inland desert. Our research activities are particularly connected to the design, construction and testing of the appropriate structures to check various humidity–intercepting materials, fog water potential assessment, drop size determination, fog collector efficiency and quality of the water collected.

Two and three–dimensional structures in various shapes have been tested to support intercepting materials: vertical nylon thread, netted nylon fabric, textiles and other materials. Several formulae have been developed to estimate fog water potential and to calculate the real and theoretical collectable water maximum. At present, our primary goal is the use of fog collected water for biodiversity conservation and recovery.

2. CAPTURE OF CREEPING FOG WATER

There are three types of fog and different types of fog collector designs to maximize the collection of fog–collected water (Tapia 1980, Soto 1985, Espejo 1992, Cereceda 1996). A detailed description of the types of fogs and fog collector designs (and their costs) is beyond the scope of this paper. However, what is important for purposes of biodiversity conservation are the prospects and problems of using fog collectors. (Additional technical information on the types of fog collectors and their applications in Chile can be obtained directly from the authors.).

3. FOG WATER AS A BIODIVERSITY SUPPORT

Life on our planet is tied to water and depends on it. In areas like the mountain coastal zone of northern Chile, where there is no rainfall, endemic vegetation and animal life are directly linked to creeping fog (Camanchaca). Over time, the coastal ecosystems have become dominated by human activities, eliminating the vegetation (especially trees) with the subsequent negative effect on biodiversity. In these areas, plants are the natural collectors of the humidity that they themselves and other related living organisms need; therefore their disappearance directly threatens the corresponding ecosystems.

Artificial water capture from Camanchaca with the fog collector allows using and applying them to conserve and restore threatened or degraded ecosystems. In the same way, it provides tap water to small isolated communities and improves their social and economic conditions.

4. USE OF FOG COLLECTED WATER

4.1 Preliminary Study for the Camanchaca Water Use for Paposo Ecosystem Recovery: Case Study 1

One of the sites mentioned by the National Forestry Corporation (CONAF) in "El Libro Rojo de los Sitios Prioritarios para la Conservacion de la Biodiversidad Biologica en Chile" is the area from Paposo (II Region) extending southward to Pan de Azucar (III Region).

Paposo is located at latitude 25° 00' S, longitude 70° 27' W and Pan de Azucar at latitude 26° 09' S, longitude 70° 45' W. The first is located 50 km to the north of Taltal City and the second at 30 km to the north of Chañaral City. The coastal zone of this area includes 300,000 hectares. Vegetation is typical of a coastal desert, with plant species highly endemic. For example, CONAF (1996) states that 134 of 259 floral species are endemic, and that 7 of 68 faunal species are endemic.

Paposo is a small village located 50 km to the north of Taltal. It has about 300 inhabitants whose main activities are fishing and small–scale mining. They face problems such as extreme poverty, poor public utility services and poor infrastructure and labor.

Paposo National Reserve (a protected area by CONAF) is located in this area, latitude 25° 00' S, longitude 70° 27' W. Due to its geographical location, the reserve has some very particular climatic characteristics which have allowed the development of an area with a unique flora (from Taltal to the North of Paposo) which has been studied by many researchers and is well known. Despite the fact that the area is protected, human activities and grazing by goats have been significant.

The presence of coastal stratocumuli and high hills very close to the coast allows the strata layer to intercept and form a plant community with endemic elements (e.g., *Griselinia carlosmunozi*, *Tillandsia tragophoba*, and *Salvia tubiflora*) with Chilean species found also in the southern zone of Valdivia. At Paposo and along the coast, there are drought–tolerant bushes and cacti that survive due to the topography which forms the Camanchaca.

Since 1992, several studies at the Cerro Mirador located near the village of Paposo at 800 m (elevation) have been done by the UCN and CONAF. The water quality was also analyzed for human and vegetable consumption.

4.1.1 Hydric Potential

The water potential of the fog was determined at Cerro Mirador (Espejo 1998) with a 1 m^2 standard collector made of Raschel net of 35 percent shade located at 1.5 m above sea level. The results in l/m^2 day ranged between a low of 0.7 in February to a high of 3.0 in August. During most months the values in l/m^2 ranged between 2.0–2.7 l/m^2.

4.1.2 Chemical Analysis of the Water Captured

The chemical analysis of water was made at the Department of Chemistry of the UCN (Espejo 1993). Samples correspond to a linear profile at 700, 750, 800 and 900 m above sea level. The following elements were analyzed: dissolved solids, solids in suspension, total solids, sulfates, chlorides, nitrates, aluminum, sodium, magnesium, potassium, ammonium, arsenic, cadmium, lead, manganese, antimony, calcium, and iron.

Samples were taken in 1995 and correspond to water captured in the last 24 hours directly from the containers. Although Chilean standards for tap water do not include potassium, sodium, and calcium, these were also considered. Table 1 shows the results for different collection heights.

According to Chilean standards for tap water (NCH 409/1), the following elements are higher than standards allow. (1)Chlorides: standard 250 mg/l. At 700 and 750 m above sea level the standard is exceeded with values of 571 and 489 mg/l, respectively. (2)Nitrates: standard 0.10 mg/l. At all levels the standard is exceeded with values of 77.51; 68.20; 55.81 and 43.4 mg/l. (3)Manganese: standard 0.10 mg/l. At all levels the standard is exceeded with values of 0.51; 0.45; 0.20 and 0.22 mg/l. (4)Copper: standard 1.0 mg/l. The standard is exceeded at 700, 750 and 800 m above sea level with values of 1.43, 1.85 and 1.94 mg/l. (5)Zinc: standard 5.0 mg/l. The standard is only exceeded at 700 m above sea level with a value of 6.49 mg/l. Water sample pH have values of about 7, which shows relative neutrality.

Table 1. Chemical analysis of the water captured and altitude

Elements	700 m	750 m	800 m	900 m
Total solids gr/l	1.4	1.2	0.69	0.6
Dissolved solids gr/l	1.28	1.16	0.56	0.56
Suspended solids gr/l	0.12	0.04	0.13	0.04
Sulfate ppm	165.00	148.00	82.00	82.00
Chlorides ppm	571.00	489.00	293.00	245.00
Nitrates ppm	77.51	68.20	55.81	43.40
Aluminum ppb	0.27	0.18	0.87	0.09
Sodium ppm	300.00	260.00	140.00	120.00
Magnesium ppm	13.34	11.54	5.39	5.64
Potassium ppm	20.75	18.09	10.11	7.45
Ammonium ppm	6.60	4.21	4.80	3.60
Arsenic ppb	15.12	6.0	25.00	9.00
Cadmium ppb	<9.0	<9.0	< 9.0	< 9.0
Lead ppm	<0.06	< 0.06	< 0.06	< 0.06
Manganese ppm	0.51	0.45	0.21	0.22
Copper ppm	1.43	1.85	1.94	0.99
Antimony ppm	< 0.26	< 0.26	< 0.26	< 0.26
Calcium ppm	45.1	45.12	40.5	27.0
Zinc ppm	6.49	4.48	2.01	1.08
Iron ppm	< 0.05	< 0.05	< 0.05	< 0.05

In the vicinity of Cerro Mirador there is little or no vegetation because of human and animal activity, combined with the fact that natural barriers for fog have been eliminated. Thus, a fog collector would contribute to the recovery of vegetation and (presumably) allow for its proper management, thereby bringing about an increase in the biodiversity of the area's ecosystem.

At present, given the characteristics and the flora and fauna of the zone and considering the potential and quality of the available water resources, UCN started a project to recover part of the Paposo endemic ecosystem which could also be applied along the coastal zone where the Camanchaca phenomenon occurs.

4.2 Horticulture and Orchards in Taltal: Case Study 2

The port of Taltal is located 375 km to the south of Antofagasta at latitude 25° 23' S and longitude 70° 26' W. A desert climate similar to

Paposo predominates. Topography and soil composition are also similar. Taltal creek has sandy soils with a good permeability and high levels of electrical conductivity.

Vegetation on the hills varies from less than 10 percent at sea level to about 50 percent in the Camanchaca area depending on humidity availability. Flora corresponds mainly to *Eulichnya, Puya, Euphorbia* and *Copiapoa.*

The coastal strip, especially the area between 300 and 800 m above sea level has a biodiversity vulnerable to human and industrial impacts because of its economic importance; some of the species are unique in the world and endemic in the area.

Taltal is very poor and has 12,000 inhabitants; economic activities are mining, fishing, tourism and agriculture. Presently, the town permanently receives tap water coming from the water reserves 55 km away at the Andes cordillera. The town is connected to the electricity system for continuous supply. A project for implementing agricultural, horticulture and cattle breeding farms by using a mixture of underground and fog–collected water is being considered for implementation.

A small–scale agriculturist association was created in December1988 in order to ask the state for 83 hectares suitable for collective experimental use.

At the end of 1999, the association became the Taltal Small–Scale Agriculturist Union, with 46 members (25 men and 21 women). Commonly, men work in small–scale mining and unskilled labor. Women are housewives and contribute to the family budget by working sporadically. The union was created in order to set up an additional productive activity and thus improve the life standard of the members.

The land has already been given to the union and is located at the El Hueso creek at 200 m above sea level, 3k m to the north and 2k m to the east of Taltal. Previous tests to determine soil fertility were successful and some products have already been produced. In this stage, water has been carried from Taltal. Three sectors have been defined for this activity, two for cattle breeding and one for horticulture. Water would be supplied by underground water from the area and with the help of Camanchaca from the nearby hills at 900 m above sea level.

4.2.1 Community Participation

Since the project began, the local community has worked cleaning the site, digging underground water wells (at low depth), cleaning old surface water currents, and in the construction, installation and control of 6 standard fog collectors with the support of the UCN. The union has successfully applied for funds to programs (public and private) related to agricultural and cattle–breeding development in order to continue the project.

In January 2000, the National Commission for the Environment (CONAMA) funded the union with US$ 8,000 for the project "Efficient Management of Water Using Complementary Water Sources: Underground and Fog–Collected Water." This project allowed the construction of two modular fog collectors with a total of 250 m^2, pipelines, a water storage dam with a capacity of 500 m^3 at El Hueso creek, and the equipment for obtaining and distributing underground water.

The community participates in all these activities on weekends and so has had the possibility to apply for more funds and obtain further support. The project's goals are to provide more job opportunities and increase agricultural and cattle breeding activities without damaging the environment and using the proper alternate technology to supply the groups with the necessary production resources.

For agricultural activity, community members have the technical assistance of an agronomy engineer and for organizational issues there is a social worker. These professionals work for the Chile Pais Program, which is a state institution created for young professionals in order to help in the development of extremely poor communities.

The project is in its last stage but its first results have been satisfactory. In fact, fruit tree growth is productive and the horticulture products (lettuce, carrot, etc.) have are of good quality and size. Also, the grass in cattle breeding areas is growing enough to meet the demands for grazing.

These activities have allowed the union to be better known and to obtain financial assistance from other sources to continue the project and to get funds for socially complementary activities (e.g., construction of the union building, implementation of training courses, etc.).

5. OTHER APPLICATIONS IN THE COUNTRY

As an illustration of the activities for endangered ecosystem recovery in other zones of the country, one other aspect of the "Chungungo" project is briefly described below.

Chungungo is a small community of fishermen, located on the shores of the Pacific Coast in the north of Chile at latitude 29° 27' S and longitude 71° 18' W. Drinking water for human consumption has been supplied to the community since 1992.

The project to supply drinking water to Chungungo began in 1987 under the joint participation of the UCN, CONAF, the University of Chile, the Meteorological Service of the Canadian Ministry for the Environment and the Atmospheric Environment Service of the Canadian Ministry for the Environment (AES). The project was financed by the International Center for Research and Development (CIID), the International Development Research Center (IDRC) and the Canadian Embassy in Chile, among others.

The system was designed to supply drinking water to 440 inhabitants, from an entrapment surface area of 4512 m^2 located in the El Tofo hills at an altitude of 78 m above sea level, using a total of 94 cloud trapping devices to extract a total of 10 m^3 per day of drinking water.

The system was under the administrative control of the Chungungo Drinking Water Committee in 1995 (CDWC). The CDWC is made up of 5 members of the local community, elected every 2 years, which has the responsibility of maintaining the system in operation, including the water treatment plant, the distribution of drinking water, and the fixing of costs for users of the systems. Maintenance and cleaning costs for the cloud trapping device is obtained from the charges paid for water by the users and a subsidy of 60 percent of the total costs from the municipal government.

However, as the local municipality was not able to deliver its financial support within the required time limits, and all the water users did not pay their water bills, this led to an inadequate maintenance of the cloud trapping devices (Edwards 2001). Problems were increased by an almost complete lack of interest by the local community in the maintenance of the system and, unfortunately, this caused the complete collapse of the system in the year 2000.

Additional simultaneous attempts were made with the project to provide drinking water to carry out two other projects in the local community: (1)the formation of farming plots, and (2)the establishment of a fish processing plant project. These projects failed due to lack of interest and participation by the members of the local community.

The viability of the project from a social point of view depends on the continuous satisfaction of the basic needs of the local people; from an ecological point of view, on the continuous productivity and protection of the biodiversity; and finally, from an economic point of view, on the production of resources without a corresponding substantial increase in costs. The use of a system to trap and get water from clouds by small local communities in arid or semi–arid zones as a complement for sustainable development requires that a balance is achieved between these three perspectives.

In the specific case of the small community of Chungungo, the water produced becomes a resource belonging to all the members of the community. In this context, the exploitation of water resources presents serious problems both in the procurement and storage due to the high demand from rural communities with limited resources. Faced with this reality, the management of water resources in local environments must be seen as a positive force promoting the ecologically sustainable use of the environment, community health, the protection of the local culture, and the local community's economic welfare.

6. CONCLUSIONS

In case study 1, the local community, as defined by Pinkerton (1993), is made up of a group of persons from the same geographical area, with common interests and concerns, who have organized themselves in a community association and as a collective organization to make decisions. Within this concept of the community it is important to take into consideration local knowledge, which in itself is an important resource in facilitating the development of profitable, sustainable and participative processes for the community (Agrawal et al. 1995)

As the water resource extracted from the clouds is one which belongs to everyone and is an essential requirement for the success of the activity by the community association and others, this means that all decision making is generally done by local persons, either public or private.

The land on which the cloud trapping devices are situated, and the land used for farming activities, is state property, which means that any proposed enlargement of the cloud trapping area, or demand for more land for other projects, must be approved by the relevant government authority.

The Taltal Association of Small Farmers has a board of Directors made up of 4 members plus 2 directors. This organizational framework has encountered many difficulties, mainly due to the fact that association members expect quick answers and solutions to their problems, which often are not forthcoming due to the time it takes state agencies to make decisions.

This tends to have a negative affect on the motivation of the board which in turn translates into a lack of support for any activities which of necessity must be dealt with in a communal manner. Anecdotally, it has been noticed that not all the members of the association are really convinced that their personal involvement in community affairs is necessary or important.

Many of the initiatives generated to help the local community are produced within the Board of Directors (Taltal Association of Small Farmers) without seeking the complete involvement and approval of all the other members of the association; this leads to the problem that many projects are approved by external bodies without the knowledge or approval of many local community members.

In the analysis of case study 2, the situation is very different from that of case study 1. In spite of the fact that the community at Paposo is geographically more closely linked, with interests and concerns that are shared by many members of the community, it has been even more difficult to accomplish the formation of a highly organized community which will permit the community to act in a united manner to face the many problems with which they are confronted. Paposo is situated 50 km to the south of Taltal, has no drinking water (all water is trucked in by water barrels), and it is considered to be an area of extreme poverty.

The Paposo Zone has a higher water potential, with respect to the capture and extraction of water from clouds, and has unlimited space of easy access which suitable for the installation of cloud trapping devices. On the other hand, in the Chungungo area the land is privately owned and because many of its surrounding zones have a high potential for development of agricultural activities there are more limitations on placement and utilization of cloud trapping devices.

The work carried out by the local community in the first case study was initiated by people with common problems who were supported by professionals of the Country Service, assigned to them by Taltal Municipality; the people have defined their own objectives and activities as the project has proceeded.

Much was been written with respect to the management of resources based in local communities, and it has been observed that when such communities are able to play a significant participatory role in the management of their resources, they often have developed methods for preventing the over exploitation of their local resources. Agreements made with the consent of the members of local communities have resulted in commitments to improve the management of woodlands, fish reserves, wildlife, water, and other local resources of common ownership in pursuit of ecological and economic sustainability (Pinkerton 1993, Pinkerton and Weinstein 1995).

The Canadian International Center for Research and Development (CIID) considers the management of natural resources by the community (MRNBC) to be a key strategy in the economic development of the communities (EDEC). The EDEC as a MRNBC is defined as a process based in and controlled by the community; it explicitly combines the benefits of economic and social development with the objective of improving the quality of life of the communities and residents of the regions, from an economic, social, ecological, and cultural perspective. The economic development strategies have arisen as an alternative to the conventional approaches to economic development. These strategies are based on the belief that the problems which communities have to deal with, such as unemployment, poverty, lack of work, environmental degradation, economic instability, and loss of control of the community by its members, must be dealt with in a holistic and participating manner (CEDC 1998).

Biodiversity conservation must consider the recovery and/or replacement of the degraded habitat, for which water resources are important. For areas like northern Chile where this resource is scarce, fog collectors (given the right environmental conditions) represent an alternative option for preserving and recovering biodiversity.

Since the water resource obtained from creeping fogs by means of fog collectors is not abundant, it is necessary to set up high efficiency systems (technical watering, economic control dispensers, etc). Complementary use as a mixture with water from other sources is also an option which needs to be considered.

The two case studies described in this paper show it is possible to use fog collectors for the conservation and recovery of biodiversity and ecosystems, as well as to provide water for local communities.

The use of Camanchaca water for biodiversity preservation and recovery in zones where water is a limiting factor also allows complementary activities that directly influence the living conditions of small towns associated with them. For instance, with enhanced water microproductive activities, tourism, energy supply, etc., can improve the social and economic conditions of the local communities.

Because the technology and implementation of fog collectors is not widely established, it is also very important to exchange data with other institutions and partners and to collaborate on possibilities for funding projects and the transfer of the technology so that it can be more widely utilized.

7. REFERENCES

Agrawal, A. 1995. Dismantling the divide between indigenous and scientific knowledge. Development and Change 26: 413–439.

CEDC, 1998. Pagina web http:www.sfu.ca/cedec/ del Centro de Desarrollo Comunitario Economico, Universidad de Simon Fraser, Columbia Britanica, Canada.

Cereceda, P., Cruzat, A., Ossandon, J., Schemenauer, R. 1996. Prospection, evaluation and construction of fog water collection systems: A new technology of simple application and large economic and social impact. IDRC–CIID.

CONAF. 1996. Libro Rojo de los sitios prioritarios para la conservacion de la diversidad biologica en Chile. Ministerio de Agricultura, Corporacion Nacional Forestal, Chile.

Edwards, M., Cereceda, P., and Schemenauer, R.S. 2001. Community involment in the fog–water collection system for Chungungo, Chile. Pages 277–279 In: 2nd International Conference on Fog and Fog Collection. St. John's, Canada, Julio 15 –20.

Espejo, R. 1992. Comportamiento de los Estratocumulos en Antofagasta y su potencial como recurso hidrico. Tesis de Grado. Universidad Austral de Chile.

Espejo, R., Burgos, C., Zuleta, R., Martinez, L., Correa, F., Ortega, P., and Vargas, L. 1993. Pages 241–246 In: Calidad del agua de los Estratocumulos Costeros. Informe preliminar. III Encuentro de Fisica Regional Norte. I Reunion Internacional Andina de Fisica, Arica.

Espejo, R., 1998. Coast fog water potencial and its applications,. Pages 341–343 In: 1st International Conference on fog and fog collection. Vancouver, Canada, Julio 19–24.

Pinkerton, E.. 1993. From Conflict to Consensus. Shared Decision–Making in British Columbia. Proceeding del Simposio Desarrollado en la Universidad de Simon Fraser, Editado por Mark Roseland, Vancouver, Canada.

Pinkerton, E., and Weistein, M. 1995. Fisheries That Work: Sustainability Through Community–Based Management. Report to the David Suzuki Foundation.

Soto, G., Cerda, J., Canto, W., Elgueta, H., and Ossandon, J. 1985. Evaluacion de las Neblinas Costeras (Camanchacas) en el Sector El Tofo. Corporacion Nacional Forestal de Chile, IV Region.

Tapia, O., and Zuleta, R. 1980. 20 Años de Camanchacas y 2 del Proyecto Mejillones. Universidad Catolica del Norte. 72 pages.

Chapter 23

GENETIC DIVERSITY AND MANAGEMENT IMPLICATIONS FOR VICUÑA POPULATIONS IN PERU

J.C. Wheeler[1], M. Fernandez[2], R. Rosadio[2], D. Hoces[3], M. Kadwell[4] and M.W. Buford[5]

[1]CONOPA, Los Cerezos 106, Lima 3, Perú; email jwheeler@conopa.org; [2]Facultad de Medicina Veterinaria, Universidad San Marcos, Lima, Perú; [3]Consejo Nacional de Camelidos Sudamericanos, CONACS, Lima, Perú; [4]Conservation Genetics Group, Institute of Zoology, Regent's Park, London, U.K.; [5]Cardiff School of Biosciences, Cardiff University, Cardiff, U.K.

1. INTRODUCTION

The scientific aims of this Darwin Initiative–funded project were to use molecular genetic markers (specifically microsatellites) to: (1)elucidate the recent evolutionary history of Peruvian vicuña populations; (2)evaluate the genetic diversity and its partitioning in those populations; (3)identify demographically independent management units within these populations for future management; and (4)assess the likely genetic effects of past and future management strategies, including the likely consequences of sustainable utilisation practices. It is important to emphasise that this is the first such study carried out on a wild South American camelid.

2. SAMPLING RATIONALE AND STRATEGY

A total of 12 populations were sampled at sites selected for the following reasons: (1)geographic locations throughout the range of habitat and reserve coverage in Peru; (2)because they were thought to have had relatively long histories of demographic isolation; and (3)because they were thought not to have been influenced by recent translocations of animals from the Pampa Galeras reserve.

Table 1 details the geographic location of each sample, sample size, type and further details. Half of each sample was deposited in the Unidad de Virología y Genética Molecular laboratory at IVITA, and the remainder was sent to the Institute of Zoology for analysis, under Peruvian CITES export permits 00240 and 00658, and UK/EU CITES import permits 67264 and 206242/01. Samples used in this study were either blood or skins. Table 2 details what is known from the 1997 and 1999 CONACS census regarding demographic composition of each population. An indication of any anecdotal evidence of recent demographic fluctuation is also given.

Table 1. Vicuña sample location details

Locality	Geographic Position	Sampling Date	Samples Collected
Tinco Cancha, Junín	75°38" W, 11°02" S	22.11.98	36
Villa Junín, Junín	75°52" W, 11°05" S	29.11.98	30
Yantac, Junín	76°18"W, 11°20" S	28.09.98	35
Tingo Paccha, Junín	75°27" W, 11°25" S	23.10.98	30
Tarmatambo, Junín	75°43" W, 11°30" S	12.09.98	27
Cachi Cachi, Junín	75°33" W, 11°38" S	27.08.97	21
Huacarpana, Ica	75°04" W, 12°50" S	15.07.97	20
Ayavi, Hunacavelica	75°15" W, 13°42" S	18.10.97	15
Lucanas (Pampa Galeras), Ayacucho (1)	74°24" W, 14°39" S	24.06.97	20
Lucanas (Pampa Galeras), Ayacucho (2)	74°24" W, 14°39" S	10.10.94	78
Lucanas (Pampa Galeras), Ayacucho (3)	74°24" W, 14°39" S	01.10.94	111
Toccra (Aguada Blanca), Arequipa	71°20" W, 16°10" S	04.11.97	3
S.A.I.S. Picotani, Puno	70°00" W, 14°55" S	26.11.97	38
Ingenio, Huacullani, Puno	69°20" W, 16°40" S	19.11.97	24

Note: all samples were blood except for Ayacucho (3)which were skin taken from hides of animals culled between 1977–1983.

Table 2. Demographic structure of Peruvian Vicuña populations. (Data from CONACS 1997 and 1999 censuses).

Locality	Total	Adult Males	Adult Females	Crías	Bachelor Bands	Solo Males
Tinco Cancha	153	17	52	33	47	4
Villa Junín	645	125	386	61	62	11
Yantac	290	29	86	43	131	1
Tingo Paccha	193	23	73	34	57	6
Tarmatambo	503	82	254	69	90	8
Cachi Cachi	497	69	221	94	108	5
Huacarpana	620	78	294	143	74	31
Ayavi	146	5	27	15	92	7
Lucanas	10,167	961	4098	2189	2575	364
	12,764	1347	5426	2215	3267	509
Toccra	129	15	60	19	34	1
Picotani	1152	113	429	242	276	92
Ingenio	45	7	27	0*	9	2

Notes: (1)crías are both sexes; (2)all populations stable, except for Ayavi (hunted), Lucanas (recovering from less than 1000 individuals in 1963–64), Toccra (recovering); (3)Ingenio data from February census; (4)1999 census data in bold.

3. METHODS

Once the samples had been collected, DNA was extracted from blood or skin using standard procedures involving Proteinase K digestion of cellular material followed by nucleic acid extraction using phenol and phenol/chloroform to remove proteins, and total DNA was precipitated in 100 percent ethanol (see Bruford et al. 1998). DNA samples were stored suspended in TE solution (10 mM Tris–HCl, 1 mM EDTA, pH 8.0) prior to analysis.

Eleven previously published South American camelid (SAC) microsatellite DNA markers were analysed (Lang et al. 1996, Penedo et al. 1998). Microsatellites are nuclear (n)DNA markers found in the chromosomes of most eukaryotes and have been found to be highly polymorphic in many species, leading to their use in applications such as paternity testing, individual profiling in forensics, population analysis and hybridisation studies (Goldstein and Schlötterer 1999). We have previously shown that microsatellites can be used effectively in hybridisation studies involving SACs (Kadwell et al. 2001) and for the purposes of this study we used these markers to measure population genetic diversity and infer recent demographic history. Table 3 shows details of the microsatellites used, their genetic diversity in terms of heterozygosity (the proportion of individuals who inherit different alleles at a given gene or locus from their parents– a robust indicator of genetic variation) and allele sizes in Peruvian vicuña.

Table 3. Summary of genetic markers.

Primer	Primer Sequence 5' – 3' (forward and reverse)	Anneal temp. (T_{ANN})	Allele size range (bp) / # alleles	H_o	ABI 373 Multiple x details
YWLL08	F-ATCAAGTTTGAGGTGCTTTCC R-CCATGGCATTGTGTTGAAGAC	58C	127 – 195 (31 alleles)	0.790	Yellow
YWLL29	F-GAAGGCAGGAGAAAAGGTAG R-CAGAGGCTTAATAACTTGCAG	58°C	216 – 226 (5 alleles)	0.513	Blue
YWLL36	F-AGTCTTGGTGTGGTGGTAGAA R-TGCCAGGATACTGACAGTGAT	60°C	106 – 180 (16 alleles)	0.699	Yellow
YWLL38	F-GGCCTAAATCCTACTAGAC R-CCTCTCACTCTTGTTCTCCTC	60°C	172 – 176 (3 alleles)	0.208	Blue
YWLL40	F-CACATGACCATGTCCCCTTAT R-CCAGTGACAGTGTGACTAAGA	59°C	180 – 192 (7 alleles)	0.390	Green
YWLL43	F-ATACCTCTCTTGCTCTCTCTC R-CCTCTACAACCATGTTAGCCA	59°C	137 – 157 (10 alleles)	0.380	Yellow
YWLL44	F-CTCAACAATGCTAGACCTTGG R-GAGAACACAGGCTGGTGAATA	58°C	105 – 141 (18 alleles)	0.591	Blue
YWLL46	F-AAGCAGAGTGATTTAACCGTG R-GGATGACTAAGACTGCTCTGA	59°C	96 – 100 (3 alleles)	0.048	Green
LCA5	F-GTGGTTTTTGCCCAAGCTC R-ACCTCCAGTCTGGGGATTTC	58°C	187 – 213 (12 alleles)	0.638	Green
LCA19	F-TAAGTCCAGCCCCACACTCA R-GGTGAAGGGGCTTGATCTTC	60°C	97 – 125 (14 alleles)	0.405	Blue
LCA22	F-TTAAGAGTCTAAAAGAGAAAGGG R-CAGATGACAGCTGGGATTGA	58°C	113 – 125 (7 alleles)	0.343	Yellow

*Notes : (1)H_0 mean observations heterozygosity ; (2)data source : YWLL08–YWLL46, Lang et al. (1996) ; LCA5–LCA22, Penedo et al. (1998).

Table 4 breaks down this analysis of genetic diversity into data for each population and compares observed heterozygosity, expected heterozygosity (assuming breeding which is random with respect to the genetic similarity of the individuals) and mean number of alleles per locus. A large difference between observed and expected heterozygosity can be caused by non–random mating (e.g., inbreeding) or selection acting against certain genotypes.

Table 4. Genetic diversity in vicuña populations of Peru.

	Expected heterozygosity	Observed heterozygosity	Mean alleles per locus	Private alleles
Tinco Cancha	0.5380	0.4924	5.0000	2
Villa Junín	0.4498	0.3811	4.7273	2
Yantac	0.4467	0.3730	4.6364	4
Tambo Paccha	0.3913	0.3848	3.2727	0
Tarmatambo	0.3770	0.3636	2.7273	0
Cachi Cachi	0.4191	0.4156	3.3636	1
Huacarpana	0.4813	0.4682	4.8182	2
Ayavi	0.4708	0.4481	3.9091	1
Lucanas (1)	0.5382	0.5364	5.1818	2
Lucanas (2)	0.5857	0.5310	6.8182	0
Lucanas (3)	0.5800	0.5279	6.3636	1
Toccra	0.4444	0.4545	2.6364	0
Picotani	0.5094	0.5072	4.5455	2
Ingenio	0.5228	0.4848	5.3636	3

Table 5 indicates pair–wise values of genetic differentiation between populations measured by the commonly-used index F_{ST}, which varies between zero (no genetic differentiation) and one (complete differentiation) and can also be regarded as the component of the total genetic variation in any two populations which is attributable to the differences between them. F_{ST} is simply calculated as:

$$F_{ST} = H_T - H_S / H_T,$$

where H_T is the expected heterozygosity in the total population and H_S is the expected heterozygosity in the sub-population in question. Table 5 also includes values of Nm, the estimated number of migrants per generation, which is calculated from F_{ST} using the formula:

$$Nm = (1 - F_{ST})/4F_{ST},$$

which is included primarily for illustration, since this calculation makes assumptions about the populations that are certainly violated in some cases. All calculations and analytical approaches were implemented using the programs GENEPOP v3.1 (Raymond and Rousset 1995) and GENETIX v4.0 (Belkhir 1999).

Table 5. Pairwise F_{ST} values (above diagonal) and Nm values (below diagonal) between populations in Peru.

	Tinco	Villa	Yantac	Tingo	Tarma	Cachi	Huaca
Tinco	--------	0.114	0.097	0.114	0.164	0.138	0.096
Villa	1.94	------	0.103	0.114	0.157	0.159	0.127
Yantac	2.33	2.19	--------	0.146	0.217	0.183	0.127
Tingo	1.95	1.95	1.46	--------	0.042	0.114	0.143
Tarma	1.28	1.34	0.9	5.77	--------	0.105	0.183
Cachi	1.56	1.32	1.12	1.95	2.14	--------	0.154
Huacar	2.38	1.71	1.71	1.5	1.11	1.38	--------
Ayavi	1.48	0.83	0.79	0.69	0.64	1.04	1.43
Luc1	4.14	2.28	2.27	1.75	1.33	2.11	9.62
Luc2	3.98	2.88	2.09	2.22	1.77	2.02	6.26
Luc3	3.39	2.97	2.25	2.3	1.65	1.88	7.42
Toccra	1.7	1.02	1.03	0.7	0.55	0.92	5.21
Picotani	1.86	1.47	1.51	0.92	0.69	0.92	1.62
Ingenio	2.93	1.66	1.82	1.07	0.93	1.31	1.83
	Ayavi	**Luc1**	**Luc2**	**Luc3**	**Toccra**	**Pico**	**Ingen**
Tinco	0.145	0.057	0.059	0.067	0.128	0.119	0.079
Villa	0.233	0.099	0.080	0.078	0.196	0.146	0.131
Yantac	0.241	0.099	0.107	0.1	0.195	0.159	0.121
Tingo	0.265	0.125	0.101	0.098	0.263	0.214	0.189
Tarma	0.282	0.158	0.106	0.132	0.313	0.267	0.212
Cachi	0.193	0.106	0.110	0.117	0.214	0.213	0.161
Huacar	0.149	0.025	0.038	0.032	0.046	0.134	0.12
Ayavi	--------	0.127	0.099	0.119	0.139	0.241	0.187
Luc1	1.72	-------	0.017	0.015	0.064	0.088	0.069
Luc2	2.27	14.33	-------	0	0.066	0.103	0.066
Luc3	1.86	16.21	9999	-------	0.049	0.085	0.072
Toccra	1.55	3.67	3.55	4.81	--------	0.159	0.145
Picotani	0.79	2.58	2.19	2.71	1.32	--------	0.079
Ingenio	1.09	3.4	3.57	3.22	1.47	2.93	--------

Figure 1 shows a dendrogram (genealogical tree of populations) generated using neighbour–joining analysis (Saitou et al. 1985) in the computer program MEGA (Kumar et al. 1993), which clusters populations according to their Nei's 1972 genetic distance (Nei 1987, which analyses genetic variation in a way analogous to F_{ST}) and enables clusters of related populations to be identified. Figures 3a–d show 2–dimensional factorial correspondence plots of four populations in northwest Junín, south Junín, central Andes (Huancavelica–Arequipa) and Puno, respectively, where the genetic diversity among populations is expressed as factors which explain the correspondence among samples in a number of dimensions (here we show the two dimensions which explain the highest proportion of the correspondence according to Benzécri 1973; using GENETIX v. 4.0– see results for a fuller explanation). Thus the spatial relationships (approximate locations) between populations can be adjudged by examining how individuals from each population cluster in 2, 3 or more dimensions.

Figure 1. Neighbour–joining tree of Nei's (1972) genetic distances between Peruvian vicuña populations.

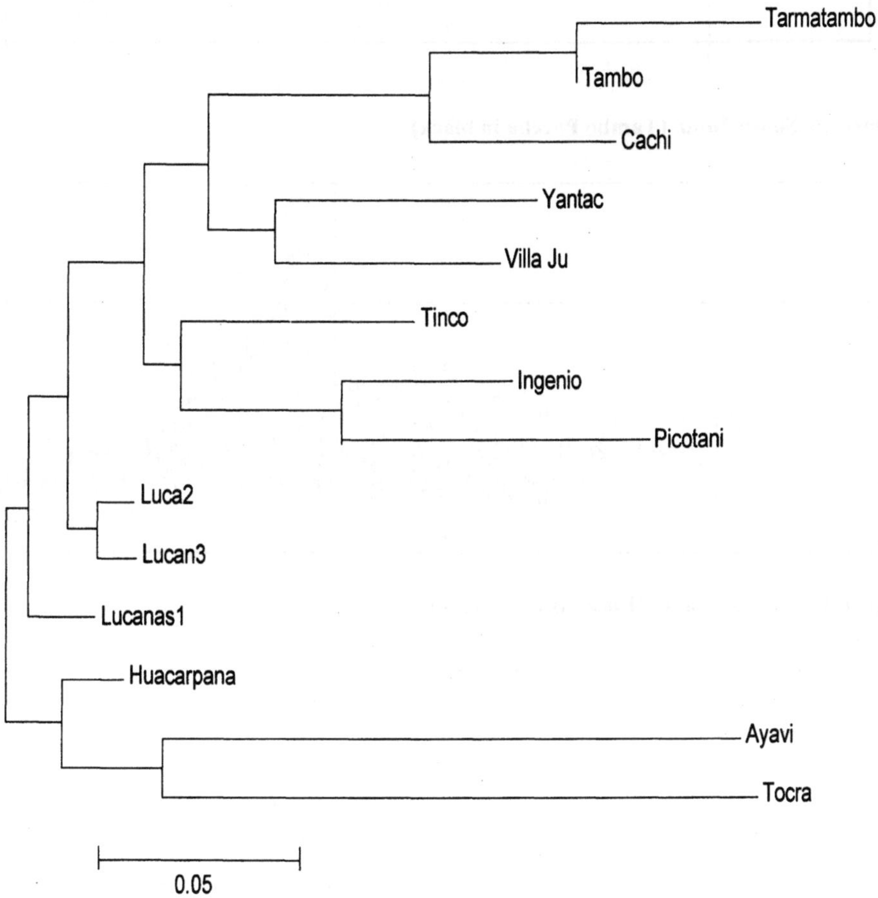

Figure 2a. Northwest Junín (Yantac in black)

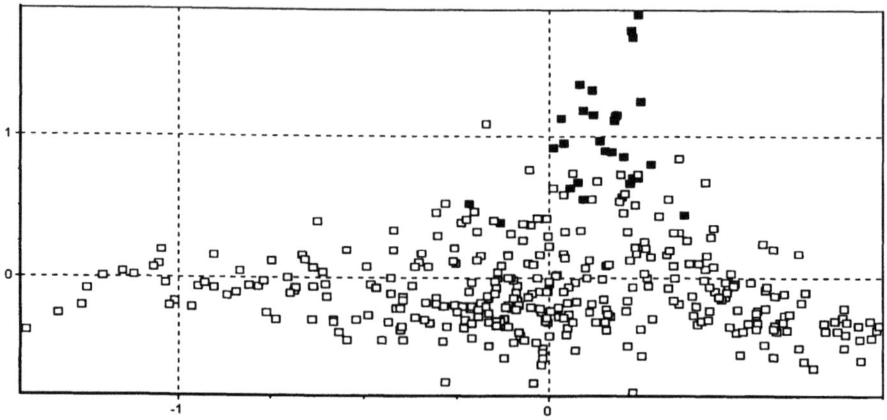

Figure 2b. South Junín (Tambo Paccha in black)

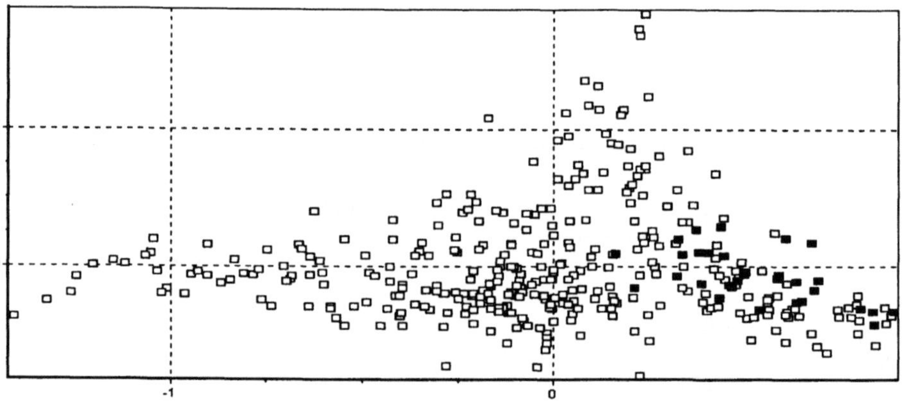

Figure 2c. Central Andes (Lucanas 2 in black)

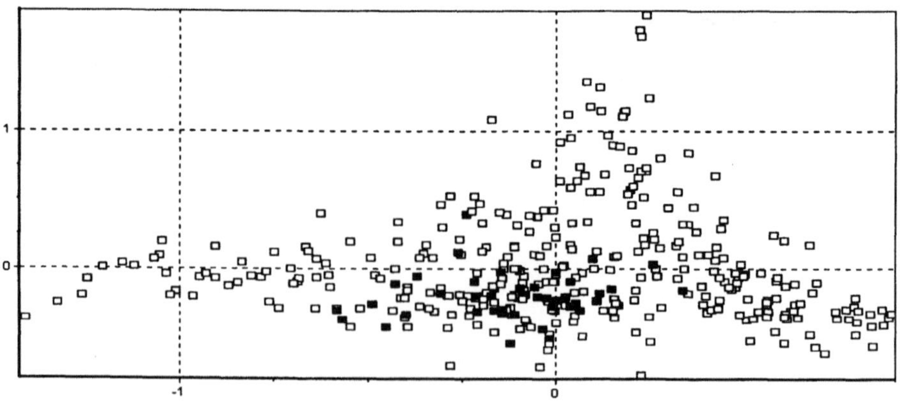

Figure 2d. Puno (Picotani in black)

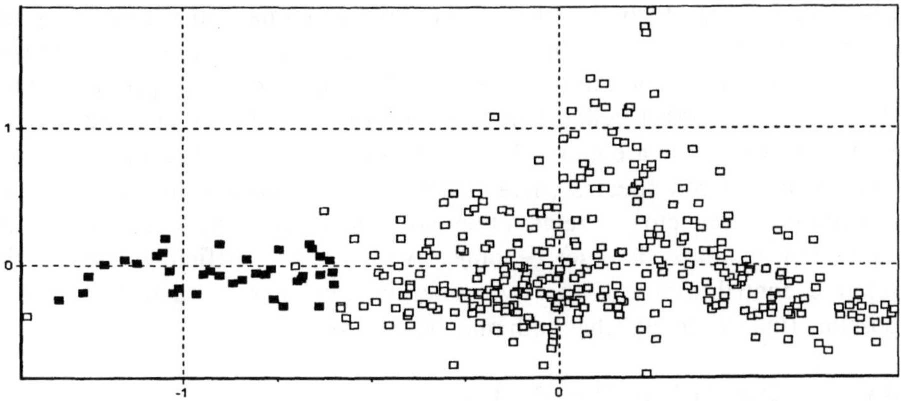

4. RESULTS AND INFERENCE

4.1 Population and Genetic Variation

4.1.1 Microsatellites

Table 3 shows that the markers used are highly polymorphic and informative in Peruvian vicuña, which means that they are ideal for study genetic diversity and differentiation. For example, YWLL08 is exceptionally informative and possesses 31 alleles and nearly 80 percent of individuals are heterozygotes. Interestingly, we have found many alleles previously unrecorded by the researchers who first developed them. For example, Lang et al. (1996), who isolated the YWLL markers used in this study from llama DNA, identified 13 alleles at YWLL 08 and 6 alleles at YWLL 36 (as opposed to 31 and 16 respectively in this study). Likewise, Penedo et al. (1998), who isolated the LCA microsatellites used here from llama DNA, found 7 alleles at LCA 5 and 10 alleles at LCA 22 (as opposed to 12 and 14 respectively in this study). However, the overall levels of heterozygosity are very much lower in this study compared with the original papers. This is perhaps surprising, given that we have sampled wild populations of vicuña and compared them to domestic camelids. This result is most likely to be due to Peruvian vicuñas having lower genetic variation within populations, although this would seem counter–intuitive, given that vicuña possess so many more alleles than their domestic counterparts. This may be explicable, however, if that majority of genetic diversity in Peruvian vicuña is found between as opposed to within populations.

An alternative explanation is the phenomenon known as "ascertainment bias" where because we have used llama–derived microsatellites (which have been chosen to be highly polymorphic in llamas) in vicuña, they are more likely to be less variable precisely because they have not been chosen because of their diversity in vicuña. Furthermore, our previous data (Kadwell et al. 2001) indicate that llamas are the domestic camelid derived from guanacos. It is a commonly observed phenomenon that microsatellites isolated in a given species (here llama) are less variable in other species, regardless of how closely related they are. However, until such a time as an equivalent guanaco data set becomes available it is difficult to assess the relative contributions of the two potential explanatory factors, and it is possible (indeed, likely) that both may be involved.

4.1.2 Populations–Genetic Diversity

Table 4 shows the levels of heterozygosity found in all populations analysed. Mean expected heterozygosity values over all loci vary between 0.377 (Tarmatambo) and 0.586 (Lucanas 2). These values are certainly lower than those commonly found in continental mammal populations (data not shown) which usually do not fall below 0.5 and thereby may reflect a recent history of population isolation with correlated genetic drift or may be a consequence of the type of microsatellites being used (see above). However, since the phenomenon seems to be found in many of the loci used, genetic drift or inbreeding may be likely. This hypothesis is supported by the observations shown in Table 4, wherein observed heterozygosity is lower than expected heterozygosity (i.e., the value expected from the observed allelic frequencies under the assumption of random mating) in all populations except one (Toccra, which suffers from a small sample size and may be biased). Such an observation is consistent with localised inbreeding or selection against heterozygotes. Furthermore, there is a significant excess of homozygosity in all populations except Huacarpana, Lucanas 1, Cachi Cachi, Toccra and Tarmatambo (GENEPOP analysis), which suggests that inbreeding and/or genetic drift is a fairly common occurrence.

Inbreeding results from mating between relatives, and can be measured by an inbreeding coefficient, F, which varies between 0 and 1, but where values above 0.05 are considered high. For example, brother/sister mating results in an F–value of 0.25. Inbreeding in vicuña, however, is highly unlikely under normal circumstances because of the social system, where both sexes disperse from their natal group. A generally elevated inbreeding coefficient can, however, follow from a situation where a population has been through a major crash or demographic "bottleneck."

Table 4 also shows a total of 20 private (population–specific) alleles which is also higher than expected for a continental population sample and is further suggestive of some level of local isolation and genetic drift. Interestingly the frequency of private alleles does not appear to correspond with population size or degree of isolation, although the northernmost population (Yantac) does possess the highest number of unique alleles.

Finally, although it is clear that the allelic diversity of the entire population is high in comparison to the domestic populations chosen by Lang et al. (1996) and Penedo et al. (1998), the mean number of alleles per locus within populations is lower than often observed in other mammals (e.g., Tarmatambo with 2.7 alleles/locus), and it is only the Lucanas 2 and 3 that show the relatively high allelic diversity often associated with such studies.

4.1.3 Populations–Genetic Differentiation

Pairwise indices of genetic differentiation between populations (F_{ST}, Nei's 1972 D) were measured and used to infer relationships between populations, calculate migration rates between them, and in combination with clustering methods, were used to graphically display inferred demographic clusters between them. Table 5 shows pairwise F_{ST} values between all populations, and the values vary between 0 in the case of Lucanas 2 and Lucanas 3 indicating complete genetic similarity (in this case not surprising since the samples were taken from the same population) and 0.31 between Toccra and Tarmatambo, a high degree of genetic differentiation between two continental mammal populations. In general, F_{ST} values are high, with 67 percent exceeding 0.1, all comparisons except two being significant at the $p < 0.05$ level and all except eight being significant $p < 0.01$. However, the values broadly reflect the known demographic relationships among the populations, with the lowest values being recorded between the Lucanas samples, and low values being found between populations in Puno and in south Junín. Such population groups are clearly demographically dependent according to these data.

Demographic dependence or independence can also be inferred from the Nm values shown below the diagonal in Table 5. Due to the generally high levels of F_{ST} found in the data set, many values of Nm are close to unity. An Nm value of one indicates that approximately one migrant is exchanged between two populations of equal size per generation (every n years), which according to seminal work of Franklin and Soulé (1980) is sufficient to prevent divergence through genetic drift. The "one migrant per generation" (or OMPG) threshold has subsequently been used as a "rule of thumb" in management of populations to maintain demographic interdependence.

However, since the study populations vary in size considerably and because many are known to have been demographically isolated for hundreds of generations, such values are meaningless without being put into the context of known population history. For instance, it is surprising biogeographically that Yantac has supposedly exchanged 2.27 migrants per generation with Lucanas 1 whereas it has only exchanged 1.12 with Cachi Cachi. Such results may partly be due to the biases mentioned above, and must be interpreted in the context of known population history. In general, the Nm values reflect apparent demographic history in the same way as the F_{ST} values, although some anomalous values appear which will be discussed later.

Genetic distances were also calculated, since they are thought to reflect the evolutionary history of populations more accurately than demographic indices such as F_{ST} (Nei 1987) and can be more effectively used by clustering algorithms such as neighbour–joining to produce dendrograms which may reflect the evolutionary relationships between the populations being analysed. Here we used pair–wise Nei's 1972 unbiased genetic distances (values not shown, although they broadly correlate with F_{ST}) to produce the neighbour–joining dendrogram shown in Figure 2 below. Branch lengths are proportional to the genetic distance but do not imply evolutionary time scale since genetic distance values can be strongly influenced by demographic fluctuations too. However, generally the results enable us to infer relationships between populations that make both biogeographical and demographic sense. For example, the three south Junín samples cluster closely together, as do two of the three northwest Junín samples and the two Puno samples. Less structure is apparent for the Lucanas, Huancavelica and Arequipa samples, however, which may be indicative of a loose demographic affiliation between those populations not apparent in the genetic distance analysis. Tinco Cancha is aberrant for reasons discussed below.

Finally, the data were subjected to 2–dimensional factorial correspondence analysis (2D–FCA) to further explore the relationships among populations. FCA is a canonical analysis particularly well adapted to describing relationships between two qualitative variables (here zero, one or two for each allele at each locus) which result in allele frequency vectors being produced for each population which are then analysed using Robertson and Hill's F_{ST} estimator to produce a series of inertial values for each individual's single locus F_{ST} estimates. Each axis of inertial values can be analysed in multidimensional space to produce a factorial correspondence plot, the most powerful of which represents the greatest proportion of the total inertial value. The result is a powerful method for recovering maximal information on the genetic relationships among individuals within and between populations using n–dimensional space. For the vicuña data, the first two axes jointly explained approximately 6.5 percent of the total inertial value, a larger than average proportion for intraspecific analyses, and a reflection on the relatively high power of the data.

Figures 2a–d show the results for the entire data set in two dimensions. What is immediately apparent is the tight clustering of many of the individuals within each population and the unique locations which several of the populations occupy in 2–dimensional space. Of particular note are the populations of Picotani (Figure 2d: the only population found between vectors –0.6 and –1.2 on the 'x' axis), and Yantac (Figure 2a: the main population found between vectors 0.9 and 1.5 on the 'y' axis). In addition, several population groups become apparent due to their strong clustering in 2D space. Four broad groups are identifiable, and Figures 2a–d highlight these and explain their origin. The populations of Yantac, Villa Junín and to a lesser extent Tinco Cancha in northwest Junín form a group occupying (x/y) 0–0.5/0.2–1.5 (see Figure 2a). Of these, Yantac, located at the northern extreme, is clearly the most different from the remainder of the population, with Villa Junín occupying 2D space intermediate with the south Junín samples. Tinco Cancha, however, has a more central distribution, similar to the Lucanas samples (see below and Figure 3).

The populations in south Junín cluster tightly, occupying predominantly the (x/y) vector coordinate range of 0–0.5/–0.5–0: the bottom right hand section of the plot (see Figure 2b). This coordinate space is also unique among Peruvian vicuña populations, and strongly indicative of an independent demographic history for this region.

A main feature of the 2D–FCA plots is a large cluster of populations in the centre of the vector distribution. This cluster comprises all the Lucanas samples, Ayavi, Huacarpana, Toccra and to a lesser extent Ingenio in Puno (x/y range –0.5 to 0.2/–0.8 to 0.5). These results recapitulate the genetic distance dendrograms in Figure 1 and suggest that the populations from Huancavelica, Ayacucho and Arequipa form a single demographically linked group of subpopulations, with some linkage to the Ingenio population south of Lake Titicaca in Puno. Of particular note is the strong relationship between the Huacarpana population and the samples from Lucanas (also evident from the F_{ST} data) and the fact that the Toccra data must be treated with caution because of the low sample size.

Finally, the Puno population of Picotani seen in Figure 2d occupies its own 2D space (x/y = –1.5 to –0.5/–0.8 to 0.2) and is very distinctive. Picotani is situated north of Lake Titicaca and not far from the Bolivian border. Its extreme distinctiveness (on a similar scale to Yantac) clearly requires particular attention.

5. ANOMALOUS POPULATIONS

During the course of this study, clear patterns have emerged from the genetic data. However, several exceptions also emerged. First, the population in Tinco Cancha in north Junín genetically resembles the central Peruvian group much more than the northwestern Junín group (Figure 3), with a similar effect evident to a lesser extent in Villa Junín. (In Figure 3, the points located between –0.3–0.2/0–0.5(x/y) correspond to descendants of animals transferred from Pampa Galeras, while those between –0.10–0.4/0.1–1.1 (x/y) represent animals with the native northwest Junín group genotype.). Second, the sample from Huacarpana in Ica is genetically more similar to the Lucanas populations than expected from its location. It is now apparent that these populations have been affected by two large translocations from the Pampa Galeras reserve (Lucanas). The translocations have severely affected the genetic structure of the local Junín populations and future management strategies should take this into account. Other translocations from Pampa Galeras have occurred during the past 35 years. These include 722 vicuña to Laccho (Huancavelica) during 1979/1980; 40 vicuña to Cañaguas (Parque Nacional Aguada Blanca–Arequipa) in 1979; 1012 vicuña to Atoxaico (Junín) in 1980/1981; 108 and 100 vicuña to Parque Nacional Huascarán–Ancash in 1980 and 1997, respectively; 25 and 170 vicuña to Cooperativa Atahualpa (Cajamarca) in 1994 and 1997, respectively; and 95 vicuña to Toccra (Parque Nacional Aguada Blanca–Arequipa) in 1997. Our study has documented the impact of the first and third in the genome of the local recipient populations, but the influence of transfers 2, 4 and 6 remains to be studied.

Figure 3. 2D–FCA plot with samples from Tinco Cancha, Junín in black

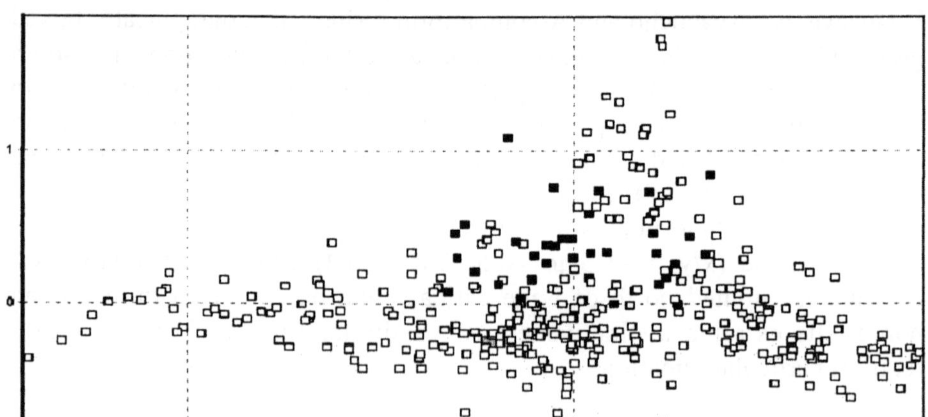

6. CONCLUSIONS AND CONSERVATION RECOMMENDATIONS

Vicuña populations in Peru seem to possess several interesting and strong genetic features, which are a result of its biology, habitat occupancy, evolutionary history and management by people in the recent past.

First, the genetic diversity of Peruvian vicuña is characterised by relatively low levels of genetic diversity within populations, but high levels of genetic differentiation between populations. Such patterns are commonly observed in threatened species with formerly large ranges which have become isolated from each other and have suffered drastic demographic contraction in recent generations (e.g., O'Ryan et al. 1998, Barratt et al. 1999). Such an effect may be predominating the genetic signal seen here in Peruvian vicuña, and should be acknowledged in future management strategies to minimise further loss of genetic diversity within individual vicuña populations.

It is especially important that future management strategies for fibre production in vicuña take account of the future genetic diversity of the populations and that the reproduction of individuals (especially males) be monitored and managed to maximise genetic diversity. Such an approach includes provision for the free movement of individuals and precludes, for example, the exclusive use of a few productive males or of active selection programs to increase fibre productivity in wild populations, since they would be likely to lead to further inbreeding and genetic drift and thereby having deleterious effects on the health of individuals and overall well–being of the population. In addition, such an approach is unlikely to lead to an increase in fibre quality either, since domestication in other fibre producing species has universally led to a significant decrease in quality. Finally, it is important to note that the vicuña has already been domesticated and is undoubtedly the ancestor of the alpaca (Kadwell et al. 2001).

However, the high numbers of private alleles and clear biogeographical structure of genetic diversity (see below) hints at an additional explanatory factor for the genetic patterns measured here. Andean animal populations are clearly occupying one of the most extreme and variable topographical environments in the world, and one where climate change, geological upheaval and changing colonisation routes must have a crucial role in shaping the genetic relationships among populations today. It may be reasonable to expect many vicuña populations to have been naturally isolated, then perhaps reconnected, and then possibly isolated again throughout much of their history. These effects, coupled with the recent anthropogenic influence on vicuña populations, make it difficult to assess how much of the genetic pattern seen here is a result of recent human exploitation and how much is a natural consequence of evolutionary history. A conservative approach to future management, however, would be to pay attention to the current patterns of genetic diversity described here, and to attempt to avoid future losses in diversity and structure.

Second, we have identified four demographically distinct groups, those in northwest Junín, south Junín, central Andes (Huancavelica to Arequipa) and Puno. Such groups could conceivably form separate management units for future demographic augmentation within and between populations. Of particular note is the extreme northwestern population of Yantac, which is genetically unique in this sample set and may represent a northern evolutionary group requiring special attention and further research. Also of note is the distinctiveness of the Puno population of Picotani, which conceivably represents a genetic group linked to the Bolivian population. It is important, therefore, to establish the genetic relationships between the Picotani population and those in Bolivia and to assess both how far this genetic group penetrates into Peru and how much it differs from the south Puno Ingenio sample. The populations in southern Junín form a clearly independent demographic unit worthy of special consideration, and the large central Peruvian population, which includes the highly diverse population at Lucanas, appears to form a large demographic management unit. However, it should be noted that the Toccra sample analysed here is not large enough to form firm conclusions.

6.1 Recommendations

1. Vicuña populations in Peru should be conservatively managed in four demographic units (northwest Junín, except Tinco Cancha; south Junín; central Andes (Huancavelica to Arequipa); Puno).

2. Genetic management should be carried out by moving individuals between populations within the same management unit to prevent further loss of genetic diversity.

3. No further translocations should be carried out from Pampa Galeras to populations outside the Central Andes management unit because of the profound effect they may have on local genetic structure (for example in Tinco Cancha).

4. Genetic management within populations requires measures to minimise further inbreeding and genetic drift and this must be taken into account management practice for fibre production. Hence the free movement of individuals within localities must be ensured. If potential barriers to free movement (e.g., fences) exist, large gaps should be kept open at all times, and these should only be closed for intensive management (i.e., chaccu).

Further, since any fencing will act as a barrier to movement, methods of intensive management should be found which will lead to the elimination of fencing in the medium term.

5. Priority should be given to obtaining samples for genetic analysis from the region of Ancash, to see if populations comprise the same "northern" genotype as exemplified in Yantac.

6. Samples should be obtained to augment the current sample from Toccra (Aguada Blanca).

7. Samples should be obtained to further investigate the genetic status of the northern Puno group, both from Bolivia and from north and west of Picotani, to establish the penetration of this group into Peru.

8. Samples should be obtained from the area south of Lake Titicaca in Puno to further investigate how closely related these animals are to the north Puno and central Andes populations.

9. Phenotypic analysis (including of fibre) should be carried out on the four sub–populations identified here.

10. A simplified, short version of this report should be given wide circulation to interested parties, especially the campesinos.

11. A similar study should be carried out when possible on the guanaco populations in Perú.

7. ACKNOWLEDGEMENTS

We would like to acknowledge the efforts of Dr. Helen F. Stanley at the Institute of Zoology for help with sampling and project design. We would also like to thank Valerie Richardson and Maria Stevens at the Darwin Initiative Secretariat, DETR London for much flexibility and help with project management. At the Faculty of Veterinary Medicine, San Marcos University we would like to acknowledge the help of Dr. Felipe San Martin, Ing. Juan Olazabal and Antonina Cano. At CONACS we would like to thank Alfonso Martinez, Jorge Herrera, Marco Antonio Zuñiga, Marco Antonio Escobar, Alex Montufar, Roberto Bombilla and all others who helped with the sampling. The research reported in this paper was carried out under Darwin Initiative for the Survival of Species grant 162/06/126.

8. REFERENCES

Barratt, E.M., Gurnell, J., Malarky, G., Deaville, R., and Bruford, M.W. 1999. Genetic Structure of fragmented populations of red squirrel (*Sciurus vulgaris*) in Britain. Molecular Ecology S12: 55–65.

Belkhir, L. 1999. GENETIX v4.0. Belkhir Biosoft, Laboratoire des Genomes et Populations, Université Montpellier II.

Benzécri, J.P. 1973. L'Analyse des Données: T. 2, I' Analyse des correspondances. Paris: Dunod.

Bruford, M.W., Hanotte O., Brookfield J.F.Y., and Burke T. 1998. Single and multilocus DNA fingerprinting. In: Hoelzel, A.R. (Ed.) Molecular Genetic Analysis of Populations: A Practical Approach, 2nd Edition, Oxford University Press, Oxford.

Franklin, I., and Soulé, M. 1980 Conservation and Evolution. Longman, New York.

Goldstein, D., and Schlötterer, C. (Eds.). 1999. Microsatellites: Evolution and Applications. Oxford University Press, Oxford.

Kadwell, M., Fernandez M., Stanley H.F., Baldi, R., Wheeler J.C., Rosadio R., and Bruford M.W. 2001. Genetic analysis reveals the wild ancestors of the llama and alpaca. Proceedings Royal Society of London B 268: 2575–1584.

Kumar, S., Tamura, K., and Nei, M. 1993. MEGA: Molecular Evolutionary Genetics Analysis, version 1.01. The Pennsylvania State University, University Park, PA 16802.

Lang, K.D.M., Wang Y., and Plante, Y. 1996. Fifteen polymorphic dinucleotide microsatellites in llamas and alpacas. Animal Genetics 27: 293.

Nei, M. 1987. Molecular Evolutionary Genetics. Columbia University Press, New York.

O'Ryan, C., Harley, E.H., Bruford, M.W., Beaumont, M.A., Wayne, R.K.Cherry, and M.I. 1998. Microsatellite analysis of genetic diversity in fragmented South African buffalo populations. Animal Conservation 1: 124–131.

Penedo M.C.T., Caetano A.R., and Cordova K.I. 1998. Microsatellite markers for South American camelids. Animal Genetics 29: 411–412.

Raymond M., and Rousset F. 1995. GENEPOP (version 1.2): Population genetics software for exact tests and ecumenicism. Journal of Heredity 86: 248–249

Saitou, N., and M. Nei. 1987. The neighbor joining method: A new method for reconstructing phylogenetic trees. Molecular Biology and Evolution 4: 406–425.

Chapter 24

SUSTAINABLE USE OF THE VICUÑA: A CRITICAL ANALYSIS AND THE MACS PROJECT

Cristian Bonacic and Jessica Gimpel
Fauna Australis, Department of Animal Sciences, Pontificia Universidad Católica de Chile, Casilla 306, Correo 22, Santiago, Chile; email bonacic@puc.cl

1. INTRODUCTION

The vicuña is a South American Camelid belonging to the infra–order Tylopoda, and family Camelidae (Osgood 1943, Wheeler 1995). Other species of South American Camelid are the guanaco (*Lama guanicoe*), which is also a wild species, the llama (*Lama glama*) and the alpaca (*Lama pacos*), both domesticated species (Bonacic 1991, Franklin and Fritz 1991, Torres, 1992, Bonacic et al. 1995). The vicuña inhabits high regions of the Andes at altitudes between 3,000 and 4,600 meters. Its range covers 9° 30' to 29° 00' latitude South (Koford 1957).

Vicuñas have been protected by the International Convention for the Trade of Endangered Species (CITES) since 1973 and by the Vicuña Convention signed by Argentina, Bolivia, Chile and Perú since 1979 (Eltringham, 1984). The Vicuña Convention was initially signed by Bolivia and Perú in 1969, and followed by Argentina and Chile in 1974. It ensured the protection of the species in national parks and private lands for future sustainable use by requiring animals to be sheared alive and returned to the wild. In Chile, the successful partnership between government agencies, local communities and international conservation agencies resulted in a rapid population recovery and the vicuña population is now classified as "out of danger of extinction" in the northern range of distribution (Parinacota Province). However, the southern populations are still in danger of extinction and little progress has been made to effectively protect them from poaching. The estimated current population size is 17–22,000 animals in Chile (99 percent of which is in the northern range) and more than 220,000 in the Andean region of Argentina, Bolivia, and Peru (Martinez, 1996, Rendon 1998, Bonacic et al. 2002). The ban on the trade of vicuna fibre has been lifted to allow shearing, and a program of sustainable use has been promoted. Since it was stated in the Vicuña Convention that protection would be followed by sustainable use for the local people's benefit, vicuña conservation is now inextricably linked to sustainable use. In 1997, the CITES authorized Perú to sell three tonnes of stockpiled vicuña fiber and to develop a trade in raw fleece (CITES 1997, Bonacic et al. 2002). As a result, the strict controls formerly imposed on vicuña wool harvest were relaxed, allowing international trade in wool sheared from live vicuñas. Currently, Perú is harvesting fiber from wild vicuñas maintained in extensive plots and Bolivia and Chile are developing capture methods and harvest systems for future large scale exploitation (Wheeler and Hoces 1997).

This paper describes the current status of the vicuña's sustainable use status in the region. In addition, we identify research needs for the future and we describe the main priorities set for the "Manejo de Camelidos Sudamericanos Silvestres" (MACS) project for the coming years. Finally, we discuss the importance of the animal welfare factor when a species is managed in a nonconsumptive way and how to include perceptions about animal welfare into concepts of sustainability.

1. METHODS

A description of the vicuña management in Chile is summarized, followed by a comparison of management in other countries. Conflicts and uncertainties for the long–term conservation of the species are analyzed based on a SWOT (Strengths, Weaknesses, Opportunities, and Threats) analysis. The SWOT analysis is used in business science to gather essential information before management decisions are taken (Miser 1995). This analysis classifies different components of a problem into standard categories to identify the critical aspects of the problem. It is the first step taken before quantitative models are developed for a particular management problem (Rea and Kerzner 1997). We used SWOT analysis by assuming that the sustainable use of the vicuña could be studied from an entrepreneurial perspective. The meaning of "business" is used here in a broad context where success can be evaluated from different points of view (ecological, economic, or social). Data and information from local communities, such as policies and ways of utilization of the species were collected between 1995 and 1998 in Bolivia and Chile through fieldwork, training courses with local communities, interviews with government officers, and literature reviews. No fieldwork was carried out in Perú, and only official information presented from papers in recent conferences was used.

1. RESULTS AND DISCUSSION

1.1 Conservation and Management

Two geographic subspecies of vicuña have been described in Chile, the Northern vicuña, *Vicugna vicugna mensalis* (18° 45'S to 19° 00'S) and the Southern vicuña, *Vicugna vicugna vicugna* (19° 00'S to 27° 30'S) (Bonacic et al. 2002). The Northern vicuña is not in danger of extinction while the Southern vicuña is endangered. The distribution between both subspecies corresponds to a continuum of scattered and fragmented groups that are less frequent from North to South. Their survival in this habitat is due in part to their highly territorial social organization and patterns of habitat use, and also to their physiological and anatomical adaptation to conditions of high altitudes and cold weather. The Vicuña Convention originally identified three main issues that needed to be considered for the protection and sustainable use of the species: (1)population recovery, (2)ecosystem conservation, and (3)social benefits from the use of the species.

The vicuña, along with other South American Camelids, is one member of a group of animals that produces rare fibres. The vicuña has a double coated fleece and is one of the most valuable and highly prized sources of animal fibre. The fine fibres are the commercially important part of the fleece and since only two percent of the fleece is made up of the coarser primary hairs, it is highly prized. Chile auctioned the first 20 kilos of fibre collected from animals captured for farming in May 2002. It is estimated that a total yield of 250 g of fine fibre could be obtained from each animal by shearing every two years. Considering a life span of ten years in the wild a potential total of one kg of fibre could be harvested from an adult vicuña.

Current management policies for vicuña include: (1)capture and shearing of wild animals, (2)captivity and farming, and (3)translocation and reintroduction. The current trend in Chile is to promote capture and extensive captive herds in the altiplano or Andean puna for commercial use. The consequences of captivity and possible future artificial selection in vicuñas are not known. Hence, from the conservation point of view three main aspects should be studied to evaluate the conservation value of in situ captive herds programs: (1)the genetic impact of artificial selection on the captive vicuñas and the positive or negative effects that this practice may cause on wild populations, (2)the risk of cross species disease transmission between livestock and enclosed wild vicuñas and the likelihood of disease transference to wild populations from accidental escapes of captive vicuñas, and (3)the impact of habitat use restriction caused by captive farming considering that vicuñas are adapted to move and select habitats in a very fragile ecosystem.

1.1 SWOT Analysis

The SWOT analysis identified five main issues that need to be studied. Ecosystem, population and socioeconomic factors were already suggested by others in previous studies (Hoffmann et al. 1983, Cueto et al. 1985, Marconi and Eyzaguirre, 1997). We also have included two new factors in our analysis: (1)economic globalization, and (2)animal welfare. Table 2 lists the main components of the SWOT analysis for the vicuña sustainable use program in the regions discussed in this case study.

The first factors to be considered were density dependent (e.g., population growth rate inverse to population size) and density independent factors (e.g., rain and primary productivity influencing carrying capacity) at the ecosystem or population levels. Both density dependent and density independent factors may regulate the current vicuña population structure (Bonacic 2000). In addition, competition with livestock can play a major role in the carrying capacity available for vicuña. Bonacic (2000) describes how the total vicuña population size in the Parinacota province of Chile began to fluctuate around an estimated carrying capacity available when domestic livestock numbers were equated to the potential primary productivity of the area.

The social and economic characteristics of each country are different owing to the different organizations of human communities (Hoffmann et al. 1983, Cueto et al. 1985, Hoces, 1996, Eyzaguirre and Velasco 1997). Local communities are well organized and in high numbers in Bolivia and Perú, while in Chile they are small and scattered with poor social organization. In addition, the impact of vicuña exploitation on a community's economy cannot be modeled uniformly for the three countries because there are differences of resource availability and levels of poverty. While vicuña population size is small in Bolivia (33,844 animals) and Chile (27,000), the total Peruvian vicuña population is estimated at over 102,000 (Eyzaguirre and Velasco 1997, Hoces 1999, Bonacic 2000). However, the levels of poverty are higher in Bolivia and Perú compared to Chile (Cueto et al. 1985,

Gonzalez et al. 1991, Eyzaguirre and Velasco 1997, Galaz and Bonacic 1998).

Table 2. SWOT (strengths, weaknesses, opportunities, and threats) analysis for the sustainable use of the Vicuña in South America.

Strengths

- Vicuña population increases within its range of distribution.
- Appropriate legal framework (CITES and Vicuña Convention).
- Humane capture and shearing methods.
- First sales with high price of the product and international market available.
- World–wide image of vicuña fiber as premium product.
- Label of ecological sustainability and non–consumptive use of wildlife.

Weaknesses

- Low production volume and lower quality of product than expected.
- Strong competition from other fine fibers (e.g. cashmere).
- No clear regional common policy about "sustainable use" of the vicuña.
- Increasing poaching.
- Few buyers and North American market is not open.
- Scientific controversy about the sustainability of management in the wild towards captive–management.
- No clear concern about animal welfare in the region.

Opportunities

- To improve income of poor local communities.
- To replace domestic livestock with wildlife species and protect the environment.
- To change attitudes of local people towards the vicuña.
- To maintain a successful conservation program.
- To provide a new fine fiber for the international market.

Threats

- Uncertainty about price stability.
- Poaching.
- Unclear legal framework to distribute the benefits.

Another primary factor is the potential market for products and a common international price for products. The commercial value of the fiber may vary when Bolivia and Chile enter the market because the amount of fibre produced might increase and potentially lower prices. The globalization of the economy could threaten the original aim of the program by increasing the demand for vicuña fiber beyond natural production limits. As a consequence, poaching may increase to meet the demand and drive the population towards extinction. Conversely, global fashion trends and interest in animal products change and prices could decline below a commercially viable threshold for exploitation. Lastly, competition from countries such as China with cashmere products and genetic improvement of domestic species (sheep, alpaca) may make vicuña fiber less desirable than in the past.

The increasing concern about animal welfare and the use of wildlife has been broadly discussed around the world. Western cultures (the main market for the vicuña fiber) could reject any exploitation of the species. The ethics behind sustainable use of wildlife is still a controversial topic (Taylor and Dunstone 1996). Therefore, the welfare of the vicuña is an understudied but very essential issue that must be added to the three main issues initially proposed by the Vicuña Convention. The aim of the Convention shifted from protection (1979–1990) to sustainable use (1990 onwards) by shearing wild vicuña in situ. However, considerations about animal welfare have diminished (Bonacic and Gimpel 1995). Although shearing seems less intrusive than an alternative program based on culling, it does involve procedures that may stress the animals and even lead to mortality. Bonacic and Macdonald (1999) has addressed some of the main issues about animal welfare by implementing experimental research looking at the effects of capture and shearing in the natural habitat of the species.

The vicuña represents a special case of nonconsumptive use of wildlife, where captured animals are sheared alive and returned to the wild with the assumption of little or no impact. However, the effect of vicuña exploitation may not just be at the level of the individual. Potentially, consequences may extend to the whole population as well as to the environment. Whereas consumptive sustainable use of wildlife such as culling affects the individual (death) causing changes to the population (i.e., changes in mortality rate), in the case of the vicuña, changes in the individual could be more subtle and a consequence of the response to stress (e.g., reproductive failure or considerable behavioral changes). Therefore, the sustainable use of the vicuña raises concerns about animal welfare as well as about conservation (Bonacic and Macdonald 1999, Bonacic 2000, Bonacic and Gimpel 2001).

The emphasis of sustainable use is on population viability, environmental protection and economic benefits, with little or no interest in the welfare of individual animals. However, it is not reasonable to neglect concern about individuals when there is a prospect of exploitation of large numbers within a population, especially when there has been a sustained effort to reach an "out of danger" status. It is thus essential to study the effects of the different management practices at an individual level in order to have the elements to assess potential population consequences. This will allow the design of a thoroughly planned management system that will be truly sustainable. In terms of vicuña management, four key aspects can be a matter of animal welfare concern: capture, transport, shearing and the overall effect of captivity.

2. THE MACS PROJECT: SUSTAINABLE ECONOMIC UTILISATION OF WILD SOUTH AMERICAN CAMELIDS

The MACS project provides information for the design and operation of the systems used for the exploitation of fibre from the wild camelids; this information is important as policy moves from being conservation orientated to allowing limited exploitation. The MACS project is funded by the European Union and began in late 2001. It will run for 4 years with two main aims: (1)to develop guides for best practices in systems of sustainable economic utilisation of wild South American camelids in Argentina, Bolivia, Chile, and Peru for the production of luxury textile fibres; and (2)to disseminate this information to producers, the European textile industry, rural communities, and regional policy makers both through publications and by establishing a network of researchers and experts.

The main project activities are organised around five semi–independent "work packages." Each work package represents a collaborative research venture between a research team in Latin America and a leading research institution in Europe, and each covers the range of scientific disciplines addressed by the project. The work packages focus on: utilisation systems, economic and social impacts, environmental impacts, biodiversity conservation, and networking and dissemination. The first work package focuses on assessing health, animal welfare and product quality as well as the impacts of management options for vicuña and guanaco at a system level, and the development of recommendations for best practices. The second work package evaluates the economic viability of management options for wild South American camelids and trade in their fibres. It also helps to identify policies to encourage appropriate management approaches and equitable distribution of benefits. The third work package evaluates significant environmental interactions and maps ecosystem indicators, such as carrying capacities. The biodiversity conservation workpackage aims to develop management strategies which will minimise the impact of sustainable utilisation on wild camelid populations. Finally, the networking work package aims to consolidate all the information arising from the project. It focuses on bibliography, networking, conferences, and training courses for teachers and community representatives on local ecology, conservation and development options. A strong emphasis within MACS is placed on adding appropriate high quality data to the existing sum of knowledge, and then using this through networking and dissemination to achieve action by consensus, education and influence on policy.

3. CONCLUSIONS

The vicuña conservation program in South America is entering a critical stage, and the direction that the program takes could conflict with the original objectives of the Vicuña Convention (i.e., to use the species in the wild for local communities' benefits). The trend for exploiting in captivity by fencing wild vicuña or captive programs is not compatible with the initial criteria of sustainable use. Captivity may promote artificial selection and natural traits could be lost in the process. Global economics and increasing concern for animal welfare threatens the long–term program viability in different ways. The impact of how global demand could stimulate poaching and how animal welfare campaigners would perceive the use of the species remains to be investigated. A sustainable use program based on capture and release of small groups of vicuña with minimal interference on the natural populations should be considered the method of choice, to which other methods should be compared. Some of the main aspects to compare between methods of management are: (1)the impact of capture, handling and shearing on different systems of use; (2)the comparative studies of carrying capacity in different regions of the altiplano and livestock competition; and (3) the social and population consequences of captive programs.

4. ACKNOWLEDGMENTS

We thank Jerry Laker, Ruth Feber, Laura Bonessi, Jay Gedir and Jenny Bryce who made useful comments on the manuscript. People from the following local communities were interviewed: Ulla–Ulla, Hichocollo, Puyo Puyo, Cañuhuma, Huacuchani, Modullani, Antaquilla, Cololo–Altarani, Ucha–Ucha y Callumpa–Medallani in Bolivia and leaders from Aymara communities from Putre and General Lagos in Chile. Park rangers, local Government officers and senior government officers were also interviewed in both countries. We would like to thank the financial collaboration to our research of the following entities: Care for The Wild, Universities Federation for Animal Welfare, People's Trust for Endangered Species, Quiborax, Johnston of Elgins. British Airways, the British Council and the Chilean Agency for International Cooperation and the European Union (ICA4-2000-10229).

5. REFERENCES

Bonacic, C. 1991. Características biológicas y productivas de los Camélidos Sudamericanos. Avances en Ciencias Veterinarias 6: 3–15.

Bonacic, C. 2000. Population dynamics of the vicuña (*Vicugna vicugna*) and estimation of carrying capacity. Pages 93–101 In: Gonzalez, B., Bas, F., Tala, C. and Iriarte, J.I. (Editors) Proceedings Sustainable Management of the Vicuña and Guanaco. Servicio Agricola y Ganadero and Pontificia Uiversidad Catolica de Chile, Santiago.

Bonacic, C., Donoso, G., and Bas, F. 1995. The guanaco in the southern Chilean Patagonia: Research needs for a sustainable use. Pages 512–515 In: Bissonnette, J. and Krausman, P. (Editors) Integrating People and Wildlife for a Sustainable Future. Proceedings of the First International Wildlife Management Congress, The Wildlife Society, San Jose.

Bonacic, C and J Gimpel, 1995. Sustainable use of South American Camelids: Theory and practice. Fine Fibre News 5: 23–26.

Bonacic, C and J. Gimpel. 2001. The response to productive management in the wild vicuna (Vicugna vicugna) and its animal welfare consequences. Pages 67–76 In: Simposio Internacional en Produccion Animal y Medioambiente, 25–27 Julio 2001. 558 pages.

Bonacic, C and D.W. Macdonald. 1999. Preliminary physiological assessment of the effects of capture for shearing on vicuna (*Vicugna vicugna*). Pages 34–39 In: Progress in South American camelids research. EAAP publication No 105, 2001. 346 pages.

Bonacic, C., Macdonald, D.W., Galaz, J. and Sibly, R.M. 2002. Density dependence in the Camelid Vicugna vicugna: The recovery of a protected population in Chile. Oryx 36: 118–125.

CITES. 1997. Convention on International Trade in Endangered Species of Wild Fauna and Flora: Appendices I and II and Interpretation (Fauna and Flora). Vol. 1997 CITES.

Cueto, L., Ponce, C., Cardich, E., and Rios, M. 1985. Management of Vicuña: Its contribution to rural development in the high Andes of Peru. Food and Agriculture Organization of the United Nations, Rome. 38 pages.

Eltringham, S.K. 1984. Wildlife resources and economic development. John Wiley, Chichester. 328 pages.

Eyzaguirre, J., Velasco, A. 1997. Censo Nacional de Vicuñas en Bolivia. Direccion Nacionald e Conservacion de la Biodiversidad, La Paz. 60 pages.

Franklin, W., and Fritz, M. 1991. Sustained harvesting of the Patagonia guanaco: Is it possible or too late? Pages 317–356 In: Robinson, J. and Redford, K. (Editors) Wildlife Sustainable Use in the Neotropics. The University of Chicago Press, Chicago.

Galaz, J. and Bonacic, C. 1998. Informe final proyecto: Plan piloto de aprovechamiento de la fibra de vicuña. Pages 24–34 In: El Altiplano de la Provincia de Parinacota Analisis de Restricciones y Condiciones de Deasrrollo del sistema de Manejo. Corporacion Norte Grande, Arica.

Gonzalez, H., Gundermann, H., and Rojas, R. 1991. Diagnostic y Estrategia de Dasorrollo Campesino en la I Region de Tarapaca, Arica, Chile. Impresos Publicitarios, Iquique. 246 pages.

Hoces, D. 1996. Ecologia de la especie. Page 74 In: Galaz, J. (Editor) Aprovechamiento de la Fibra de Vicuña en Los Andes de Argentina, Bolivia, Chile y Peru. Corporacion Norte Grande and CONAF, Arica.

Hoces, D. 1999. Uso sustentable actual de la vicuña. Page 79 In: Gerken, M., Reinieri, C., and Bransmann, S. (Editors) 3rd European Symposium on South American Camelids and Supreme European Seminar. Universitat Gottingen, Gottingen.

Hoffmann, R., Otte, R., Prado, K.C., Rios, and F.P.D. 1983. El Manejo de la Vicuña Silvestre. GTZ, Eschborn. 705 pages.

Koford, C. 1957. The Vicuña and The Puna. Ecological Monographs 27: 152–219.

Marconi, M., and Eyzaguirre, J. 1997. Reglamento Para La Conservacion y Manejo De La Vicuña: Guia Practica Para Su Aplicacion. Direccion Nacionald e Conservacion de la Biodiversidad, La Paz. 40 pages.

Martinez, A. 1996. Legislacion en Peru. Page 74 In: Galaz, J. (Editor) Aprovechamiento de la Fibra de Vicuña en Los Andes de Argentina, Bolivia, Chile y Peru. Corporacion Norte Grande and CONAF, Arica.

Miser, H. 1995. Handbook of system analysis. John Wiley & Sons Ltd., New York. 361 pages.

Osgood, W.H. 1943. The mammals of Chile, Vol. 30. University Chicago, Chicago. 268 pages.

Rea, P..J., and Kerzner, H. 1997. Strategic planning: A practical guide. John Wiley & Sons, Ltd, New York. 326 pages.

Rendon, O. 1998. Experiencia Boliviana en el manejo comunal de la vicuña. Page 38 In: Gonzalez, B. and Zapata A. (Editors) Concha Manejo Sustentable De La Vicuña y El Guanaco. Pontificia Universidad Catolica de Chile, Santiago.

Taylor, V. and Dunstone, N. 1996. The exploitation of mammal populations. Chapman & Hall, London. 415 pages.

Torres, H. 1992. South American wild Camelids: An action plan for their conservation. IUCN, Glandz.

Wheeler, J., and Hoces, D. 1997. Community participation, sustainable use, and vicuña conservation in Peru. Mountain Research and Development 17: 283–287.

Wheeler, J.C. 1995. Evolution and present situation of the South–American Camelidae. Biological Journal of the Linnean Society 54: 271–295.

Chapter 25

SUSTAINABLE USE OF ANDEAN WILDLIFE AND LOCAL DEVELOPMENT OF RURAL COMMUNITIES IN DRY AREAS OF LATIN AMERICA: A COMMENTARY

Bernardo Peredo Videa
Consultant Horizon Solutions Site, Bolivia, IPRAVI, Braulio Vera 6648, La Paz, Bolivia; email berbolivia@yahoo.com

1. INTRODUCTION

Due to its relatively sparse population, lack of development and diverse geography, Bolivia is one of the best places in South America for conserving wildlife. However, the use and commercialization of various native wildlife species without proper management results in diminishing the prospects of conserving those species.

Local communities with high rates of poverty living in the arid and semiarid regions in Bolivia often over–utilize wildlife species because of lack of valid alternatives for local development, i.e., few opportunities for satisfying basic needs and improvement of their quality of life. It is crucial to build bridges and promote best practices that will allow conservation and sustainable use of natural resources as well as development of local communities in arid and semiarid regions in Bolivia.

The vicuña (*Vicugna vicugna*) is the smallest representative of the South American Camelids that inhabit the Andean altiplano or puna ecosystem. Estimates of total vicuña population in the 16th century exceeded several million; this number had declined steeply during the 20th century as a result of hunting. A census in the late 1960s showed that there were about 10,000 vicuñas in Perú, but fewer than 2000 vicuñas in Bolivia and Chile. However, recent successful population recovery programs have begun to shift the emphasis away from conservation towards the sustainable use of the species. For example, increasingly studies are demonstrating the benefits of sustainable use of the vicuña in order to maintain the high quality of its fibre.

The quirquincho (*Chaetopractus nationi*) is a species endemic to Bolivia and northern Chile and is an important part of the culture of the Andean zone of Bolivia. Currently, the high use of the quirquincho has increased the threat of extinction for the species. The Andean Ostrich (*Pterocnemia pennata*), extensively used for traditional and cultural dances and costumes, also runs the risk of being over–utilized.

Because all of these three species are used by local communities in traditional ways, their management represents both challenges and opportunities for multidisciplinary and local participatory approaches to achieve sustainable conservation. This paper discusses how a combined ecological and socioeconomic approach involving three sectors is necessary to develop sustainable use programs for the three species. The first sector comprises rural communities who utilize the species for traditional costumes and cultural activities involving medicinal, handicraft, and religious purposes. The second sector utilizes the species for making some of the Andean region's musical instruments, such as charangos and matracas and costumes. The third sector uses the species to make souvenirs, amulets, and other materials and instruments for tourists and for export to other nations.

Despite the fact that Bolivia is a poor country, it is culturally and ecologically diverse. The Bolivian Highlands (Puna and inter–Andean Valleys), with an average elevation of 3,500m. occupies approximately a fourth of the country and stretches over areas of La Paz, Oruro, Potosí, and parts of Tarija, Chuquisaca and Cochabamba. The indigenous population consists of the two linguistic groups, Quechua and Aymara who for the most part practice communal land tenure and cultivation methods with a heavy dependence on livestock production and rainfed crop production. Although crop production is the most important source of livelihood it is also a risky endeavor because of frequent droughts. Accordingly, livestock plays a crucial role in these systems because they are the only perennial crop that farmers have.

However, the sectoral uses of vicuña, quirquincho, and the Andean Ostrich lack adequate formal planning or regulatory control. As a result, there is a need for higher government to work jointly with the local communities to find sustainable management alternatives, particularly in the case of the most utilized wildlife species. This will allow local people to derive the cultural and economic benefits of using the species without depleting them.

2. QUIRQUINCHO

The quirquincho or hairy armadillo (*Chaetopractus nationi*) is endemic to Bolivia and northern Chile and is an important part of the cultural folklore of the Andean zone of Bolivia. The quirquincho is threatened by the expansion of agriculture, habitat modified for human activities, and the use and commerce of the species for the fabrication of musical instruments (e.g., matracas, charangos).

Approximately 1950 specimens of quirquincho are removed from the wild each year for use in the manufacture of traditional instruments, handicrafts and souvenirs (including stuffed specimens), charms, traditional medicine, and food (Caceres 1995). The Ministry of Sustainable Development estimates that without regulation of use the quirquincho will face extinction in Bolivia (DNCB 1997). There also is evidence of increasing export of quirquincho to the USA and Europe. For example, groups of dancers in local festivities related to the Carnival of Oruro and other similar traditional festivities use matracas with quirquincho shells. Those groups have 150_200 dancers, and each one carries a matraca as part of the traditional dances and clothes. Based on conversations with local artisans, they affirm a trade of musical instruments made with quirquincho shells to border countries. This trade is an important source of income for local artisans but it also is illegal and because of this sometimes the instruments are seized by customs. When this happens the local artisans suffer a loss of important income.

3. ANDEAN OSTRICH

The Andean Ostrich or Suri (*Pterocnemia pennata*) occurs only in South America from northern Chile, Bolivia and Peru to southern Argentina. Several factors contribute to the decrease in Suri populations: hunting for feathers for use in traditional handicrafts, costumes, and for souvenirs for tourists; gathering of eggs for local consumption; predation by cougars and dogs; and overgrazing of habitat by other domestic animals (DNCB 1997).

4. VICUNA

Vicuñas inhabit the Andean altiplano or puna, a high altitude arid grassland ecosystem between 3000 and 4600 metres above sea level. Their current geographical distribution extends from 9° to 29° LS in South

America and covers parts of four different countries: Argentina, Bolivia, Chile and Perú.

The vicuña has a social organization with family groups consisting of a territorial male, several females (2 to 7), and the young of the year. Vicuñas inhabit year–round territories that are defended vigorously by the male. The excess males form large bachelor groups that are not territorial. Solitary animals, usually males, are also found. Vicuñas are well adapted to their habitat, which is characterized by large daily temperature fluctuations with most nights below freezing, intense solar radiation and low levels of oxygen. The fine fleece of vicuñas protects them from the extremes of temperature and solar radiation, while their oval erythrocytes are adapted to pick up oxygen at low pressures.

The wool of vicuñas is much sought after because of its fine fibers (13 to 15 micron in diameter) and because it has long been used in high fashion markets. The demand for this wool has led to mass killings of the vicuñas because of the difficulty of capturing them alive.

Despite increasing demand for the vicuña, their population numbers have increased in the past 20 years going from endangered status in 1969 to vulnerable in 1972. This change is a direct result of conservation programs (Wheeler 1995). According to Bonacic et al. (in press), the present population levels for vicuñas (total of 223,850 individuals) are estimated at: Argentina, 23,000; Bolivia, 39,000; Chile, 19,850; and Peru, 142,000.

In the early stages of vicuña conservation programs, sustainable use was encouraged as a means for conservation of the species. The vicuña is a unique example of a long–term conservation strategy where the government, local communities and international agencies co–coordinated efforts and funding to reduce the chances of extinction as well as to increase population numbers. Consequently, the vicuña is now out of danger of extinction and can be managed in a sustainable use program. While vicuña population size is small in Bolivia (approximately 33,000–34,000 animals) and Chile (approximately 27,000 animals), the total Peruvian vicuña population is estimated at over 142,000 animals (Bonacic 2000).

In Bolivia, vicuña were initially under Appendix II of CITES with a zero harvest quota but now have a commercial quota. Bolivia is starting a research program and developing a mobile capture system that could be used in different communities around the high plains. The capture system is following strict animal welfare standards and is committed to maintaining the species as wild animals with the minimum of human intervention.

The Bolivian government is organizing local communities to capture vicuña on their common land. Bolivia successfully appealed for a change of status in the last CITES Convention (2000) to allow the country to enter the fiber market. Currently, the Bolivian government is planning more experimental captures throughout the range of the vicuña and training for people in each community. Poaching is still frequent and an illegal market for vicuña products still exists, and public involvement in the management of vicuña is in its early stages. The Bolivian government has not yet made commercial contacts abroad in support of a regulated export market. However, there is a strong internal pressure to produce local vicuña products and to involve local knitters in manufacturing the fiber.

Local communities are numerous and well organized in Bolivia and Perú, while in Chile they are small and scattered with less social organization. Levels of poverty are higher in Bolivia and Peru and this can increase the exploitation of the species. Consequently, the impact of vicuña exploitation cannot be modeled uniformly for the three countries.

The globalization of the economy could threaten the original aim of vicuña conservation or sustainable use programs. Globalization could increase the demand for vicuña fiber beyond natural production limits. As a consequence, poaching may increase to meet the demand and drive the population towards extinction. Conversely, global fashion trends and interest in animal product change and prices could cause populations to decline below a commercially viable threshold for exploitation. Lastly, competition from countries such as China with cashmere products and genetic improvement of domestic species (sheep, alpaca) may make vicuña fiber less desirable than in the past.

5. SOCIAL, ECONOMIC, AND CULTURAL CONSIDERATIONS

Most of the consumers of vicuña, quirquincho, and Andean Ostrich wildlife are local people who share many of the same customs and traditional costumes and geographical territory, but who also share a common vulnerability to their welfare because of high rates of poverty. Consequently, these people are target groups insofar as government programs to improve opportunities for local development based on the sustainable use of Andean wildlife.

There are also other groups of people that migrated to urban cities but who share many of the same customs and traditional costumes of people in rural communities. Still other groups of people who have migrated to urban areas do not use traditional costumes but use the species for different commercial activities, especially in handicrafts and musical instruments, e.g., in Bolivian folklore. Finally, there are tourists who visit local markets and purchase typical handicrafts and souvenirs thereby providing a source of income for local artisans. All of these groups are important in the formal and informal economy of utilized wildlife species.

6. CONCLUSION

There are important socioeconomic needs that have to be met in order to promote best practices for conservation of biodiversity in arid and semiarid zones. These include: (1)improving income sources and benefits to local inhabitants; (2)developing pilot projects with short–term results and result–based management approaches; (3)implementing multidisciplinary approaches and programs for sustainable use of wildlife; (4)improving local capacity building and participation in project management and decision making processes; (5)creating sustainable use rather than restrictive policies for local communities that will generate income for local people; (6)developing communication and consensus among stakeholders to reduce poor coordination; and (7)implementing development plans involving sustainable use of natural resources and particularly wildlife management programs for local development of rural communities in arid and semiarid zones. In particular, given the similar concerns and issues shared by Bolivia, Chile, and Perú regarding the three Andean species discussed in this paper, it would appear that prospects for South–South cooperation could be developed.

7. REFERENCES

Bonacic, C., Gimpel, J. and Galaz, J.L. In press. Current advances in research and management of the vicuña (*Vicugna vicugna*) in Chile. Ciencia e Investigación Agraria.

Bonacic, C. 2000. Population dynamics of the vicuña (*Vicugna vicugna*) and estimation of carrying capacity. Pages 93–101 In: Gonzalez, B., Bas, F., Tala, C., and Iriarte, J.I. (Editors) Proceedings Sustainable Management of the Vicuña and Guanaco. Servicio Agricola y Ganadero and Pontificia Uiversidad Catolica de Chile, Santiago.

Cáceres, F. 1995. Actual Status of Quirquincho *Chaetopractus nationi*, Andean Species Endangered. Consultant for DNCB, La Paz, Bolivia.

(DNCB) Ministry of Sustainable Development and Environment. 1997. Proposal for Inclusion of Quirquincho *Chaetopractus nationi* into Appendix II de CITES. National Conservation Directorate of Biodiversity, La Paz.

Wheeler, J.C. 1995. Evolution and present situation of the South American Camelidae. Biological Journals of the Linnean Society 54: 271–295.

Chapter 26

CONFLICTS AND DILEMMAS BETWEEN POVERTY AND BIODIVERSITY IN THE SEMI–ARID SERIDÓ OF NORTHEAST BRAZIL

Eleonora Tinoco Beaugrand
Department of Social Sciences, Universidad Federal do Rio Grande do Norte, Natal, Brasil; email nora@natalnet.com.br

1. INTRODUCTION

The following article, which is based on the Sustainable Development Plan of Seridó (SSDP), analyzes parts of the environmental and economic aspects of the plan. The SSDP was a pioneering and innovative experience of strategic participative planning in the State of Rio Grande do Norte, Brazil. The SSDP, which began in September 1999 and concluded in September 2000, involved various state agencies, organizations and representatives from civil society during its preparation. The SSDP was prepared by the Inter–American Institute of Agricultural Cooperation and the Rio Grande do Norte State Government. The initial impetus for the SSDP came from the requests of the regional political, entrepreneurial, unionist and religious leaderships made to representatives from the federal and state governments for decision making on the economic crisis which had been going on in the region since the 1980s. This crisis entailed severe environmental effects apart from the economic and social consequences, and these effects were intensified by the prolonged dry seasons which occur cyclically in the northeastern semi–arid region, including the Seridó region.

Conceptually speaking, the SSDP adhered to the Brundtland Commission's definition of sustainable development: "...the one that fulfills the needs of the present without jeopardizing the capacity of the future generations to fulfill their own needs...." (CMMAD 1987). According to Buarque (1999) in a wider sense: "...local sustainable development is the process of social change and growth of opportunities of the society, rendering compatible in time...and in space, the economic growth and efficiency, environmental conservation, life quality and social equity, starting from a clear commitment towards the future and solidarity between the generations." It was with this view of local sustainable development, where popular participation is an educational process and development promoter, that the SSDP was conceived. The methodology of (situational) strategic planning for the SSDP was based on techniques of Matus (1996) and Buarque (1999).

This case study presents a descriptive analysis of the problems and prospects of the Seridó which ultimately will have to be dealt with prior to implementation of an effective SSDP.

2. DESCRIPTION OF THE REGION

The Seridó region of Rio Grande do Norte is 12,965 square kilometers in area, which represents 24.3 percent of the total area of the state of Rio Grande do Norte, whose area is 53,306 square kilometers. The Seridó is situated in the semi–arid meridional region of Rio Grande do Norte (90 percent of the state territories are situated where the soils are shallow and stony and have a low fertility. The semi–arid climate, low rainfall, and periodic droughts, combined with human activities in the region which cause deforestation, forest fires, the discharge of pollutants into area reservoirs, soil erosion, and environmental degradation has been increasing.

Although it represents a fourth of the state's area, the Seridó has an approximate population of 289,700 people, representing 11.2 percent of the state population. Demographically, there was a small population growth between 1991–1996 on the order of 0.33 percent per year.

The economy of the Seridó is based on cotton, cattle raising and mining (e.g., extraction of scheelite). Two of these sectors have been economically depressed since the 1980s as a consequence of various factors. The cultivation of arboreal cotton has been affected by the opening of trade for cotton imports, by the "bicudo" (cotton) pest, and by the growing substitution of cotton by synthetic fibers. Further, other regions from the state have started producing herbaceous varieties of cotton whose fiber is smaller; however, they are highly productive when cultivated with modern technology (Beaugrand et al. 2001).

The mining economy has been depressed due to increasing competitiveness of other producers, especially the Chinese. This has destabilized one of the major economic sectors of the region which is responsible for the growth and development of the main regional urban centers. In part, to compensate for economic problems of the region some governmental programs have been implemented and have contributed to the increased production of other agricultural products such as milk, sheep, goat, fish (farming), corn, manioc, cashew, and fruits. In addition, programs to foster urban activities have been developed. These include small industries based on craftwork, the ceramics industry, bakeries, caps and textile industries, shoes industries, and small establishments in the food, beverages, and services sectors.

Despite whatever economic gains which have been made, many of these economic activities have contributed to the environmental degradation of the area: deforestation, extensive agriculture and cattle raising, firewood for domestic and industrial use, mining activities, forest fires, use of agrochemicals and fertilizers, the lack of management practices and soil conservation in economic exploitation, soil degradation caused by the removal of materials for ceramics, and extensive cattle raising.

In 1991, the Seridó went through a severe social and economic decline compared to the state as a whole. The average Human Development Index (HDI) of the Seridó was 0.481, while the average of HDI for Rio Grande do Norte was approximately 0.62. The family income of the Seridó is low (about $USD 70): 72 percent of the population had a per capita family income which is equal or lower than half a minimum wage. Income levels were the worst in the Serras Centrais geographic zone, where low income levels affected 86.24 percent of the population. The issue of unemployment also prevails in all areas of the Seridó. Despite the social and economic hardships of the Seridó, the region has a cultural identity which is receptive to the development of new economic options and, above all, to gather the social energies required for a regional development project.

2.1 The Seridó Region and the Environment

Environmental quality and prospects of sustainable development in the Seridó are threatened by the development of some economic sectors, most importantly the ceramics industry. Following, the Seridó's ecosystems, water resources, soil resources, forest resources, demands for forest–based fuels, mineral resources and environmental changes (especially desertification) are described.

2.2 The Ecosystems of the Caatinga Region and the Northeast Deciduous Forests

Caatinga is a natural region which integrates the northeastern semi–arid region, together with Sertão, Seridó, Agreste, Curimataú, Carrasco and

Serras. Within the caatinga, there are locations that can be used for traditional or technological agricultural production, as well as areas suitable for small size irrigation projects. Identifying suitable locations requires scientific and appropriate management of soil and water resources in order to avoid degradation and desertification. With respect to promoting the SSDP, because much of the caatinga already has been exploited due to mining, the ceramics industry, and other activities within extractive patterns, careful management is required. For example, the SSDP concluded that there are increasing pressures on use of the caatinga lands and forests as a source of energy.

Some of this pressure to make greater use of caatinga lands and forests stems from the need for jobs and the pressure from urban and rural entrepreneurs who deal with timber cutting and coal production (Mendes 1997). The economic exploitation of forest resources is responsible for 29 percent of the income generated by the small rural producers of Rio Grande do Norte, therefore fostering the continuity of those activities (Ibama 1998). Currently, the industries that are causing the greatest risks to the preservation of biodiversity of the Seridó are the ceramics, vegetable oil, soap, and bakery industries. The red ceramics industries are situated in the valleys of intermittent rivers and their wood harvesting contributes to the depletion of the woods surrounding these watercourses. The ceramics industries also use as raw material the alluvial soils from the riverbanks, which are the richest soils in the Seridó (Mendes 1997).

2.3 Water Resources

One of the characteristics of the Seridó is that resources are renewable at slow rates. The amount of available water resources of Rio Grande do Norte State and the Seridó is the same as the capacity of the dams in the region. The state's dam capacity used to be 3.495 billion cubic meters of water. More recent studies carried out by the Department of Water Resources of Rio Grande do Norte reveal that the volume was increased by 3.592 billion cubic meters by 1995 (Serhid 1999).

2.4 Soil Resources

Productive soils of the Seridó also are becoming scarce due to over–utilization and soil erosion (Carvalho 1988). Due to the decrease of the production of cotton which occurred in the 1980s, the soils of Seridó (which typically are not very fertile) started being used solely for the extensive development of caatinga whose species were intended for timber extraction for the production of firewood and charcoal, used mainly in the ceramics industries and bakeries.

2.5 Native Vegetation and Forest Types

The last survey of the forest vegetation of Seridó was carried out in 1988, using aerial photographs taken in 1985 (scale 1:40.000). The results were part of the Plan of Forest Management of Seridó the Region (UFIB 1992). The survey identified three vegetation types of the Seridó: (1)thin natural vegetation, (2)semi–dense natural vegetation, and (3)dense natural vegetation.

The natural vegetation covered 83.40 percent of the mapped area, out of which 14.94 percent of the area was covered by the thin vegetation, 52.30 percent by the semi–dense vegetation, and 16.15 percent by the dense vegetation. The total forest biomass in the 670,063 ha of native woods of Seridó is 31.57 million steres, out of which 23.06 million steres, or 73.0 percent, represent the exploitable reserve, while the areas which are restricted in use because of conservation or preservation mandates equal 8.51 million steres, or 26.96 percent of the total volume of biomass. A number of local industries have been using the algaroba firewood (*Prosopis juliflora*) as fuel, thereby using less of the native forest for energy purposes.

The thin natural vegetation has two predominant species, Aspidosperma pirifolium and Mimosa hostilis, both in terms of number of trees (71 percent) as well as basal area (82 percent). This vegetation has been heavily exploited by humans and this has led to a high level of habitat degradation. In terms of volume, the natural forest biomass is relatively low. It has a reserve of 0.97 million cubic meters or 3.22 million steres, and it covers an area of 120,064 ha out of which 107,609 ha. Approximately 12,455 ha of natural vegetation have been set aside for permanent conservation or preservation.

The most common species in the semi–dense natural vegetation are *Aspidosperma pirifolium*, *Croton hemiargyreus*, *Mimosa hostilis*, *Pithecolobium dumosum*, *Cobretum leprosum* and *Caesalpinia pyramidalis*. The semi–dense vegetation has the greatest volume of biomass in the Seridó with a reserve of 5.05 million cubic meters (16.73 million steres). It covers an area of 420,242 ha, out of which 336,147 ha are being exploited or liable to exploitation. Approximately 84,095 ha have conservation or preservation status.

Basically, two species are dominant in the dense natural vegetation of the Seridó: *Croton hemiargyreus* and *Caesalpinia pyramidalis*. In this kind of forest, there is a larger floristic diversity. The dense natural vegetation of the Seridó has the smallest exploitable area compared to the other two vegetation types and ranks second in terms of volume of biomass, with a reserve of 3.49 million cubic meters (11.62 million steres). It covers an area of 129,757 ha, out of which 75,806 ha are exploitable and 53,951 ha have some protected status.

2.6 Forest–Based Energy Demand and Consumption

In the Seridó, ceramics production places huge demands on the forests for fuel wood. According to UFIB (1992), the consumption of fuel reached a volume that is equivalent to 501,500 steres per year, which represented approximately 12,000 ha of deforested area yearly. In the household sector, the consumption reached approximately 350,300 steres per year, out of which 146,400 steres per year (42 percent) corresponded to urban household consumption and 203,900 steres per year (58 percent) corresponded to rural household consumption. As for the industrial sector, the consumption had been approximately 151,200 steres per year. However, at the time when the sampling was carried out the sector was going through a period of relatively high idle capacity (34 percent) due to the fact that 1897 was a year of declining economic activity in the country. The potential consumption of the industrial sector working at full capacity is approximately 227,600 steres per year (50 percent more than the capacity in 1987) (IICA 2000).

Some of the recommendations of the SSDP were that a new forest survey should be carried out to provide additional knowledge on reasons for the increased consumption of fuel wood over the past years. For example, the number of ceramics industries increased from 7 to 71 between 1987 and 1999. At the same time, other industrial sectors which also consume forest wood for energy needs also have increased. These include soap, vegetable oil, coffee, confectioners and bakery industries.

2.7 Desertification Process

The semi–arid environmental conditions of climate, soil, water and vegetation associated the heavy human use have contributed significantly to the increasing desertification processes in the Seridó. This region has been assessed to be one of the most affected by desertification in Rio Grande do Norte.

Desertification in the Seridó covers an area of approximately 2,341 square kilometers (18.06 percent of the region's area), where there are 244,000 inhabitants (84 percent of the population of the Seridó). The main consequences of desertification are: soil destruction (through erosive processes), depletion of the native vegetation, silting of water reservoirs, decrease of groundwater capacity, loss of biological diversity, reduction of alternatives to development in the region, increased social and economic impoverishment, and migration of people from rural areas.

2.8 The Importance of the Ceramic Industries

The ceramics industry is very importance for the economy of the Seridó and the state of Rio Grande do Norte. It is the primary and often the sole industrial activity in some municipalities. This sector generates approximately 4000 jobs, with special emphasis on child labor, where the wages are lower than those earned by adults. Out of the 71 ceramics industries in operation in the region, 46 (65 percent) are distributed in 5 municipalities only: Jardim do Seridó (6), Carnaúba dos Dantas (16), Parelhas (21), Santana do Seridó (2) and Equador (1). The remaining 25 industries are distributed in the other municipalities of the region.

The ceramics industry is a heavily mechanized industry which utilizes large amounts of sand, clay, kaolin, diatomite, feldspar, mica, quartz and talc. Fuel for the industries comes from the forest trees that are burnt and used as firewood in kilns of the ceramic industries. However, economically the ceramics industry produces a large variety of products: bricks, roof tiles, earthenware, and pottery handcraft, among other things. Ceramics industries represent the major economic activity of the Seridó, but at the same time they cause irreparable damage to the environment by intensifying deforestation, soil degradation, desertification, and biodiversity loss. Despite short–term economic benefits of the ceramics industry, over the long–term the rural population is mostly affected by the adverse effects of environmental degradation and is mostly incapable of mitigating damages; therefore, over the long–term the rural poor are forced to migrate to other cities, often being forced to live in more miserable conditions than the ones they were living in before. Certainly, if industrial activities adhered to the goals of sustainable development the people of the Seridó would not be currently facing the dilemma of having to find, at any cost, alternatives to present development activities.

A complicating factor with respect to the ceramics industry is that it is has a low technological and managerial efficiency. Many kilns are rudimentary and self–made, use inefficient combustion systems, lack methods and tools for product and combustion control, lack laboratory and technical support, lack sufficient methods and tools for drying of material, lack necessary mechanization, lack a more highly qualified labor force, manufacture products with little uniformity, and have low levels of investment sufficient to yield capital return for improvements.

According to the Brazilian Institute of Environment (Ibama 1998), the average wood consumption of the 71 ceramics industries of the region is

39,760 steres per year. If divided by the average annual wood production (42 steres per hectare), there will be a total deforested area equal to 947 hectares per year. According to estimates made by the SSDP, the deforested area to supply the annual demand of the ceramics sector of the Seridó is approximately 3,800 ha. This is a much higher than the 947 ha per year estimate of consumption officially stated to Ibama (1998).

2.9 Dangers and Environmental Impacts

The Serido's mining economy is based on extraction and manufacturing processes which date back to more than half a century. It has not been carried out according to the modern requirements of environmental legislation. Nevertheless, some enterprises have adopted the norms required by the state and federal environmental legislation.

Economically stronger and more highly organized industrial enterprises are the ones that manage to better comply with the environmental legislation, while the opposite occurs in relation to the smaller businesses of mining production and to the prospectors. This fact does not indicate that the large and well–organized mining enterprises do not cause significant environmental impacts. In fact, they have caused environmental degradation in many areas of the Seridó. The use of non–renewable natural resources that provide short–term economic efficiency and benefits to the rural poor constitutes a challenge which needs to be overcome but because of the poor socioeconomic conditions of rural people in the area the solutions must directly or indirectly (but tangibly) benefit these people. This is, of course, one of the primary tasks of making the transition to sustainable development, especially where the poor are concerned.

The SSDP has revealed other environmental impacts from mining, such as waste from the mines and the use of alluvial soil, previously used for growing crops. Thus, there are other environmental impacts deriving from the activities of the ceramics industry. Apart from the loss of fertile soils used for farming and cattle raising, the ceramics sector still contributes to desertification through the burning of scarce forest resources and the resulting erosion of the soils.

3. FINAL CONSIDERATIONS

Semi–arid environments are harsh and can be fragile, and care must be taken in the utilization of their resources. In the Seridó, degradation of the local water resources is due to the destruction of forest vegetation caused by deforestation and forest fires, to the storage and uncontrolled use of water and by the discharge of pollutants in the water reservoirs. Another important factor whose effects are felt in the quantity and quality of water resources is soil erosion resulting from the depletion of the vegetation and the development of farming and mining sectors, without any conservation measures. The native vegetation of the region, especially the caatinga, has been decreasing in terms of area and quantity due to the process of environmental degradation caused by deforestation, extensive farming and cattle raising, extraction of firewood for domestic and industrial use. These impacts are separate from the direct impacts of forest fires and fuel wood utilization caused directly by mining activities.

The traditional cutting practice of the Caatinga has been contributing to the reduction of biodiversity, caused mainly by the destruction of the habitat and predatory hunting, which supply the illegal trade of wild animals.

As for the development of farming and cattle raising, there are other factors which generate environmental impacts and should be highlighted: (1)the forest fires (which destroy the vegetation and microorganisms); (2)the widespread use of agrochemicals and fertilizers which contribute to soil and water pollution, with damaging effects for human beings; (3)the lack of management and soil conservation practices while the economic sectors are being developed; (4)the soil degradation resulting from the removal of material for the ceramics sector; and (5)extensive cattle raising.

Another indicator used to evaluate the effects of human activities in the region was the relationship between "forest areas and people employed in rural activities." The greater the ratio between these two variables, the lower will be the occurrences of deforestation. This relationship in the Seridó region, which used to be 18.76 hectares per employed person in 1985, increased to 19.72 in 1995. It represents a growth of less than 1 percent. For the State, the indicator nearly doubled in the 10 years of analyses, increasing from 7.83 hectares per person employed in rural activities in 1985 to 13.17 hectares per employed person in 1995. The data reveals that the risk of deforestation in the Seridó is nearly twofold lower than the one in the state, bearing in mind that in the Seridó this risk presently is stabilized due to a poor economy, while in the state this risk showed a relative growth of 68 percent in 10 years.

The apparently more favorable situation for the Seridó, shown through the indicator "forest areas and people employed in rural activities" should be considered very carefully, as it can be a result of a reduction of the number of people employed in the rural area of Seridó, as shown by the results on the increase of migrations from rural areas. It can also derive from the decrease of cotton–cultivated areas, whose category changed to "thin vegetation" after 1985.

Through the preparation of the SSDP, it became obvious that Rio Grande do Norte is one of the northeastern states that is mostly affected by the harm done to the environment. The actual resources of Rio Grande do Norte are being used without the necessary precaution with regards to the norms and standards of sustainability, environmental conservation, and economic rationality. Within this perspective, the deforestation of the native vegetation areas, the endangerment and flow reduction of the rivers and creeks, and expansion of the areas undergoing desertification can be identified as major problems.

However, the short–term human socioeconomic needs confronting rural people in the Seridó may become worse in the short run due to their problems of trying to live without occupational prospects and other basic amenities. Until these problems are overcome, exploitation of the environment is likely to continue and sustainable development will be more difficult to obtain.

4. REFERENCES

Beaugrand E., Tinoco, D., Matos, J., 2001. O plano de desenvolvimento sustentável do Seridó: Uma experiência de planejamento participativo e de implantação de um sistema de gestão no Rio Grande do Norte – ENAPAD.

Buarque, S. C. 1999. Metodologia de planejamento do desenvolvimento local e municipal sustentável. Brasília, IICA. 196 pages.

Carvalho, O. 1988. A economia política do Nordeste: Secas, irrigação e desenvolvimento. Rio de Janeiro. 505 pages.

Duque, J. 1964. O Nordeste e as lavouras xerófilas. 1. BNB, Fortaleza.

FAO. 1990. La descentralización de los sistemas de planificación agropecuaria en America Latina. Roma.

(Ibama) 1998. Ministério do meio ambiente dos recursos hídricos e da Amazônia legal MMA. Instituto Brasileiro do Meio Ambiente e dos Recursos Renováveis–Ibama (1999) Manejo florestal sustentado da caatinga. 2 ed. Brasília.

Matus, C. 1999. Adeus senhor presidente, governantes governados, FUNDHAP. São Paulo. 381 pages.

Mendes, B. 1997. Importância social, econômica e ecológica da caatinga. Anais do I Simpósio Brasileiro sobre o meio–ambiente e Desenvolvimento Sustentável do Semi–Árido Coleção Mossoroense, Serie E, vol CMXLVIII – Mossoró.

(SERHID) 1995. Secretaria de recursos hídricos–caracterização das bacias hidrográficas do Rio Grande do Norte, Natal.

(SERHID) 1999. Secretaria de recursos hídricos–caracterização das bacias hidrográficas do Rio Grande do Norte, Natal.

(SSDP) 2000. Estado do Rio Grande do Norte, Secretaria de Planejamento e Finanças, Instituto Interamericano de Cooperação com a Agricultura IICA, Conselho de Desenvolvimento Sustentável do Seridó, Plano de Desenvolvimento Sustentável do Seridó, Volume 1. 316 pages.

(SSDP) 2000. Estado do Rio Grande do Norte, Secretaria de Planejamento e Finanças, Instituto Interamericano de Cooperação com a Agricultura IICA, Conselho de

Desenvolvimento Sustentável do Seridó, Plano de Desenvolvimento Sustentável do Seridó, Volume 2. 272 pages.

(UFIB 1992). Projecto UNDP, FAO, Ibama BRA/87/007. Plano de manejo florestal para a Região do Seridó do Rio Grande do Norte. Natal RN, Ibama, Volume 1.

Chapter 27

LATIN AMERICAN PLANT SCIENCES NETWORK: A PROGRAM FOR THE DEVELOPMENT OF PLANT SCIENCES AND CONSERVATION OF BIODIVERSITY IN LATIN AMERICA

Susan Maldonado[1], Osvaldo Sala[2], and Gloria Montenegro[3]
[1]Red Latinoamericana de Botánic a, Instituto de Ecología–UNAM, A.P. 70–275. D/F/ 04510, Mexico, email rib@ecologia.unam.m; [2]Laboratorio de Ecologia, IFEVA, Universidad de Buenos Aires, Argentina; [3]Facultad de Agronomía, Pontificia Universidad Católica de Chle, Santiago, Chile; email rib@ecologia.unam.mx

1. INTRODUCTION

Several factors contribute to biodiversity loss: poverty, economic marginalization, undervaluing of natural resources, poor social participation in decisions concerning development and resource utilization, scientific and other ignorance about resources, and where resources are present in each country, etc. Many of the factors stem, in part, from constraints and unequal access by different groups of people to educational opportunities. Although the majority of the world's human population and terrestrial biological diversity is located in developing countries, the countries have only six percent of the world's scientists. Within large geographical regions there are uneven distributions of scientific expertise and capacity. For example, an analysis of the number of papers presented by plant scientists during the last four Latin American Botanical Congresses (1986, 1990, 1994, 1998) shows that an overwhelming majority of the papers presented at those meetings come from only five countries: Argentina, Brazil, Chile, Mexico and Venezuela despite participation of between 12–19 countries in each meeting (Maldonado, unpublished). Several countries of the region have not contributed a single paper in those meetings in a time period spanning almost 20 years, while others have increased their presence during this period. This case study describes the experiences of the Red Latinoamericana de Botanica (RLB) to increase scientific capacity in plant sciences in Latin America. The RLB is a consortium of academically prestigious institutions in six Latin American countries (Argentina, Brazil, Chile, Costa Rica, Mexico, and Venezuela). Since 1988, these institutions have worked together collaboratively to offer graduate level training to students from and in Latin American countries. The RLB also serves as a focal point for organizing scientific meetings, workshops, and collaborative research projects. The training centers of the RLB's participating institutions are contributing significantly to increase the number of qualified plant scientists in Latin America by receiving students from countries lacking graduate level programs of particular fields of competence. This training is helping not only to increase the critical mass of well–trained people, but also to alleviate the isolation among scientists by providing opportunities for cooperative, comparative, and multidisciplinary research. More specifically, the RLB has emphasized training in the fields of biology, ecology and food production and conservation of natural resources, thereby helping to increase the scientific knowledge needed for informed policy and decision making for conservation of biodiversity throughout Latin America.

2. DEVELOPMENT OF THE RLB

Partly as a result of sharing concerns about the above situations, a group of plant scientists from several countries in the region held a meeting more than fifteen years ago to discuss ways of improving the opportunities for young students to pursue either graduate degrees or specialized courses in the plant sciences. In September 1987, a proposal entitled "A Network of Latin American Universities in the Plant Sciences," prepared by Drs. Mary Kalin de Arroyo, Liliana Cardemil, and Sonia Dietrich, was submitted to the Jessie Smith Noyes Foundation, New York (Kalin 1988). Subsequently, a consortium of several institutions was established to offer first class graduate training to students in Latin America. The first group of participating institutions was from Argentina (Universidad de Buenos Aires), Brazil (Instituto de Botánica, Universidade de São Paulo), Chile (Universidad de Chile, Pontificia Universidad Católica de Chile), Costa Rica (Universidad de Costa Rica), Mexico (Universidad Nacional Autónoma de México–UNAM), and Venezuela (Instituto Venezolano de Investigaciones Científicas, Universidad de Los Andes). A select group of internationally recognized plant sciences scholars collaborated to offer courses and accept graduate students from countries other than their own and to help students carry out their research programs leading to their degrees. These institutions also became foci for the organization of bi– and multinational research projects and regional specialized courses and scientific meetings; all supported wholly or in part by the RLB.

3. GOALS OF THE RLB

The RLB's goals are to: (1)increase the number of plant scientists trained to conduct research on conservation of the region's biodiversity, (2)increase collaboration and professional contacts between scientists of Latin America interested in problems of conservation and resource management, and (3)promote productive international relations between the countries of Latin America regarding the conservation of biodiversity. In order to accomplish these goals, the RLB offers programs that provide an opportunity to young men and women to obtain first–class academic training. The opportunities include short–term specialized courses for mid–career professionals working on conservation and management of natural resources, support for students pursuing doctoral degrees or postdoctoral work including individual training at highly specialized laboratories in the region, and support to attend high–level workshops or seminars and congresses. The RLB's programs also have fostered collaborative research projects between established scientists of two or three institutions in different Latin American countries.

4. OUTCOMES OF RLB PROGRAMS

A partial list of RLB activities and accomplishments from 1989 through 2001 includes: doctoral and masters level fellowships (269); research on binational projects, databases, thesis grants, and research grants–in–aid (151); contributions to infrastructure in centers and institutions (23); graduate courses (42); and scientific events (80). These activities have led to the: (1)training of over 150 men and women from 16 Latin American countries at the doctoral, masters and short–term levels; (2)full or partial funding of over 40 graduate courses attended by over 500 students of the region; (3)full or partial funding of over 80 scientific events including four Latin American Botanical Congresses; and (4)over 150 small grants for research throughout Latin America. A more specific measurement of the success of the RLB is through the scientific productivity of its beneficiaries. During the period of the RLB's operation its scholars have published approximately 100 scientific articles and book chapters. Former graduate students (at the doctoral and master's level) have written many publications, either as products of their theses research or as a direct product of their training supported by the RLB. In addition, over 120 papers have been presented at national and international congresses, with support from the RLB. More importantly, a large number of former RLB fellows have been promoted in their academic or administrative posts upon their return to their institutions, and many are actively involved in the training of new human resources in their countries by contributing to raising the academic level of their institutions and countries. In a survey conducted by the RLB in 1998, the majority of the former fellows stated that the improvement in their professional positions was a direct result of the training received with the support of the RLB. On the other hand, the RLB has been promoting the sharing of expertise and know–how already existent in the regions through increased scientific interactions at all levels. The President, the Executive Director, and members of the RLB's Executive or Scientific Committee have been promoting activities by attending regular national, regional, and international scientific meetings. Further, the RLB Scientific Committee and Executive Committee meetings have been held periodically in different countries of the region. Occasionally, scientific activities have been organized simultaneously with these meetings so that younger people and researchers can benefit from listening to Scientific Committee and Executive Committee scientists. Members of these committees have provided conferences in academic institutions about the role of the RLB in the development of plant sciences or of their individual and institutional expertise areas, have been willing to meet with educational authorities, and have interviewed promising young students and distributed written material about the RLB's activities.

The RLB has also expanded and improved the services it uses to disseminate information to better meet the increasing needs for information. Since 1994, the biannual newsletter "RLB Boletín Informativo" has produced and covered RLB news as well as served in the disseminating of information, which arrives at the RLB headquarters to the Latin American botanical community at large. The continued expansion of the RLB's Web site (http://www.rlb-botanica.org) has increased the RLB's ability to receive and disseminate information from a great number of institutions. The site contains links to the information on the Network's collaborating institutions, and on Latin American researchers and academic staff in plant sciences (e.g., summary of their lines of research, principal publications, etc.) and any other information relevant for the botanical community in Latin America. We believe that through these kinds of cooperative actions the RLB will continue to contribute to expanding botanical knowledge and fostering a greater sense of identity of professional plant scientists throughout the region. Continuing these efforts into the next years should further strengthen Latin American institutions by supporting the local scientific communities.

Funds invested until now have been over US$ 3,100,000 from a total budget of more than US$ 3,600,000 financed from grants from several private foundations and by various aids provided by institutions from the RLB training centers. More than 53 percent of the RLB funds spent until now have been dedicated to training activities (fellowships and graduate courses), 12 percent to research grants, approximately eight percent to scientific events, approximately three percent to infrastructure at collaborative institutions, approximately six percent to committee meetings and regional coordination, and approximately 17 percent to administration. These resources have been available from grants from a number of private foundations as well as from in–kind support offered generously by several of the institutions constituting the RLB, such as reduction of tuition fees for scholarship holders (Universidad de Chile and Universidad de Costa Rica), lodging expenses for students and professors attending graduate courses, office space and communications infrastructure (Universidad de Chile and UNAM), secretarial assistance (UNAM), travel and expenses for representatives of the institutions attending scientific committee meetings, etc.

5. ORGANIZATION OF RLB

In terms of its basic structure and scope of activities, RLB has always been an integrated network with a high degree of South–South connections (Kalin et al. 1994). One of the critical problems of developing an integrated network focused on concern on how to reach an agreement to distribute the coordination of activities and the administrative duties (Kalin 1988). In this sense, the coordinators and committee members are selected for fixed terms and their responsibilities are shared. It is important to emphasize that the organizational structure of the RLB is dynamic and includes the participation of many people of diverse countries. In each decision making opportunity a potential RLB participant and the opinion of all the committee members are consulted. Furthermore, the Executive Committee and the Scientific Committee convene once a year and every two years, respectively, in a Latin American country to discuss policy, budgetary matters, program major graduate courses and symposia, and review applications for fellowships and financial help for courses and meetings. The RLB is governed by an Executive Committee, presently constituted by the President, Osvaldo Sala (Universidad de Buenos Aires); assisted by the Executive Director, Susana Maldonado; one Vice President, Daniel Piñero (Universidad Nacional Autónoma de México); the Regional Coordinator for Mexico and the Caribbean, Victoria Sosa; the Regional Coordinator for Central America,

Sonia Lagos–Witte; the Regional Coordinator for South American Andean Countries, Enrique Forero; the Regional Coordinator for South American Southern Cone, Gloria Montenegro; and three of the Institutional Representatives, Lilian Zaidan, Michele Ataroff, and Javier Simonetti. Furthermore, the RLB operates with a Scientific Committee, composed of the members of the Executive Committee (excepting the Executive Director) plus the Institutional Representatives of some of the collaborative Latin American training Network centers. The Network's graduate training centers or the collaborating institutions are listed in Table 1.

6. LINKS TO ARID AND SEMIARID ZONES

The RLB focuses on helping to solve problems faced by countries with high biodiversity (hotspots), including those in arid or semiarid zones. The majority of the collaborative institutions of the RLB are situated in countries with important arid and semiarid zones: Argentina, Brazil, Chile, Mexico, and Venezuela. For example, in these centers the RLB fellows have worked to develop theses in different subjects related to arid and semiarid regions and problems. According the objectives of this book's project "Promoting Best Practices for Conservation and Sustainable Use of Biodiversity of Global Significance in Arid and Semiarid Zones," an organization like the RLB could share its experiences in developing and working with centers of excellence, facilitating the exchange of information and expertise inside the region, coordinating training and research programs, and providing training of scientists, etc. One of the most important objectives of the RLB is the conservation and sustainable use of biodiversity in the Latin American region through capacity building, partnership, and networking. These objectives have much in common with those of the biodiversity project of which this case study is a part. Our experience indicates that it is very important to know and to use the available infrastructure in our countries to foster partnerships and capacity building. In other words, it is not always necessary to create something (entirely) new. For example, it might be more useful to develop an adequate database of institutions dealing with arid and semiarid zones, either at the Latin American region level or more globally.

Table 1. RLB Network Graduate Training Centers and Institutions

Country	Institutions	Areas of Expertise
Argentina	*CIRN-INTA and CICV-INTA, Castelar, Buenos Aires *Universidad de Buenos Aires, Buenos Aires *Universidad de La Plata, La Plata *Instituto de Botánica Darwinion, Buenos Aires *Instituto de Botánica del Nordeste, Corrientes *IMBIV, Universidad Nacional de Córdoba, Córdoba	*Plant Systematics *Plant Physiology *Biochemistry *Systematics of Genetic Resources *Paleobotany *Biogeography *Ecology
Brazil	*Instituto de Botánica, São Paulo *Universidade de São Paulo, São Paulo *Universidade de Campinas, Campinas *Universidade Federal de Rio Grande do Sul, Porto Alegre *Universidad Federal de Viçosa, Minas Gerais *Universidade Federal de Pernambuco, Recife *Universidade Federal de Brasilia, Brasilia.	*Plant Physiology *Biochemistry *Plant Systematics *Tropical Rainforest *Phytosociology *Coastal Plant Resources *Plant Anatomy *Paleontology
Chile	*Universidad de Chile, Santiago *Pontificia Universidad Católica de Chile, Santiago *Universidad de Concepción, Concepción *Universidad Austral de Chile, Valdivia *Universidad de La Serena, La Serena	*Plant Ecology *Plant Physiology *Plant/Animal Interactions *Plant Taxonomy *Marine Ecology *Phycology *Biogeography *Paleontology *Conservation and Sustainable Use
Costa Rica	*Universidad de Costa Rica, San José	*Molecular Biology *Biochemical and Molecular Characterization of Plant Genomes *Plant Biology
Mexico	*Universidad Nacional Autónoma de México (UNAM), Mexico, D.F. *Colegio de Posgraduados, Montecillo *Centro de Investigación y de Estudios Avanzados del IPN, Irapuato *Instituto de Ecología, A.C., Xalapa.	*Tropical Plant Ecology *Plant/Animal Interactions *Genetics of Plant Populations *Plant Physiology *Ethnobotany *Plant Systematics *Conservation Biology
Venezuela	*Instituto Venezolano de Investigaciones Científicas (IVIC), Caracas *Instituto de Ciencias Ambientales y Ecológicas (ICAE, ex CIELAT), Universidad de Los Andes, *Universidad Central, Caracas	*Tropical Plant Ecology *Ecophysiology *Dynamic of Ecosystems *Tropical Agroecosystems

7. LESSONS LEARNED

We think that there are at least four important lessons learned by the RLB that could be useful for developing best practices for the conservation and sustainable use of biodiversity: (1)use of existing infrastructure and expertise is very important and should be utilized whenever feasible, (2)the coordination and interchange at different levels of government needs to be fostered for both the establishment of networks and their operational efficacy and maintenance, (3)the dissemination of information needs to be transparent and as wide as possible, and (4)networks such as RLB with a proven record of accomplishments should be supported by funding sources obtainable with more difficulty than experienced by traditional academic institutions.

8. ACKNOWLEDGMENTS

Sincere appreciation is extended to all foundations and institutions, without whose help the RLB would not exist. We personally thank Dr. John Lemons (TWNSO project consultant and Professor, Department of Environmental Studies at the University of New England, Biddeford, ME, USA) for his invitation to present this study case during the "Latin America and Caribbean Regional Workshop" held 18–22 March 2002 at the Facultad de Agronomía e Ingeniería Forestal, Pontificia Universidad Católica de Chile, Santiago, Chile. We also thank the GEF/UNEP funding agencies for their support of the project "Promoting Best Practices for Conservation and Sustainable Use of Biodiversity of Global Significance in Arid and Semiarid Zones," which is implemented by the Third World Network of Scientific Organizations in Trieste, Italy.

9. REFERENCES

Kalin de Arroyo, M. 1988. Towards the Establishment of a Latin American Plant Sciences Network. Report presented to The Jessie Smith Noyes Foundation, New York. 206 pages

Kalin de Arroyo, M., Dietrich, S., Forero, E., and Maldonado, S. 1994. The Latin American Plant Sciences Network: A collaborative regional effort in science training. Agroforestry Systems 28: 45–62.

Chapter 28

THE BRAZILIAN BIODIVERSITY VIRTUAL INSTITUTE

Érica Speglich[1] and Carlos Alfredo Joly[2]
[1]Centro de Refêrencia em Informação Ambiental/CRIA, Av. Romeu Tórtima, 388, 13084–520, Campinas/SP, Brasil, email erica@cria.orb.br; [2]Department de Botânica/IB, Universidad Estadual de Campinas/UNICAMP C.P. 6109, 13083–970, Campinas/SP, Brasil, email cjoly@unicamp.br

1. INTRODUCTION

São Paulo state, located in the Southeastern region of Brazil, is the most industrialized state of the country, and has a population of over 35 million people. The two major biomes of the state, Atlantic Forest and Cerrado (Savannah), have been reduced to 12 percent and 2 percent of their pre–Columbian land area, respectively. With the exception of the coastal mountains (Serra do Mar), which are still covered with large extensions of remnants of native Tropical Rain Forest, inland forest and cerrado remnants are highly fragmented. Although forest clearing started in early 1800's, it grew rapidly in the last half century. From 1962 to 1992 the state lost more than 50 percent of its native cerrado cover (Governo do Estado de São Paulo 1993; http 1).

The relevance of biodiversity conservation in these two biomes, Atlantic Forest and Cerrado, has been recently recognized with their inclusion in the list of "hotspots" (Myers et al. 2000). Therefore, it is not surprising that the biodiversity numbers of the state are extremely high, around 8,000 species of higher plants, 5,500 of algae, more than 2,000 of vertebrates, more than 500,000 of invertebrates (the number of microorganisms can only be speculated). At least 30 percent of these species are endemic, which makes even more urgent the development of tools to increase our knowledge and establish policies for sustainability.

A major problem confronting scientists and decision makers is the fact that available information regarding biological resources is fragmented, dispersed, difficult to access and underused. As a consequence of the lack of an updated cartographic base, key information such as the location of sampling sites is usually inaccurate. Further, there has been a need to systematize sampling, using GPS to locate the sampling sites and areas, to develop an integrated databank for storing this information, and to produce accurate and reliable maps for plotting the spatial distribution of species within the state.

In March 1999, the "Research Program on Conservation and Sustainable Use of the Biodiversity of the state of São Paulo," named BIOTA/FAPESP (BFP), or "The Biodiversity Virtual Institute" (http 2) was officially created by the state of São Paulo Research Foundation/FAPESP (http 3). In this paper, we describe the creation process and the strategies being applied for connecting research projects, researchers, research students, the data they produce as well as how to qualify people for working with biodiversity conservation. In particular, we pay close attention to the "lessons learned on creating partnerships and capacity building."

2. THE CREATION OF BIOTA/FAPESP

The creation of BF was the result of three years of collaboration of numerous biodiversity scientists and organizations in Brazil. It is important to point out that from the beginning we started discussing the importance of connecting research and researchers, therefore a homepage (http 2) and discussion list (http 4) were developed. Through the discussion list information was obtained about the importance of making the information on biodiversity knowledge available to public access via the Internet. One of the most important discussion topics concerned copyrights of, for example, a list of birds, or fishes or plants of São Paulo state published only in the Internet. Once this was solved, by tagging to the "on–line" publication a metadata label with the copyright information, we started publishing the available species lists for the state.

These lists were a starting point for a more thorough inventory of the available biodiversity knowledge. Taking into account that since the large European expeditions to Brazil in the 1800's, species (mainly of vertebrates and higher plants) present in the state of São Paulo have been recorded, we decided that it was important to evaluate the existing knowledge about different taxonomic groups, ranging from viruses to mammals and angiosperms, as well as the list of personnel and institutions working with each taxonomic group, and the state ex situ and in situ infrastructure for its conservation. At this stage there were approximately 70 researchers involved.

In order to consolidate these inventories and begin a collaborative effort to study the biodiversity of the state, in July 1997 we organized a workshop with over 100 participants from many research areas and institutions. The quality of the documents prepared for that meeting encouraged us to publish them in a series of seven volumes ("Biodiversidade do Estado de São Paulo: Síntese do Conhecimento ao Final do Século XX" [Biodiversity of the state of São Paulo: A Synthesis of Knowledge at the End of the 20th Century]), and to make them fully available through the Internet (http 5). During this workshop we defined the following long–term objectives for all the BFP research projects, (using the broadest definition of biodiversity as stated in the "Convention on Biological Diversity" [CBD]): (1)to understand the processes that generate and maintain biodiversity, as well as those that can result in its deleterious reduction; (2)to standardize sampling, making the use of GPS mandatory; (3)to make information relevant to conservation and sustainable use of biodiversity available to decision makers; (4)to ensure fast and free public access to this information; and (5)to improve teaching standards on subjects related to conservation and sustainable use of biodiversity.

In accordance with the rules of the state of São Paulo Research Foundation/FAPESP, researchers interested in submitting proposals for the BFP organized themselves according to "thematic research projects," which were then submitted to an international panel of experts during 1998. Every "ad hoc" referee received a document defining the basic points relating the specific project with the proposed research program on characterization, conservation and sustainable use of biodiversity we were planning to create which was attached to each proposed research project.

The individual research projects and the proposed BFP were highly recommended by more than 90 percent of the "ad hoc" referees who participated in this evaluation process. Once the quality of our proposal has been recognized, the Scientific Director of FAPESP, Professor José Fernando Perez, submitted the idea of the Program to the Superior Council of FAPESP. The Council approved the Program during its meeting on February 10, 1999 and on March 26 it was officially launched.

3. BIOTA/FAPESP PROJECTS

The initial 25 proposals were merged into 18 thematic projects, connected in such a way that each of them would be feasible individually, but when combined with other projects would promote interaction between research teams and, therefore, result in broadened approaches and a gain in data quality. A "Coordination Group" (http 6) organized several meetings among project leaders to ensure that all the 25 proposals were integrated within the 18 research projects, and to seek possible points of connectivity and guarantee they were present in the final version of the projects submitted to BFP. The experience of this "Coordination Group" was maintained by the BFP through the establishment of a board composed by four ex–members of the "Coordination Group." The task of this board is to analyze not only the compatibility between new proposals and aims of BFP, but also to ensure their connectivity with on–going projects.

After three years, there are 36 major research projects within BFP's framework (http 7). The research projects linked to BFP have been conducted in order to increase the academic knowledge on the biodiversity of São Paulo state and simultaneously produce data potentially useful for improving state policies on the conservation and sustainable use of its biodiversity. For this purpose, projects aiming at improving biological knowledge were associated with projects seeking the economic use of species.

All major public universities (USP, UNICAMP, UNESP, UFSCAR), some private universities (such as PUC, UNAERP and UNISANTOS), research institutes (such as the Instituto de Botânica, Instituto Florestal, Instituto Geológico, INPE), EMBRAPA Centers, and NGOs (e.g., Instituto Socioambiental, Fundação SOS Mata Atlântica, Conservation International and Reference Center on Environmental Information-CRIA) are taking part in the BFP. The number of researchers linked to those institutions within the state of São Paulo is approximately 350 Ph.Ds and 200 graduate students. In addition, there are approximately 80 collaborators from other Brazilian states and approximately 50 from abroad. An important feature of BFP is that the researchers have added common goals to their individual research projects. Furthermore, they are using a set of common tools that have been developed for integrating data within the BFP. The current projects are related to a wide variety of subjects, including: (1)microorganisms; (2)flora; (3)fauna; (4)species interactions; (5)secondary compounds of plants species; (6)carbon sequestration; (7)landscape ecology and fragmentation; and (8)watershed functions.

4. THE ENVIRONMENTAL INFORMATION SYSTEM

The establishment of a standard record form to register sampling data also enhanced connectivity among projects. Sampling data forms include nine mandatory fields: (1)sampling author; (2)date; (3)locality, including the geographical coordinates obtained by GPS; (4)municipality; (5)watershed; (6)taxa (including standard form for species lists); (7)methods; (8)ecosystem; and (9)conservation unit (if applicable). Besides these nine mandatory items, there are more than forty supplementary fields that can be used to give more detailed information, if required, regarding the specific taxonomic group or research.

These tools have proved to be suitable to all new research projects and are strongholds of the BFP. They were also essential to the construction of a databank for registering all samples collected by researchers participating in the BFP. All data produced is included in the Environmental Information System (http 8) developed by the Reference Center on Environmental Information/CRIA (http 9) in collaboration with the Instituto de Computação (http 10) of the State University of Campinas/UNICAMP. This system uses free computational languages and software and therefore can be applied for developing, at low cost, similar systems in other states of Brazil or elsewhere.

5. DIGITAL ATLAS

Along with the development of the databank and its interface with Internet, a digital map of the state of São Paulo (http 11), in a 1:50.000 scale, was produced in collaboration with the Instituto Florestal (http 12), UNICAMP (Instituto de Geociências) (http 13), Faculdade de Engenharia Agricola (http 14) and CEPAGRI (http 15). The map has detailed information about such things as urban areas, roads, county boundaries, rivers, areas covered by *Eucalyptus* spp and *Pinnus* spp, wood exploitation forests, conservation units, and remnants of native vegetation. The natural vegetation is divided into primary and secondary Atlantic forests, all three physiognomies (open grassland, grassland with shrubs and trees, predominantly trees and shrubs) of Cerrado, riparian forests, floodplain vegetation, and coastal vegetation. The digital atlas is an assembly of the 416 cartographic charts from the 1972 IBGE (Instituto Brasileiro de Geografia e Estatística (http 16) map of São Paulo state, updated with Landsat 5 or 7 satellite images from 1998/99.

The geographic coordinates, one of the mandatory fields from the standard sampling form, connect the databank and the digital map, allowing, in this stage a display of the spatial distribution of occurrence sites of species registered in the SinBiota. The system allows zooms, besides the connection with the standard sampling form related to the sites plotted on the map, and the visualization of the registered information concerning that species.

6. BIOTA NEOTROPICA

The BFP launched in 2001 the on–line only journal BIOTA NEOTROPICA (http 17) to publish results of original research, associated or not to the program, concerned with characterization, conservation and sustainable use of biodiversity within the Neotropical region. Papers may be submitted within the following categories: articles, inventories, thematic revisions, taxonomic revisions and short communications, in English, Portuguese or Spanish. We hope that BIOTA NEOTROPICA will become an important connection tool for everyone working with biodiversity in Latin America.

7. REFERENCES

Governo do Estado de São Paulo, 1993. Inventário Florestal do Estado de São Paulo. Secretaria do Meio Ambiente, Coordenadoria de Informações Técnicas, Documentação e Pesquisa Ambiental, Instituto Florestal, São Paulo.
(http 1) http://www.biota.org.br/info/saopaulo/index)
(http 2) http://www.biota.org.br)
(http 3) (http://www.fapesp.br)
(http 4) (http://www.biota.org.br/mailman/listinfo/biotasp-1)
(http 5) (http://www.biota.org.br/publi)
(http6) http://www.biota.org.br/info/historico/coordenadores
(http 7) (http://www.biota.org.br/projeto/index?search)
(http 8) (http://sinbiota.cria.org.br)
(http 9) (http://www.cria.org.br)
(http 10) (http://www.ic.unicamp.br)
(http 11) (http://sinbiota.cria.org.br/atlas)
(http 12) (http://www.iflorestsp.br)
(http 13) (http://www.ige.unicamp.br)
(http 14) (http://www.agr.unicamp.br)
(http 15) (http://www.cpa.unicamp.br)
(http 16) (http://www.ibge.gov.br)
(http 17) (http://www.biotraneotropica.org.br)
Myers, N., Mittermeier, R.A, Mittermeier, C.G., Fonseca, G.A.B., and Kent, J. 2000. Biodiversity hotspots for conservation priorities. Nature 403: 852–858.

Chapter 29

CAPACITY BUILDING TO SUSTAINABLY USE BIODIVERSITY IN DRYLAND REGIONS OF LATIN AMERICA AND THE CARIBBEAN

Leonard Berry
Director, Florida Center for Environmental Studies, Florida Atlantic University, 3932 RCA Blvd, Suite 3210, Palm Beach Gardens, FL 33410, USA; email berry@fau.edu

1. INTRODUCTION

Latin America and the Caribbean are well known as the site of great biodiversity, but the focus of attention is most often on the admittedly important world heritage of the tropical and mountain rainforests and wetlands of the region. This paper directs attention to the globally significant biodiversity of the arid and semiarid parts of the region. The area includes some of the driest deserts in the world in Chile and large areas of semiarid subtropical conditions, some of which are influenced by the proximity to, and the influence of, the Andean mountain chain. The paper reviews the current funding priorities in the region, attempts in a general way to identify biodiversity conservation needs in terms of networks and institutions and addresses the most effective processes to build on and expand the already sound capacity in the region.

2. CONSERVATION INVESTMENTS IN LATIN AMERICA AND THE CARIBBEAN

Castro et al. (1999) have provided an important service by analyzing the patterns of conservation investment in the region and this section draws on that database in relation to the drier areas. In the period of analysis

(1990–1997) biodiversity funding for the region is identified as $3.26 billion, though some funding sources were not reported in this survey. The funding includes $180.4 million for the Caribbean, $113 billion dedicated to Central America and Mexico, $1.78 billion dedicated to South America, and $164 billion not dedicated to any specific region. This funding was from international donors and does not reflect government contributions within the region. A further breakdown of the funding shows that less than 15 percent has been spent on drylands and a similar percentage on semiarid grasslands and savannas. While spending on any ecosystem could reasonably be said to be inadequate, it is clear that in terms of area a large percentage (two–thirds) of the funding is directed to the tropical and subtropical forests. The arid and semiarid areas, which even in this region make up 50–60 percent of the total area, are relatively underserved.

The recommendations of the Castro review include reviewing levels of funding to priority ecoregions and consider the need for increased funding to those currently receiving modest amounts. This should include:

1. Ensure that all ecosystem types are sufficiently represented in investment portfolios. The majority of funding attributed to the ecoregion level was directed at tropical and subtropical broadleaf forests.
2. Explore additional avenues for collaboration among and between organizations
3. Work within the same aggregate region. There may be potential for multilaterals and bilaterals to cooperate on projects in the Andes.
4. Ensure that countries needing additional funding receive it. Currently, 28 countries account for only 9.7 percent of total funding.
5. Investigate whether low levels of funding directed toward capacity building and outreach are sufficient to encourage long–term in–country stewardship of natural resources.
6. Improve communication within all levels of the conservation community (large and small donors, national governments, NGOs, etc.) through more frequent contact, collaboration on projects and sharing of funding information.
7. Improve record keeping of projects supported to increase transparency of information and facilitate information exchange.

There is a clear need to obtain additional financial and human resources to better identify and conserve biodiversity in the Latin America and Caribbean region. The rest of this paper suggests ways in which this process can be most efficient and most effective.

3. THE NEED FOR NETWORKS

Ecosystems are seldom the unit of study in academic or government programs. It is hard to create and sustain the team approach needed and it is usually even harder to obtain the financing. In most universities and countries there are seldom critical masses of researchers on dryland issues. A great part of the investment funding outlined above was used for the protection and conservation of critical parts of ecosystems and only a small part in the analysis and recording of biodiversity within those ecosystems, a vital part of the establishment of baselines and for the understanding of ecosystem dynamics. Research was only 5 percent of the total funding. Additional funding for research and analysis is a high priority. At the same time the development of networks to share existing information is important. The types of networks that might be useful. Include: (1)networks between scientists in different regions with similar interests; (2)networks between scientists of different disciplines interested in the same or similar ecosystems; (3)networks between scientists and ethno–scientists interested in the same topic region; (4)networks between scientists and local educational and community organizations; and national and regional networks.

4. NETWORKS BETWEEN SCIENTIST

As events in the region have shown there is a critical need for the establishment of networks, linking scientists (and nonscientists) with parallel or overlapping interests. There is already a well–established network linking botanists across the region. An excellent example of a scientific network is the Red Latinoamericana de Botanica (RLB). The RLB is a consortium of 23 academically prestigious institutions in six Latin America Countries (Mexico, Costa Rica, Venezuela, Brazil, Chile and Argentina) already operating for 14 years (see Maldonado et al. this volume). The network has raised over $13 million dollars, trained 152 individuals from 16 countries, funded graduate courses and provided over 150 small grants throughout Latin America. This network is of special relevance in the understanding of taxonomy of plants and this has also been successful in capacity building. Although the network links scientists with expertise in drylands, its scope is much broader than that, covering all ecosystems in the area.

A national state network, which can be used as a model, is BIOTA/FAPESP the Biodiversity Virtual Institute. This is a Brazilian initiative in the State of São Paulo. It links all state major universities, research institutes and the local EMBRAPA centers totaling over 400 researchers and an equal number of graduate students. Over 32 major research projects are linked under the network and biodiversity information gained through the network is stored on one integrated database. (The BIOTA/FAPESP website is http:www.biota.org.br).

These two networks and others are excellent examples of scientific collaboration that could well be emulated in the region with a focus on arid and semiarid regions and their biodiversity.

5. NETWORKS ON PARTICULAR ECOSYSTEMS

Networks that focus on particular ecosystems are harder to form because of the traditional communication problem across disciplines and the structures of universities. Where they occur they are usually in the form of interdisciplinary centers where scientists can spend part or even all of their time. As we seek to understand how ecosystems work in an integrated way there is a need for incentives to develop these structures and to develop mechanisms for improving cross disciplinary communication and integration. An integrated information base is one effective way of encouraging cross linkages and technologies such as remote sensing imagery, GIS databases and GPS location technology are also important. Such techniques allow this integration to be effective and provide the basis for good cross disciplinary science and a better understanding of the dynamics of the biological system of the area.

Such cross disciplinary structures can be part of universities or university systems, or in some cases be part of government funded centers. There are examples of long existing institutes or centers that effectively combine a university setting with active state or federal government participation through advisory boards, exchange of personal, joint projects and other mechanisms. If structured effectively, these combine the independence of a university setting wit the practical orientation of operational agencies. National Parks, agricultural agencies, water resource agencies, forestry departments, and non–university educational ministries have proved to be effective and supportive members of biodiversity oriented centers.

6. NETWORKS INVOLVING ETHNOSCIENCE

Networks between scientists and ethno–scientists are generally recognized to be important, but are not as widely utilized, as they should be. It is clear that in some specialized areas, indigenous medicinal materials for example, local knowledge is better and more discriminating than scientific knowledge, though in most cases it is not as well structured and formalized.

However, individual scientists have gained important insights into the ecology of endangered species through cooperation with local people and the linking of ethno–science with traditional ecology and taxonomy (Lemons and Montenegro 2002). A vital side effect of involving local people in the search for understanding of the components and dynamics of the biodiversity within ecosystems is the pay off in local cooperation in conserving and sustaining those systems.

7. LINKS WITH LOCAL EDUCATIONAL SYSTEMS

An extension of the ethnoscience linkage is the involvement of local and national education systems. At its most effective, important biodiversity hot spots would be characterized by local information systems linked with the schools, and school and community participation in bird, plant, mammal and other identification and with educational material developed with the assistance of involved scientists. Appropriate material and insights would be available for local schools and community organizations and involved scientists would periodically be involved in locally based presentations. Components of this scenario have worked in parts of South America and the Caribbean, but funds and insights have rarely, if ever, come together to activate the whole network concept. It would be a valuable contribution to biodiversity conservation.

8. REGIONAL, NATIONAL, GLOBAL NETWORKS

Arid and semiarid biodiversity conservation at national and regional (cross national and global levels) is another vehicle for cooperation. At the national level there is a need for cooperation between government departments, universities and research institutes on biodiversity conservation and analysis. Governments have in general gained a new awareness and focus on the problem of biodiversity through their participation in and agreement to the international conventions on biodiversity and desertification and by involvement in various international journals. Taking this awareness and translating it into a focus on arid and semiarid biodiversity is a new challenge, especially in resource poor countries. However, major advances are possible through the coordination of involved government departments, (often agriculture, forestry, water) with universities and NGOs. International funding can be an incentive, but this can only be fully effective if there is a product involving local capacity building and valuable local products. These can include economic products and materials, which fit into the national educational system. In fact, a major benefit to biodiversity conservation can be awareness and knowledge building through education at all levels.

9. COMMUNICATION, INSTITUTIONAL DEVELOPMENT, CAPACITY BUILDING

This is characterized as the information age, but for many of us the problem is finding ways to sift and define information networks focused on particular topics and issues. In the context of this paper, the links between research and scientific information on biodiversity in an ecosystem, the added value of local and ethno–science and the incorporation of this understanding of one ecosystem to the broader global audience of specialist and non–specialists means that more and more effort needs to be applied to communicate both electronically and through more traditional modes. There is a special need in the area of arid and semiarid biodiversity, within regions such as the Caribbean and Latin America and also between continents and regions. This GEF/UNEP funded project implemented by the Third World Network of Scientific Organizations (Trieste, Italy) project is an important start, which needs to be followed by ongoing networking. In the water field the Inter American Water Resources Network, now a decade old, has grown in its water information work into a global network attracting a specialized audience in a productive set of face–to–face and virtual exchanges. Such a network is an attractive possibility for arid and semiarid biodiversity specialists.

Communication is obviously important, but building the institutional base to provide the knowledge to communicate is an important prior step. Universities provide the base for much of a country's research and higher–level education, though there is a growing role for government based research in fields important to biodiversity. University Departments funded for research are a powerful tool for knowledge development, though departments tend to serve one discipline and often have strengths in one part of that discipline.

Research and information centers have the benefit of being able to bring different disciplines under one roof either physical or virtual, but sometime have problems of establishing career paths for their researchers and long term funding. Centers that link government and universities can combine the long–term scientific work with the sustainability provided by the government support and practical application. Examples of such institutions in Africa with a 30–40 year history of sustained activity provide potential models.

Capacity building can be an important general goal and the support of specialized education and training in subjects related to biodiversity conservation is a priority for the Latin America and Caribbean region and for the world. However, capacity building in relation to institutional development is an important specific goal. Centers of excellence in arid and semiarid biodiversity should be identified and strengthened through the combined efforts of national and international agencies. In this context the combination of skills needed can be given special attention and long–term ecosystem research can be encouraged in an interdisciplinary setting. The set of skills needed include the traditional scientific, and analytic disciplines, but also information technology and capability is also important. Models from around the world need to be examined and key components transferred.

Arid and semiarid biodiversity is an important component of the biodiversity wealth of the Latin American and Caribbean region. Within the region there are important centers of excellence in components of ecosystem analysis, but the level of research and analysis needed is handicapped by lack of resources and the character of the institutions involved. They were generally inherited from a previous set of traditions and priorities. Now with ecosystem assessment a focus, research, information sharing, network promotion, and institutional development all need to be part of a coherent regional and global strategy. This Third World Academy activity is playing an important role in defining that strategy.

10. REFERENCES

Castro., G., and Locker, L. 2000. Mapping Conservation Investments. Biodiversity Support Program, Washington, D.C.

Joly, C.A., and Speglich, E. 2002. BIO/FAPESP– The Biodiversity Virtual Institute: A Brazilian Experience That Can Be Used as a Model for Biodiversity Conservation and Sustainable Use Research Programmes. Pages 51–53 In: Lemons, J. and G. Montenegro (eds.) Proceedings of the Latin America and Caribbean Regional Workshop Promoting Best Practices for Conservation and Sustainable Use of Biodiversity of Global Significance in Arid and Semiarid Zones. Pontificia Universidad Catolica de Chile, Santiago, Chile. 72 pages.

Lemons, J., and Montenegro, G. (eds.). Proceedings of the Latin America and Caribbean Regional Workshop Promoting Best Practices for Conservation and Sustainable Use of Biodiversity of Global Significance in Arid and Semiarid Zones. Pontificia Universidad Catolica de Chile, Santiago, Chile. 72 pages.

Chapter 30

THE IMPORTANCE OF NATIVE TREES IN SUSTAINING BIODIVERSITY IN ARID LANDS

Michael D. Robinson
Department of Biology, Sultan Qaboos University, Muscat, Sultanate of Oman; email mrobins@squ.edu.om

1. INTRODUCTION

The value of science is greatly increased when it is used to benefit society. Too frequently those who do and understand basic science are not involved in its applications, and those charged with resolving societal concerns, such as the management and sustainability of natural resources, are not aware of research that can assist them in making decisions. The objective of this paper is to provide professionals, who are not ecologists but who are involved in conservation and management of natural resources, a summary of ecological research that indicates desert trees are an important resource. I have chosen trees as the focus of this paper because I believe that they alone harbor a large proportion of the biodiversity in dryland ecosystems, and the sustainability and function of these ecosystems depends on their welfare. This view is not to discredit other types of vegetation, for they are important and share several biological processes and some of the biodiversity associated with trees. However, generally less is known of their ecological and functional properties. When possible I have incorporated ecological information on *Acacia tortilis* (samr in Arabic) from various regions of its range, including recent studies from northern Oman. Samr is a ubiquitous tree throughout the arid and semiarid regions of North Africa and the Middle East. Its dominance in many of the varied ecosystems in the region suggests it is a keystone species, one that is vital for maintaining productive ecosystems.

The discussion proceeds by considering how trees develop microclimates, their interacts with other plants and animals, and their role in sustaining below–ground soil fauna and flora. This latter aspect is given special emphasis because of its central importance in ecosystem function. The soil biota has received far less attention by ecologists, managers and planners than the more visible species above the ground. The well–being of the soil biota must be taken into account in development and management plans or serious impairment of vital ecosystem processes could result. Understanding the interdependency of the above and below ground biota is the key to developing rational plans of resource conservation and sustainability. A vital question for which currently there is no clear answer is to what extent can natural systems loose component species and still remain sustainable? Some views of ecologists on the relationship between biodiversity and the efficiency of ecosystem function are discussed. The paper ends with a review of the possible benefits arid ecosystems can offer societies.

2. INTERACTIONS WITH CLIMATE

During the past half–century ecologists have investigated the many ways that desert trees alter the extreme climates in which they grow. On a hot summer day one can easily appreciate the respite their foliage offers from the intense solar radiation. Beneath the trees, summer surface temperatures are normally 5 to 15 degrees centigrade less than the surrounding open desert. Depending on the tree's size, branch architecture and foliage density, radiation decreases 40 to 70 percent by the time it reaches the ground, creating a strong vertical thermal gradient. This aspect of microclimate is used by numerous invertebrates, small ectothermic vertebrates and a few birds to adjust body temperatures and various physiological processes. The variety of thermal conditions provided by trees greatly exceeds that in low shrubs and open desert.

In deserts where winter nocturnal temperatures occasionally fall below freezing, tree canopies significantly reduce the amount of radiative heat lost to the cold night sky (Lowe and Hinds 1971). In the Sonoran Desert of North America, the warmer microclimate provided by tree "nurse plants" is critical to the survival of young cactus (Noble and Loik 1999), and possibly other organisms during their first years of life.

Trees also influence the distribution of precipitation to their immediate surroundings. A significant amount of water is intercepted by foliage during rain events of low intensity (< 25 mm). Belsky et al. (1989) estimated that during light rain showers *Acacia tortilis* canopies reduce the water reaching the soil by 50 to 100 percent. Rain interception becomes negligible during heavier (>25 mm) rains. Rain that is intercepted by the leaves and branches flows downward along small channels in the bark, forming hundreds of tributaries onto the main trunk. The stem flow, as it is called, is 2 to 10 times richer in ions from minerals in the dust collected by the leaves and branches. Stem flow courses down the bole and into the rhizosphere of the tree. Whitford et al. (1997) estimate that canopies of the microphyllous shrub, *Larrea divaricata*, intercept 35 percent of the incoming rain, of which approximately 17 percent is diverted into stem flow. The disadvantages of reduced soil moisture beneath the canopy may be somewhat offset by the cooler temperatures, lower transpiration rates and greater concentrations of nutrients.

Significant thermal differences exist within the subcanopy environment of *A. tortilis* (Al Harthi 2000). Soils of the northern sector are commonly several degrees cooler than the southern aspect. However, the photosynthetically active radiation (PAR) reaching the surface is uniform. In contrast, the distribution of the dominant annual plant species is not, suggesting other ecological factors vary significantly within this microenvironment (Robinson 1999). Ecological specialization at such a small scale has not been reported in arid environments and may contribute to the overall species richness of Oman's desert flora.

3. INTERACTIONS WITH SOILS AND MICROORGANISMS

Most desert soils are coarse–grained, low in soil moisture and organic matter, and deficient in several plant nutrients (Whitford 2002). But soils that develop in the microclimate surrounding desert trees have chemical and physical properties suitable for limited agriculture. The maturation of these soils requires decades, but loss of their fertility occurs quickly once the tree dies or is removed.

During the life of a tree soil fertility increases as organic litter and finer–grained soil particles accumulate around base. This “macrolitter” is reduced to smaller particles by mechanical breakage and by invertebrate feeding. The addition of detrital organic matter decreases the soil bulk density and improves moisture retention. Together with cooler soil temperatures an environment is created that is more suitable for decomposition and nitrogen fixation by symbiotic and free–living bacteria. The development of soils in the rhizosphere of desert trees probably requires 30 to 60 years, with maximum development occurring after a century. During this period a succession of plants, fungi, invertebrates and other organisms colonize the environment, increasing biodiversity and the complexity of the trophic relations among the soil biota. Simultaneously, biological processes (e.g., herbivory, predation, nitrogen fixation, death and excretion, hydraulic lift by roots) concentrate nitrogen and other essential elements forming isolated "fertile islands" among the inhospitable soils of open desert (Barth and Klemmedson 1982, Garner and Steinberger 1989).

Most research on desert nutrient cycles has focused on nitrogen, which along with water most limit desert plant productivity. Most relevant to North Africa and the Middle East is the research on *Acacia tortilis* and several other tree species common to African savannas (Belsky et al. 1989, 1993; Weltzin and Coughnour 1990, Vetaas 1992, Belsky 1994). This research clearly shows that soil carbon and nitrogen are significantly greater at shallow depths near the bole and concentrations decrease with depth and distance from the tree. Carbon–to–nitrogen ratios, which indicate potential microbial activity, are lowest under trees, and microbial biomass is 35 to 60 percent greater than surrounding grassland. Nearly 2.5 times more nitrogen is mineralized from subcanopy soils than those away from the trees (Belsky et al. 1993). These findings accord well with a general pattern of nitrogen distribution derived from research in various North American deserts (West and Klemmedsen 1978, Whitford 2002).

Approximately 70 to 98 percent of the ecosystem nitrogen resides in the soil, which is not an unusually high proportion and is comparable to grassland and forest biomes. Nitrogen concentrations are quite high in many desert plants. Leaflets of *A. tortilis* contain 16 percent crude protein (equivalent to 2.7 gN/100 g ash–free dry weight), and *Prosopis cineraria* (ghaf tree) has between 14–18 percent (Brown 1988). Surface litter beneath samr trees contain 11.5 percent crude protein, providing a potentially rich source of nitrogen for the soil biota (Robinson unpublished).

Nitrogen exists in a variety of inorganic and organic forms, but it is difficult to assess how much of the nutrient pool is available for plant growth (West and Klemmedson 1978). The conversion and release of mineralized nitrogen involves a diverse biota and appropriate soil moisture conditions which may only be present for brief periods. In hot deserts a considerable amount of the inorganic nitrogen is lost to the atmosphere through denitrification and volatization (West and Klemmedsen 1978, Whitford 2002).

Whitford (2002), Whitford and Parker (1989) and Zak and Freckman (1991) provide an idea of the variety of organisms involved in various processes that regulate the rates of decomposition and mineralization in the North America deserts. The main results of their research and that of Bengtsson et al. (1996), Crawford (1991) and Coleman (1996) are summarized below.

Surface litter is first broken down by a variety of large invertebrate detritivores. Crawford (1991) records six orders of arthropods known to feed on the surface litter, and 14 species occur in the Mohave Desert of North American (Rundel and Gibson 1996). Although some groups (e.g., Collembola and Psycoptera) may be seasonally abundant, in terms of biomass, termites are the dominant surface detritivores. Much of the surface litter is decomposed and mineralized by their symbiotic gut fauna. Similar symbionts have been identified in several desert beetles and orthropterans. Nematodes are common predators feeding on bacteriovores, fungivores and smaller predators. Omnivorous taxa eat yeast, bacteria and protozoans. Forty eight percent of the total nematode biomass is from bacteriophagous species. The Acari (mites) are the most abundant group of microarthropods in surface litter. Six orders are common in North American deserts, and 8 subfamilies have been recorded in a single order. They feed on other mites, nematodes, fungus, cyanobacteria and soil algae.

Buried litter is the base resource for a diverse assemblage of subsurface decomposers. Soil fungi are one of the primary groups. Mandeel (2002) isolated 28 taxa of fungi from the rhizosphere of *Zygophyllum qatarense*, a South Arabian succulent shrub. In North America a succession of different microarthropod groups utilize the litter as it ages. Ameoboid and ciliate protists feed mainly on bacteria and yeast, which ingest microliter and root exudates. Species–level taxonomy is uncertain for many members of the soil fauna, but they can be placed in trophic groups defined by their feeding habits. Links between the functional groups define the main pathways of energy and materials. In forest ecosystems each group may have 100 or more species (Bengtsson et al. 1996), but fewer should be expected in arid environments. The main trophic groups in the soil biota are those feeding on organic detritus (macro and microdetritivores), on fungi (fungivores), on microbes (microbivores) and predaceous species. Omnivory is quite common. Food chains in desert soils may have 5–7 links, and taxonomically distinct chains can occur in different communities (Belnap and Phillips 2001), as well as at different sites in the rhizosphere. Apparently, the numerous energy transfers are possible because the organisms are ecotothermic and have relatively high production efficiencies (Coleman 1996).

In soil experiments, the removal of mites results in a marked decrease in the rate of mass loss from litter buried in soil. Without the microarthropods, the nematode populations increase rapidly and soon the bacteria and yeast are overgrazed. Thus the mites regulate predaceous nematodes, allowing bacteria populations to efficiently decompose litter. Mineralization is a critical soil process that converts organic compounds to inorganic forms that can then be utilized by plants. Carbon is primarily mineralized by yeast and fungi, whereas predaceous protozoans, nematodes and some mites convert organic nitrogen to inorganic compounds. Fungi occupy a considerable area around the decomposing plant debris, and in the process they immobilize several essential plant nutrients that they assimilate. Turnover of fungal biomass through death and grazing by microarthropods is the most important process in regulating the release of inorganic nutrients to the soil. Whitford and Parker (1989) conclude that immobilization of nitrogen by the saprophagic soil microflora controls the rate of primary production, and maintenance of healthy microarthropod populations is essential for the release of nitrogen and the continuity of the cycle. Such groups are examples of keystone species and are likely to have important roles in ecosystem stability and function. Considerable research effort is necessary to gain a first impression of which functional groups are present and the basic interactions among them; but an elementary understanding of food webs is essential for formulating rational solutions to land management problems, solutions that will not create future problems.

The tree rhizosphere is an arena of consumer activity and biological diversity. The soil of the rhizosphere is only about 10 percent of the total soil volume but contains 90 percent of the biological activity (Coleman 1996). Without trees, biodiversity and the level of nutrient cycling will decline. Conserving them is the most practical and cost–effective way to sustain the resources they provide and the biodiversity associated with them.

4. INTERACTIONS WITH PLANTS

Trees interact with other plants by either competing with them for resources or by facilitating their germination and growth. The two contradictory associations can change through time in response to various biotic and abiotic factors. However, a discussion of these complex relationships is not within the scope of this paper. Instead, I will concentrate on how desert trees promote biodiversity by providing favorable conditions for the establishment and growth of certain perennial and annual plants. In some cases the association is so strong that future survival of the dependent species may depend on the future existence of the host tree; and paradoxically, the tree may be overtaken and eventually killed by the plant it helped to establish.

The fact that some perennial plants are highly dependent on "nurse trees" has been noted in the literature for many years (Noble and Liok 1999). Their dependence appears mainly related to the moderate microclimate and greater water and nutrient resources surrounding the trees. In summer, plants under the canopy are able to conserve water because transpiration rates are 2–3 times lower (Abd el Rhaman and Batanouny 1965, Smith et al. 1987). The photosynthetic adaptations of many perennial savanna forbs reflect the different demands of their environments. C_3 species occur in shaded areas under trees and C_4 plants, which have greater water use efficiency, live mainly in open grassland (Scholes and Archer 1997).

In northern Oman two perennials frequently grow under the canopy of samr. One, *Lycium shawi*, also occurs in open desert, especially on the margins of wadis, and its presence under trees is probably due to seed dispersal by birds roosting in the trees. The other, *Ephedra ciliata*, is a climber that grows only in the subcanopy environment.

Annual plants frequently have greater productivity beneath tree canopies compared to open desert (see e.g., Smith et al. 1987, Robinson 1999). Species richness can also be greater under perennial plants, and in some arid ecosystems the species composition is distinct from that of open desert. The diversity of annuals is great in drylands, where they constitute around 50 percent of the flora. In northern Oman approximately 50 species of annuals occur in the low coastal desert (Robinson 1999), and about 25 of these are most abundant in the woodland areas. Six annual species are strongly restricted to the subcanopy environment. Two of them, *Plantago ovata* and *Paronychia arabica*, and a third more ubiquitous species, *Aizoon canariense*, produce about 90 percent of the ephemeral biomass under the trees (Robinson 1999). Ephemerals are important seasonal forage for a variety of herbivores and contribute considerable detritus material which helps sustain the soil fauna. Large seed banks develop beneath the trees and because erosion is less there, these refuge sites may contain the only viable gene pools of some taxa, especially in badly degraded deserts where populations of some annuals no longer exist in open spaces (Brown and Porembski 1997).

5. INTERACTIONS WITH ANIMALS

Animals use trees for food, nesting and resting sites and refuge from predators. In Oman, and throughout North Africa and the Middle East trees provide large herbivorous mammals sites for feeding, digesting and resting and waiting out the intense midday summer heat. Recent research by Williams et al. (2000) suggests that without trees the Arabian Oryx (*Oryx leucoryx*) would probably not survive the summer period. By remaining completely inactive and shaded by trees (*Maerua crassifolia*) during the day, oryx are able to radiate excess body heat to the cooler microclimate and also reduce their daily energy requirement by half. This occurs during a period when fresh forage is very rare, and may partially account for their survival in extreme desert habitats. The oryx is a critically endangered species in Oman numbering less than 250 individuals (Fisher 1999). Camels, gazelles, goats, sheep and donkeys probably receive similar physiological benefits.

While feeding in trees, insects continuously drop partially eaten leaves and twigs and excrement onto the soil below. Dung excreted by larger animals decomposes more rapidly than undigested organic matter and contains appreciable amounts of nitrogen. Locusts are famous for the extensive damage they wreak upon crops, but the possible benefits they may contribute to ecosystems during their feeding bouts have not been discussed. Robinson (2001) quantified the amount of fecal nitrogen deposited by Arabian tree locusts (*Anacridium melanorhodon*) during a moderate population outbreak. The mean standing crop of feces under the roost trees (*A. tortilis*) was 65.3 gm/m^{-2}. The feces contained a nitrogen equivalent of 3.2 gN/100g (ash–free dry weight). On one hectare, the locusts deposited about 1 kg of fecal nitrogen under samr trees. This equals about one–third of the total standing crop of detritus nitrogen under perennial vegetation in the Mohave Desert (Rundel and Gibson 1996). However, without knowing the rate at which fecal nitrogen is mineralized, it is not possible to evaluate the impact of such large and irregular inputs on plant productivity.

Deserts are rich in reptile species, especially lizards. About 39 taxa inhabit the low desert regions of northern Oman (A. Gardner, personnel communication), and of these, 7 lizards and 2 snakes commonly utilize trees. Thirty birds breed in the same area. Ten nest on the ground, and 20 (66 percent) raise their young in trees (J. Ericksen, personnel communication). I am not aware of research that enumerates the arthropod species living solely or mainly on trees in the Middle East. Wisdom (1991) summarizes several studies of the herbivorous insects living on mesquite trees (*Prosopis glandulosa*) in North American deserts. Mesquite appears to have a similar ecological role as *Acacia tortilis*, and can serve as an estimator of the potential insect community on samur. Wisdom lists 6 orders representing 40 families and 200 species of herbivorous insects. Mesquite trees have a geographical range that is approximately one–fifth, the area inhabited by *A. tortilis*. It seems reasonable to assume that at least a similar number of species occur on samr trees over its much greater geographical range but generally more arid climates. In the Mohave Desert, which in some respects is climatically similar to Oman's northern deserts, the number of foliage arthropods collected on 5 shrub species varied annually from 150 to 197 (Rundel and Gibson 1996). Perhaps 75 to 100 species utilize samr trees in northern Oman.

6. TREE LOSS IN NORTHERN OMAN

Four trees are common in the low deserts of northern Oman: *Acacia tortilis* (samr), *Prosopis cineraria* (ghaf), *Ziziphus spina–christi* (sidr) and *Maerua crassifolia* (sarh). Samr is by far the most abundant and widespread species. All four may occur together as scattered individuals, often bordering wadi channels, but only samr forms extensive monospecific stands. It appears to tolerate drought and salinity more than the others, and in the arid interior of central Oman it occurs virtually alone. Ghaf trees are more localized, occurring as woodlands on the eastern fringes of the Wahiba Sands and in isolated drainage basins throughout the country. Various aspects of the ecology and conservation of this tree have been reported in Dutton (1988). A comprehensive management plan for the ghaf woodlands is needed along with the infrastructure to implement it, if the trees are to continue to provide animals and the bedouin tribes living there the resources they have used for many years.

A second area where tree conservation is urgently needed is the Batinah, a narrow coastal strip that extends west from Muscat between the Western Hajar Mountains and the Gulf of Oman. The tree formations in this area vary from very open communities of *A. tortilis* and a few shrubs, to more heavily vegetated stands of mixed tree and shrub species. In a few areas, samr trees form open woodlands, which represent the remnants of an arid tropical savanna that was previously more widespread through the lowlands of southern Arabia (Kürschner 1998). This particular woodland is badly degraded from overgrazing and human activities, and the native trees in this region face various threats.

In 1992 the Batinah was the most populous region of Oman with an annual growth rate in excess of 3 percent (MOD 1993). Today, ten years later, probably more than half of the country's people live in the relatively small area of Muscat and the Batinah coast. To keep pace with Oman's accelerated development and population growth, considerable expansion of housing, agriculture and various light industries is underway in the Batinah, compressing the natural landscapes into a narrowing band against the Hajar mountains.

Gravels are being quarried on the alluvial fans to meet the demands for regional building projects. Surface mining is the most severe form of ecosystem disturbance. Gravel quarries have removed about 5 percent of the surface strata from a 250 km^2 area of the eastern Batinah. This activity accelerates desertification (Al Awadhi 2001). Removal of the protective lag gravels increases wind erosion, and extracting the deeper layers destroys soils and their biotic communities. Heavy haul vehicles compact the soil, reducing water infiltration and increasing runoff. Wind transport of the fine–grained tailings that are left behind, and other quarry activities increase dust pollution 16–17 times (Al Awadhi 2001). Generally, regional governments do not consider quarrying to be an environmental problem so there is relatively little regulation of the industry's activities.

The Batinah is Oman's largest and most productive agricultural area. Nearly all of the best agricultural land has been put under cultivation, but as fields become salinized and lose productivity, new land is being cleared at the expense of the tree and shrub communities. Most of this newly cleared land is classified as unsuitable for agriculture (MAF 1990). Irrigation water is supplied by ground water pumping, and as a result seawater is entering the aquifers and increasing the salinity. Plantations of dead date trees and dying woodlands of *Prosopis cineraria* attest to effects of the poor water quality. Near Sur, farms 4 km inland from the sea are being irrigated with well water that exceeds international standards for chloride, sodium and electrical conductivity (Victor and Al Farsi 2001). In addition, pesticides and other chemicals are reaching the ground water.

The human food crops grown in Oman provide for about 75 percent of the total consumption of those species, the remainder is imported (MAF 1995). Will expanding agriculture onto substandard land provide Oman the increased food production it is seeking; and will this be of more value to society than the existing resources and services provided by the natural communities? This question requires careful study.

Many trees in the Batinah show signs of moderate to heavy grazing. Two species, *P. cineraria* and *M. crassifolia*, are favored browse by livestock. In some areas the latter species, which lacks defensive spines, is overgrazed and nearly unrecognizable. *A. tortilis* and *Z. spin–christi* are also grazed, but their heavier, denser spines keep browsers from causing too much damage to the foliage. While browsing, animals may consume reproductive structures (flowers, seed pods) along with new foliage, reducing the plant's reproductive potential. Pods that fall to the ground are eagerly consumed by all livestock, and if seedlings appear they, are soon taken along with nitrogen–rich detritus leaves, which livestock consume when fresh forage is scarce. Removal of the litter increases soil temperatures, which possibly affects seed germination. Seed production may also be decreasing from a scarcity of pollinators. The virtual absence of detritus beneath the trees in many areas of the Batinah may also have profound effects on the soil biota. A distinct characteristic of Oman's rangelands is young trees are very rare.

Finally, the future of several native trees may be compromised by exotic species that have been introduced into Oman during the past 25–35 years. The possible effects invasive species may have on a native flora is illustrated by *Bromus tectorum*, a grass native to the Middle East, North Africa, and the Mediterranean and Central Europe, which has invaded several North

American deserts. Several studies document the rapid and fundamental changes that occur as alien plants become established. Two years after colonizing an undisturbed grassland community, the biomass of *Bromus* litter was twice that of the native grasses. *Bromus* litter has significantly greater C:N and lignin:N ratios than the native species. Inorganic N was 50 percent lower. The change in litter quality and chemistry decreased the nitrogen available to microbes and the rates of net N mineralization, reducing the productivity of the native forbs. *Bromus* increases fire frequency, which in turn results in nitrogen loss through volatilization (Evans et al. 2001).

Concurrent changes occurred in the soil biota. Invaded sites generally had lower species richness and abundance of fungi and invertebrates, but a greater abundance of active bacteria. The soil food web species composition varied significantly among different types of *Bromus*–invaded grasslands and the indigenous grasslands (Belnap and Phillips 2001). Fundamental changes to the structure of the food webs are likely to have notable effects on the nutrient cycles and are possibly related to those reported by Evans et al. (2001).

Alien trees can have pronounced impacts on woodlands and the vertebrates that inhabit them. In the Kalahari Desert, Dean et al. (2002) report that bird communities are richer in species and more diverse in the native *Acacia* karoo woodland compared to woodlands consisting introduced *Prosopis* trees. Differences in the vegetative structure and plant species in the *Prosopis* woodland have affected the composition of the bird assemblages. In the alien woodlands raptors are absent, frugivores are rare and only about half of the insectivore species are present.

Velvet Mesquite (*Prosopis juliflora*), is a leguminous tree from the arid southwestern United States and northwestern Mexico. It was introduced into Oman about 25 years ago and has been cultivated as an ornamental species. As in other countries where it was introduced, it has escaped from cultivation and is quickly spreading into natural habitats. Velvet Mesquite is fast growing and probably more salt–tolerant than *P. cineraria*, which it is replacing in several coastal areas of Oman. It is known to possess allelopathic chemicals. Unlike native legumes, animals are reluctant to eat the seedpods, and its seeds may increasing in seed banks. The subcanopy environment of mesquite is usually vegetation less, suggesting that the soil has become chemically unsuitable for germination of native plants. Various countries have tried numerous methods to remove *P. juliflora*, but none have been very successful. Documenting the impacts of this tree to the biodiversity of Oman's lowland desert communities is urgently needed.

7. BIODIVERSITY AND ECOSYSTEM FUNCTION

The primary goal of this workshop is to identify practices that will allow sustainable use of biodiversity. While many understand what biodiversity is and various methods to conserve it, fewer realize that, more than being a descriptive statistic of natural communities, biodiversity appears to be a prime regulator of ecosystem function. Until we understand how fluctuations in species' numbers and abundances alter the efficiency of a system and the output of its products, we cannot hope to have a very realistic picture of how to maintain sustainability. Ecologists are now trying to identify the components of biodiversity that most affect ecosystem function (Naeem 2002).

In theory, an ecosystem's capacity for some process, such as productivity or decomposition, is enhanced as biodiversity increases. More species allow alternative pathways to develop and to disseminate energy and materials, giving greater stability to the system. Critical to the issue of sustainability is an understanding of how many species can be lost before function is seriously impaired. Short–term experiments in grassland communities (Hector et al. 2001) indicate that productivity reaches a maximum level when plant species number about half the maximum richness that occurs in nature. One might hastily conclude that no serious loss of function would occur until half the plant community was eliminated. However, Hector et al. (2001) and others (see Naeem 2002) caution that what is important is the sequence of species extinctions from the community, and which species or functional group is lost, not simply the number of taxa. Based on functionality, some species (keystone species) are more essential than others and their loss could be catastrophic, while the removal of others would have little effect. Even if all the functional groups are present, they may depend on a diverse pool of species that biologists may not consider essential to the system.

The experiments beg the questions: Which species or functional groups are essential for maintaining the system and its functions? Is it possible to predict the order in which species are lost as the system becomes degraded? Our present knowledge suggests that we may be able to single out species that are crucial to the function of these artificial systems during short–term experiments. But will this subset of taxa continue to maintain the system through long–term environmental fluctuations? We do not know, and it seems doubtful we can predict the outcome. Many biological systems do not have linear responses, instead they behave like complex systems best described by power law distributions, especially during brief phase transitions between ordered and chaotic states (Lewin 1992, Kauffman 1995). This alone should urge us to be cautious. Given the unpredictable nature of future environmental change it seems reckless to assume we can get by with a smaller assemblage of core species. In fact, greater species diversity and the functional redundancy it may bring are positive attributes to buffer the system against change. In arid ecosystems trees are biological attractors and a large proportion of the community biodiversity interacts with them. Tree loss is certain to have adverse effects. Conserving them will benefit society in unexpected ways.

8. SOCIAL BENEFITS PROVIDED BY WOODLANDS

Soil is an important part of a nation's assets, an asset that is being lost at an alarming rate. An estimated 17 percent of the earth's vegetated land surface has been degraded by human activities since 1945 (Oldeman et al. 1990). Human–induced land degradation in arid regions is costing societies a direct loss of income estimated at 42.3 billion dollars per year (UNEP 1991, Dregne et al. 1992). Woodlands and other vegetation help protect soil from the direct destructive effects of rain. Vegetation and the soils that develop over the centuries are the most effective regulators of the hydrological cycle.

Loss of trees and shrubs from the watersheds of the Batinah will increase runoff, potential for flooding, and economic loss to agricultural, urban and industrial areas. Most of the nutrients in dryland ecosystems are concentrated in the uppermost 5 cm of soil. With increased runoff and soil erosion, nutrients will be transported down slope to the sea, or deposited in reservoirs behind retention dams. Increased sedimentation in the reservoirs reduces the volume of water captured and the functional lifetime of the dam. Nutrients trapped in reservoirs percolate into the deep groundwater and are unavailable to plants and other organisms. Although recharge dams divert runoff water into aquifers, they also decrease the natural inflow of sediments to the coastal areas, resulting in abnormal beach erosion.

Retaining areas of natural vegetation in the watersheds will reduce nutrient loss and soil water evaporation and ultimately the rate of soil salinization. Preserving woodlands also conserves the soil biotic communities that are critical for the production of food, fibers, wood, biomass fuels, waste removal and the organisms that have produced important pharmaceuticals (Daily et al. 1997). Their role in recycling these and other materials is fundamental to achieving sustainability, and their functions cannot be replaced by any human technology. In arid lands especially, their value to our societies is yet to be explored.

Natural biotic communities provide a large proportion of our food crop pollinators. This fact is little appreciated by scientists and non–scientists alike. Approximately 90 percent of the world's flowering plant species depend on animals for effective pollination (Nabhan and Buchmann 1997). Studies suggest that plant populations in habitats fragmented by human activities (habitat degradation, agrochemicals, introduced pests, and development) are no longer producing enough seeds to regenerate themselves, due mainly to inadequate pollinators. Burd (1994) concludes that even in ecosystems unaffected by man the reproductive success of nearly half the world's plants may be more limited by pollinator scarcity than by other often cited limiting factors (e.g., nutrient availability, climatic change, seed parasites). As a consequence of general biodiversity loss, the variety of pollinators of wild and domestic plants is diminishing.

In most agricultural systems it has been assumed that crops were pollinated principally by honey bees (*Apis mellifera*). Nabhan and Buchmann (1997) estimate that 12 genera of bees pollinate about 73 percent of the world's crop species, of which *Apis* accounts for only 15.5 percent. In addition to *Apis*, 37 genera of invertebrates and 7 general of vertebrates are know to pollinate 1332 cultivated plants. In the United States agricultural economy the value of non–*Apis* pollinators is between 4.1 and 6.7 billion dollars per year.

Relevant to Oman are the studies of Southwick and Southwick (1992) on alfalfa, which is the primary fodder crop of the Sultanate. They estimate that without native pollinators, which will be needed to supplement declining honey bee colonies in the USA, the alfalfa seed crop production would decrease 70 percent, equivalent to 315 million dollars per year. This argues strongly for protecting pollinators living in adjacent natural habitats of the Batinah from pesticides and land clearing. These woodlands are also home to insectivorous birds that help regulate insect populations.

The fact that about 20 percent of the world's arid lands are currently degraded (Dregne 1983) underlines the urgency for developing cost–effective methods of land reclamation. Various technologies will be needed to accelerate the natural soil processes that have evolved over centuries. Our richest source of biological materials for these technologies is the natural communities that are still productive.

Because of their low and variable rainfall and frequent winds that remove topsoils, deserts present a special challenge to land restoration science. For the past 20 years scientists have been using fungal mycorrhizae as a means of accelerating the revegetation of degraded lands. Mycorrhizae form mutualistic associations with roots of certain plants. Plants with mycorrhizae grow faster, have greater rates of nutrient and water uptake and tolerate drought better (Allen 1989). Fungal propagules are lost as the topsoil erodes, and to re–establish them requires the presence of mycotrophic plants. In North America, severely disturbed land is commonly colonized by non–mycotrophic annual plants such as Chenopodiaceae, Brassicaceae, Amaranthaaceae, Zygophyllaceae, and other families that are also common in Oman. As long as these plants persist the mycorrhizal fungi will not reestablish in the community (Allen and Allen 1988). As the first step to restoration, cultivated seedlings of mycotrophic species are planted with inocula of fungi and nurtured until self sufficient (Allen 1989). This process should be done on above average rainfall years to increase the probability that the plants survive. A next step, which has not been tried, would be to inoculate the soil with biota representing functional groups known to be essential to litter decomposition and mineralizataion. This will require a sound knowledge of the food webs and the technology to culture the indigenous organisms. Recreating the natural soil environment is probably the most efficient way to revérse desertification over the long term.

9. ACKNOWLEDGEMENTS

I wish to thank the organizers for inviting me to participate in the Biodiversity Workshop, "Promoting Best Practices for Conservation and Sustainable Use of Biodiversity of Global Significance in Arid and Semi–Arid Zones," under the auspices of the Global Environmental Facility (GEF) and the Third World Network of Scientific Organizations (TWNSO). Rahma Al Siyabi assisted me during the workshop.

10. REFERENCES

Abd el Rhaman, A., and Betanouny, K. 1965. The water output of the desert vegetation in different microhabitats of Wadi Hoff. Journal of Ecology 53: 139–145.

Al Awadhi, J. 2001. Impact of gravel quarrying on the desert environment of Kuwait. Environmental Geology 41: 365–371.

Al Harthi, L. 2000. Effect of the canopy morphology on the microclimate beneath Acacia tortilis. BSc (Honours) Thesis, Sultan Qaboos University, Oman.

Allen, E. 1989. The restoration of distrubed arid landscapes with special reference to mycorrhizal fungi. Journal of Arid Environments 17: 279–286.

Allen, E. and Allen, M. 1988. Facilitation of succession by the non–mycotrophic colonizer Salsola kali (Chenopodiaceae) on a harsh site: Effects of mycorrhizal fungi. American Journal of Botany 75: 257–266.

Barth, R and Klemmedson, J. 1982. Amount and distribution of dry matter, nitrogen and organic carbon in soil–plant systems of mesquite and paloverde. Journal of Range Management 35: 412–418.

Belnap, J. and Phillips, S 2001. Soil biota in an ungrazed grassland: Response to annual grass (Bromus tectorum) invasion. Ecological Applications 11: 1261–1275.

Belsky, J. 1994. Influences of trees on savanna productivity: tests of shade, nutrients and tree–grass competition. Ecology 75: 922–932.

Belsky, J., Amundson, R., Duxbury, R., Riha, J., Ali, A., and Mwonga, S. 1989. The effects of trees on their physical, chemical and biological environments in a semi–arid savanna in Kenya. Journal of Applied Ecology 26: 1005–1024.

Belsky, J., Mwonga, S., Amundson, R., Duxbury, R., and Ali, A.1993. Comparative effects of isolated trees on their undercanopy environments in high and low–rainfall savannas. Journal of Applied Ecology 30: 143–155.

Bengtsson, J., Setala, H. and Zheng, D.1996. Food webs and nutrient cycling in soils: Interactions and positive feedbacks. Pages 30–38 In: Polis, G. and Winemiller, K. (Editors) Food Webs. Integration of Patterns and Dynamics. Chapman and Hall, N.Y. 472 pages.

Brown, K. 1988. Ecophysiology of Prosopis cineraria in the Wahiba Sands, with reference to its reafforestation potential in Oman. Pages 257–270 In: Dutton, R.W. (Editor) The Scientific Results of the Royal Geographic Society's Oman Wahiba Sands Project 1985–1987. Journal of Oman Studies, Special Report No. 3. Office of the Advisor for Conservation of the Environment, Diwan of the Royal Court, Muscat. 576 pages.

Brown, G. and Porembski, S.1997. The maintenance of species diversity by minature dunes in a sand–depleted Haloxylon salicomicum community in the Kuwait desert. Journal of Arid Environments 37: 461–473.

Burd, M.1994. Bateman's principle and plant reproduction: The role of pollen limitation in fruit and seed set. Botanical Review 60: 81–109.

Coleman, D. 1996. Energetics of detritivory and microbivory in soil in theory and practice. Pages 39–50 In: Polis, G and Winemiller, K. (Editors) Food Webs. Integration of Patterns and Dynamics. Chapman and Hall, N.Y. 472 pages.

Crawford, C. 1991. The community ecology of macroarthropod detritivores. Pages 89–112 In: Polis, G (Editor) The Ecology of Desert Communities. University of Arizona Press, Tucson. 456 pages.

Daily, G., Matson, P. and Vitousek, P. 1997. Ecosystem services supplied by soil. Pages 113–132 In: Daily, G. (Editor) Nature's Services. Societal Dependence on Natural Ecosystems. Island Press, Washington, D.C. 392 pages.

Dean, W., Anderson, M., Milton, S., and Anderson, T. 2002. Avian assemblages in native Acacia and alien Prosopis drainage line woodland in the Kalahari, South Africa. Journal of Arid Environments 51: 1–19.

Dregne, H. 1983. Desertification of Arid Lands. Advances in Desert and Arid Land Technology. Vol 13. Harwood Academic, Chur, Switzerland. 242 pages.

Dregne, H., Kassas, M. and Rozanov, B.1992. A new assessment of the world status of desertification. Desertification Control Bulletin 20: 16–18.

Dutton, R.W. (Editor) 1988. The Scientific Results of the Royal Geographical Society's Oman Wahiba Sands Project (1985–1987). The Journal of Oman Studies, Special Report No. 3. Office of the Advisor for Conservation of the Environment, Diwan of the Royal court, Muscat, Oman. 575 pages.

Evans, R. Rimer, R., Sperry, L. and Belnap, J. 2001. Exotic plant invasion alters nitrogen dynamics in an arid grassland. Ecological Applications 11: 1301–1310.

Fisher, M. 1999. The conservation status of the terrestrial mammals of Oman: A preliminary Red List. Pages 109–129 In: Fisher, M., Ghazanfar, S. and Spalton, A. (Editors). The Natural History of Oman. A Festschrift for Michael Gallagher. Bakhuys Publishers, Leiden. 206 pages.

Garner, W. and Steinbuger, Y. 1989. A proposed mechanism for the formation of 'Fertile Islands' in the desert ecosystem. Journal of Arid Environments 16: 257–262.

Hector, A., Joshi, J., Lawler, S., Spehn, E. and Wilby, A. 2001. Conservation implications of the link between biodiversity and ecosystem functioning. Oecologia 129: 624–628.

Lewin, R. 1992. Complexity. Life at the Edge of Chaos. Macmillen Publishing Co., N.Y. 208 pages.

Lowe, C. and Hinds, D. 1971. Effect of paloverde (Cercidium) trees on the radiation flux at ground level in the Sonoran Desert in winter. Ecology 52: 916–922.

Kauffman, S. 1995. At Home in the Universe. The Search for the Laws of Self–Organization and Complexity. Oxford University Press, Oxford. 321 pages.

Kürschner, H. 1998. Biogeography and introduction to vegetation. Pages 63–98 In: Ghazsanfar, S., and Fisher, M. (Editors) Vegetation of the Arabian Peninsula. Kluwer Academic Publishers, London. 362 pages.

MAF 1990. General Soil Map of the Sultanate of Oman. Ministry of Argiculture and Fisheries. Mazoon Printing, Muscat, Oman.

MAF 1995. The Process of Agriculture and Fisheries Development (1970–1995). Ministry of Agriculture and Fisheries, Muscat, Oman.

Mandeel, Q.A. 2002. Microfungal community associated with rhizosphere soil of Zygophyllum qatarense in arid habitats of Bahrain. Journal of Arid Environments 50: 665–681.

MOD 1993. General Census of Population, Housing and Establishments 1993. Ministry of Development, Muscat, Oman. 334 pages.

Nabhan, G. and Buchman, S.1997. Services provided by pollinators. Pages 133–150 In: Daily, G. (Editor) Natures Services. Societal Dependence on Natural Ecosystems. Island Press, Washington, D.C. 392 pages.

Naeem, S. 2002. Ecosystem consequences of biodiversity loss: The evolution of a paradigm. Ecology 83: 1537–1552.

Noble, P. and M. Loik. 1999. Form and function of cacti. Pages 143–163 In: Robichaux, R.H. (Editor) Ecology of Sonoran Desert Plants and Plant Communities. University of Arizona Press, Tucson. 303 pages.

Oldeman, L., van Englen, V. and Pulles, J.1990. The extent of human–induced soil degradation, Annex 5. In: Oldeman, L, Hakkeling, R. and Sombroek, W. (Editors). World Map of the Status of Human–Induced Soil Degradation: An Explanatory Note. Wageningen International Soil Reference and Information Center.

Robinson, M. 1999. Desert annuals under samur trees in Oman: biomass, density and distribution. Journal of Tropical Ecology 40: 11–17.

Robinson, M. 2001. Desert nitrogen cycles: Fecal nitrogen from a population of the Sahelian Tree Locust in Oman. Journal for Scientific Research–Science and Technology, Sultan Qaboos University 6: 33–38.

Rundel, P. and Gibson, A. 1996. Ecological communities and processes in a Mohave Desert ecosystem. Cambridge University Press, Cambridge. 369 pages.

Scholes, R. and Archer, S. 1997. Tree–grass interactions in savannas. Annual Review of Ecology and Systematics 28: 517–544.

Smith, S., Patten, D. and Monson, R. 1987. Effects of artifically imposed shade on a Sonoran Desert ecosystem: Microclimate and vegetation. Journal of Arid Enivironments 13: 65–82.

Southwick, E. and Southwick, L. 1992. Estimating the economic value of honey bees (Hymenoptera: Apis) as agricultural pollinators in the United States. Economic Entomology 85: 621–633.

UNEP (United Nations Environmental Programme). 1991. Status of Desertification and Implementation of the United Nations Plan of Action to Combat Desertification. Nairobi, UNEP.

Vetaas, O. 1992. Micro–site effects of trees and shrubs in dry savannas. Journal of Vegetation Science 3: 337–344.

Victor, R. and Al Farsi, A. 2001. Water quality and invertebrate fauna on farm wells in an area affected by salinization in Oman. Journal of Arid Environments 48: 419–428.

Weltzin, J. and Coughnour, M. 1990. Savanna tree influence on understorey vegetation and soil nutrients in northwestern Kenya. Journal of Vegetation Science 1: 325–334.

West, N. and Klemmedson, J. 1978. Structural distribution of nitrogen in desert ecosystems. Pages 1–16 In: West, N. and Skujins, J. (Editors) Nitrogen in Desert Ecosystems. Dowden, Hutchinson and Ross, Stroudsberg, Pa. 314 pages.

Wisdom, C. 1991. Patterns of heterogeneity in desert herbivorous insect communities. Pages 151–179 In: Polis, G. (Editor) The Ecology of Desert Communities. University of Arizona Press, Tucson. 456 pages.

Whitford, W. 2002. Ecology of Desert Systems. Academic Press, N.Y. 343 pages.

Whitford, W. and Parker, L. 1989. Contributions of soil fauna to decomposition and mineralization processes in semiarid and arid ecosystems. Arid Soil Research and Rehabilitation 3: 199–215.

Whitford, W., Anderson, J., and Rice, P. 1997. Stemflow contribution to the fertile island effect in creosotebush, Larrea tridentata. Journal of Arid Environments 35: 451–457.

Williams, J., Ostrowski, S., Bedin, E., and Ismail, K. 2000. Seasonal variation in energy expenditure, water flux and food consumption of Arabian Oryx Oryx leucoryx. Journal of Experimental Biology 204: 2301–2311.

Zak, J. and Freckman, D. 1991. Soil communities in deserts: Microarthropods and nematodes. Pages 55–88 In: Polis, G. (Editor) The Ecology of Desert Communities. University of Arizona Press, Tucson. 456 pages.

Chapter 31

SITE SELECTION CRITERIA FOR MANGROVE AFFORESTATION PROJECTS IN OMAN

Peter Cookson[1] and Tomoo Shoju[2]

[1]*Department of Soil and Water Sciences, Sultan Qaboos University, Al–Khod 123, Muscat, Sultanate of Oman; email peterc@squ.edu.om;* [2]*Japan International Cooperation Agency and Ministry of Regional Municipalities, Environment and Water Resources, Sultanate of Oman*

1. INTRODUCTION

The ability of *Avicennia marina*, the only indigenous mangrove species in Oman, to tolerate both an arid climate and high salinity allows this tree to form an integral part of the coastal habitat. Fouda and Al–Muharrami (1996) estimated the area of mangroves to be 1003 ha, scattered around the coast in no less than 30 sites. The factors that allow mangroves to flourish in some sites and not others are not fully understood. It has been speculated that over grazing in isolated areas eradicated trees and dispersal from distant forests has not occurred. Other possible causes for the decline in mangrove forests are the climate shift to more arid conditions some 6000 years ago, excessive exposure to wind and tide action, and unsuitable soil properties at many locations for germination and seedling development. In an attempt to arrest the decline in mangrove areas, a comprehensive afforestation program has been initiated.

Afforesting sheltered kwars and lagoons with mangroves is expected to provide habitats for a host of marine and bird life. Al–Muharrami (1994) reported that 51 species of fish, 31 species of crustaceans and 51 species of mollusks were found in the mangrove forests at Shinas, Qurm and Mahout Island. Mangrove forests have also been long exploited by humans for shelter, firewood, seafood and forage for grazing animals (Constable and Stiffe 1883). Most mangrove forests are grazed, either by goats or camels, to some degree, despite a Ministerial Decree banning the use of natural trees for this purpose in 1993.

A study of mature trees of *Avicennia marina* in the Qurm Nature Reserve in Muscat (Cookson and Lepiece 1997) found that soils of loamy sand texture, being aerobic in the surface horizons and anaerobic below, appeared optimal for growth. However, the success of an afforestation program depends on selecting transplanting areas where seedling development into young trees is most favorable. The aim of the present study was to investigate soil properties at four recently transplanted sites. Seedlings at these sites exhibit a range in development from early senescence to attaining a maximum height of 0.78 m. Sand particle size distribution, texture, bulk density and salinity were measured at each site.

1. TRANSPLANTING PROCEDURES

Six month old mangrove seedlings were transplanted to Suwadi in February 2002, and four month old seedlings to Sur in November 2002. Seedlings were transported to each site from the Al–Qurm nursery in a shaded lorry, the plastic bags holding the seedlings were removed, and seedlings carefully placed in 20 cm deep holes at a spacing of one meter. No fertilizer or chemical amendments were applied.

1.1 Description of Transplanting Sites

Approximately 12,000 seedlings were transplanted to an estuary at Suwadi, in three adjacent blocks of land that extended from low to high tide level. The intertidal distance was approximately 50 meters. Approximately 8,000 seedlings have been transplanted around the shores of an extensive tidal lagoon at Sur, at three sites, designated A, B and C. Site A was near a small, muddy port where the soil was anaerobic. Site B was located on sandy material some distance from the port, which attracts attention from the local football players. Site C represented a relatively open sandy beach where some mature mangrove trees survive.

1.1 Sampling Procedure

At Sur, soil samples were taken at random from different locations in each site from two depths, 0–20 cm and 20–40 cm. The colour of the soil was described at the time of sampling. At Suwadi, soils were sampled from beside individual seedlings in a transect stretching inland from most seaward point for 24 meters. Samples for bulk density determination were taken using a rigid walled cylinder 10 cm in depth.

1.1 Soil Analysis

After drying soils at 80 °C, texture was determined by dispersing samples in calgon solution in 1 liter cylinders and measuring suspension density, with a hydrometer. The proportions of sand, silt and clay particles were calculated using the method of Klute (1965) Afterwards, soil samples were washed and the material passing passed through a 2mm sieve discarded. Each sample was dried and passed through a series of sieves, 0.053, 0.106, 0.250, 0.500 and 1.000 mm in diameter. The amount of sand retained by each sieve was weighed and expressed as a percentage of total sand. The bulk density of samples from Suwadi was calculated from the internal dimensions of the sampling cylinder and the dry weight of the sample. Values of soil bulk density were expressed in g/mL. Saturated hydraulic conductivity (K) of soil was measured by determining the time taken for a known volume of water at a set head to pass through a repacked saturated soil column of known cross sectional area and depth (Klute 1965). Values of K were in cm/min.

1. RESULTS AND DISCUSSION

1.1 Soil Properties

Percentages of sand fractions as determined by wet sieving in the soil samples are given in Table 1a–b. The sand fractions of soils from most sites were dominated by particles of either fine or very fine diameter. However, soils from location 1 at site C contained relatively larger proportions of coarser sands than in other sampling sites. Much of the coarser material at this location appeared to consist of a large proportion of crushed shell fragments. An interesting relationship appeared to exist between the relative proportion of fine to very fine sand and the degree of aerobism found in the soils. The degree of aerobism was assessed from soil colour, i.e., reddish soils were judged to be aerobic and grayish soils anaerobic. If the ratio of fine to very fine sand is less than one then the soil tended to be anaerobic, and if greater than one, then aerobic. Soil textures were mostly sandy or loamy sand. However, at Site A silt contents in excess of 35 percent produced fine textured sandy loam soils, which were strongly anaerobic.

The physical characteristics of soil influence the amount of water a soil can hold (i.e., the available water) and the rate at which water drains from the soil. According to Landon (1984) the available water held by a fine sand is 100 mm/m, whereas for a very fine sand is amount is 180 mm/m. The figure of 180 mm/m is actually larger than the figure given for finer textures, including Silt Loam, i.e.170 mm/m.

Saturated hydraulic conductivity values of soils ranged from 0.01–0.54 cm/min, and varied considerably between sampling locations. Conductivities of water through saturated sandy loam soils were often at rates of less than one tenth of those through saturated sands. These figures are consistent with the observation that. anaerobic conditions were associated with soils of low saturated hydraulic conductivity having a ratio of fine to very fine sand of less than unity, and silt contents of over 20 percent.

The significance of relationships between various sand fractions and saturated hydraulic conductivity were assessed by using Spearman's ranked correlation coefficients as given in Table 2. The proportions of larger sand particles were not correlated to saturated hydraulic conductivity of the soils. In contrast, the proportions of fine sand were positively related to hydraulic conductivity, while the proportions of very fine sand were negatively related. The proportions of fine relative to very fine sand were significantly related to increased hydraulic conductivity. High silt and clay tended to significantly reduce hydraulic conductivity. Brown and McLachlan (1990) accounted for the lower hydraulic conductivities in fine than coarse beach sands by the fact that smaller pore sizes of the finer material restrict the flow of water more than larger pores of coarse material. The reduced pore sizes, in effect, reduce the rate at which seawater flushes through the surface layer of fine beach sands.

Soil properties at Sur and Suwadi can be interpreted from a wider perspective of estuary development. Clifton (1992) considered that tides shape estuaries into a series of tidal flats and drainage channels. In the upper reaches of estuaries, sediments tend to be 'muddier' than downstream because of discharge from rivers. Sands often dominate downstream, where coarse material is deposited by tides. However, depositional patterns in different estuaries are subject to a range of site specific factors, including complex combinations of tides, tidal currents, wave action, river discharge, precipitation temperature and local flora and fauna. Using the criteria listed by Brown and McLachlan (1990) for assessing the degree of exposure experienced by intertidal zones then the beaches at Sur and Suwadi would fall in the sheltered range. Specifically, these beaches are characterized by a slight wave height, rarely more than 1m, very wide surf zone, high amount of very fine sand, a median particle size diameter of between 0.180–0.250

mm on a low intertidal slope, reduced zones either at or near the soil surface, and burrowing animals present.

Table 1a. Soil properties of Mangrove transplanting sites

Site	Depth (cm)	vcS %	cS %	mS %	fS %	VfS %	fS:vfS ratio
Site A -Sur							
Location 1	0-20	1.2	2.6	3.3	39.3	53.6	0.73
	20-40	1.1	3.2	3.6	23.7	68.3	0.35
Location 1	0-20	1.4	3.8	5.1	23.1	66.5	0.35
	20-40	0.8	2.7	4.3	20.3	71.9	0.28
Site B -Sur							
Location 1	0-20	1.2	3.3	11.7	67.8	16.0	4.24
	20-40	0.9	2.8	6.8	72.1	17.4	4.14
Location 2	0-20	1.0	4.0	10.6	64.6	19.7	3.28
	20-40	0.4	3.0	8.9	69.7	18.0	3.87
Location 3	0-20	0.6	2.0	6.8	60.6	30.0	2.02
	20-40	1.0	2.8	8.8	70.4	17.0	4.14
Location 4	0-20	0.2	0.7	3.5	78.8	16.8	4.69
	20-40	0.5	1.4	4.9	28.4	6.7	4.24
Site C -Sur							
Location 1	0-20	6.5	11.4	15.8	41.1	25.2	1.63
	20-40	8.8	13.5	16.1	34.3	27.2	1.26
Location 2	0-20	4.1	7.0	8.1	30.1	50.7	0.59
	20-40	4.4	5.2	7.5	32.1	50.9	0.63
Location 3	0-20	0.1	1.1	5.8	38.6	54.4	0.71
	20-40	1.4	2.9	7.0	50.2	38.6	1.30
Site D -Suwadi							
Location 1	0-20	0.3	0.7	4.1	85.3	9.0	0.81
Location 2	0-20	0.2	0.8	2.0	77.5	17.6	1.17
Location 3	0-20	0.2	0.4	1.4	76.8	19.2	1.74

soil size vc: 2.0–1.0mm, c: 1.0–0.50mm, m: 0.50–0.25, f: 0.250–0.106, vf: 0.106–0.053mm

Table 1b. Soil properties of Mangrove transplanting sites

Site	Sand %	Silt %	Clay %	Texture	Sal dS/m	SHC cm/min	Aeration status
Site A -Sur							
Location 1	56.2	35.3	8.5	SL	17.4	0.0320	anaerobic
	55.7	37.4	6.9	SL	20.3	nd	anaerobic
Location 1	47.3	44.6	8.1	SL	17.1	0.0102	anaerobic
	50.9	41.8	7.3	SL	18.8	nd	anaerobic

Table 1–b (contined). Soil properties of Mangrove transplanting sites

Site	Sand %	Silt %	Clay %	Texture	Sal dS/m	SHC cm/min	Aeration status
Site B -Sur							
Location 1	83.7	11.0	5.4	LS	19.6	0.2947	aerobic
	85.3	7.2	7.5	LS	20.5	nd	aerobic
Location 2	90.8	5.9	3.3	S	19.9	0.5476	aerobic
	88.6	6.6	7.8	S	20.4	nd	aerobic
Location 3	90.7	5.0	4.3	S	20.9	0.5393	aerobic
	89.5	6.1	4.4	S	19.1	nd	aerobic
Location 4	95.5	3.9	0.6	S	19.8	0.5089	aerobic
	90.6	5.0	4.4	S	23.3	nd	aerobic
Site C -Sur							
Location 1	76.4	16.5	7.1	LS	18.7	0.1225	aerobic
	77.4	21.2	1.4	LS	22.0	nd	aerobic
Location 2	68.9	26.0	5.2	SL	19.5	0.0559	anaerobic
	68.8	24.1	7.1	SL	20.6	0.0592	anaerobic
Location 3	76.4	20.5	3.1	LS	26.1	0.0625	anaerobic
	70.9	24.2	5.0	SL	24.0	nd	aerobic
Site D-Suwadi							
Location 1	97.1	2.6	0.3	S	18.6	nd	aerobic
Location 2	98.1	1.2	0.7	S	9.6	nd	aerobic
Location 3	96.8	3.2	0-20	S	18.2	nd	aerobic

S:sandy, LS:loamy sand, SL:sandy loam; sal: salinity; SHC: saturated hydraulic conductivity

Table 2. Spearman ranked correlation coefficients between soil properties

	Percent Sand	Percent Silt	Percent Clay	Saturated Hydraulic Conductivity
Vc sand	-.49*	.52*	.22 ns	-.46 ns
c sand	-.30 ns	.36 ns	.18 ns	-.23 ns
M sand	.26 ns	-.21 ns	-.18 ns	38 ns
f sand	.74***	-.71**	-.23 ns	. 83**
Vf sand	-.83***	.84***	.34 ns	-.77**
f:vf ratio	.87***	-.89***	-.35 ns	.83**
percent sand	-	-.98***	-.61**	.94***
percent silt	-.98***	-	.55*	-.95***
percent clay	-.62**	.55 ns	-	-.69*

Note: ***, **, * indicate correlation coefficients significant at the 0.001,0.01 and 0.05 level, respectively. ns indicates a non–significant correlation coefficient.

1.2 Seedling Growth Performance

Mangrove seedlings were four months old when transplanted to the sites at Sur. Within two months obvious differences in seedling survival rates were recorded. Virtually all seedlings failed to grow in site A, where soil was anaerobic. At site B, again nearly all seedlings failed to develop where soils were anaerobic, but in areas of aerobic soil between 25 and 30 percent of seedlings survived transplanting. At Site C, where soils were sandiest around 40 percent of mangrove seedlings survived.

The low rates of seedling survival found at Sur are in contrast with the more than 80 percent survival rate recorded at Suwadi. Even at Suwadi, however, the rates of seedling growth differed within the transplanting site. The effect of variations in soil bulk density on the growth of one year old mangrove seedlings is shown in Figure 1. Bordering the estuary channel, soils were compact with bulk densities of over 1.5 g/cm^3, and mangrove growth height was above 0.50 m. Soil bulk density initially decreased inland and then increased. Mangrove seedlings tended to be taller in soils with bulk densities of less than 1.3 g/cm^3. The lower bulk density may have allowed deeper root penetration either through reducing soil resistance to root extension or a greater degree of soil aeration between tides. Variations in bulk density of beach sands has been long recognized (Chapman and Newell 1947), and is often considered to result from wave action producing a rippled surface. The crests of sand ripples at both Suwadi and Sur appeared to have lower bulk densities than of the troughs between them.

Figure 1. Relationships Between Soil Bulk Density and Mangrove Seedling Height Eight Months After Transplanting

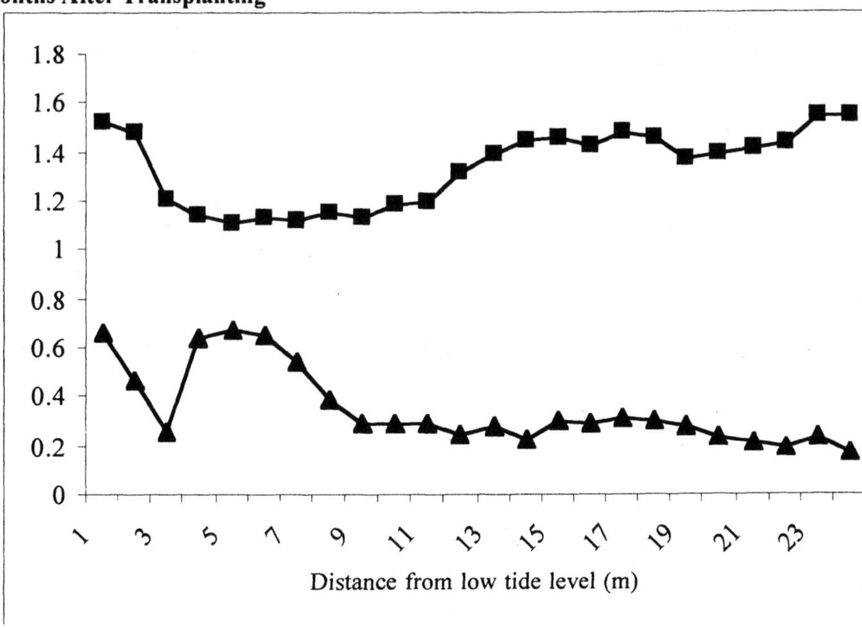

Salt did not accumulate in excess of 25 dS/m in soil to a depth of 20 cm, when regularly flushed with seawater. Soil salinity in the transplanting sites under study did not appear to be a limiting factor to *A. marina* establishment. However, soil salinity at depths greater than 40 cm has been found to exceed seawater salinity in mangrove soils at Qurm (Cookson and Lepiece 1997). Excessive salinity in subsurface soil horizons may limit rooting depth of established mangrove trees. However, increased depth of rooting may also help to reduce salinity through improving subsurface drainage and, thus, allow more effective flushing by seawater.

1.3 Criteria for Selecting Potential Transplanting Sites

Successful establishment of rigorously growing seedlings of *A. marina* is essential for afforestation of coastal Oman with mangroves. Potential afforestation sites need to be carefully selected to avoid the waste of effort of producing healthy seedlings in a nursery and, then, subsequently suffering nearly 100 percent seedling loss following transplantation, as was the case at Site A in Sur. This study concentrated on assessing only the soil characteristics of proposed afforestation sites. Site selection factors not investigated in this study, included degree of beach exposure to waves, i.e. whether beaches are dissipative or reflective, length of exposure of seedlings to air between tides, and the influences of climate. Afforestation of open beaches and the influence of climate (for example in Salalah) on seedling development are issues requiring future investigation.

Since *A. marina* have been transplanted only to sheltered coastal areas of Northern Oman, the site selection criteria listed below are applicable only to lagoons, and estuaries:

1. Soil surface should be free of a salt crust, and salinity should not exceed 30 dS/m
2. Soils should be aerobic and sandy in texture, (with fine to very fine sand ratios of more than one, and silt and clay contents below 10 and 5 percent, respectively).
3. Saturated hydraulic conductivity of surface soil layers should exceed 0.1 cm/min
4. Soil bulk density should less than 1.3 g/cm^3 (unless bordering tidal channels).

2. CONCLUSIONS

The study reported on the variable survival rates of over 20,000 mangrove seedlings transplanted to four coastal sites in Oman. Seedling survival rates varied from 0.1 to over 80 percent, and the main reason for this variation was the varying degrees of anaerobic conditions in surface soil horizons that, in effect, suffocated growth. It was concluded that poor aeration was related to a restriction of seawater flow, either because of a low fine to very fine particle size ratio in sandy textured soils or relatively high amounts of silt and clay in sandy loam textured soils. An additional factor influencing seawater drainage was, also, high bulk density of soil. Soil salinity did not appear to be a serious factor in causing seedling development to differ between sites.

3. REFERENCES

Al–Muharrami M.B.A. 1994. Resource Management of Mangroves in the Arid Environment in the Sultanate of Oman. Unpublished M.Phil. Thesis, Department of Marine Sciences and Coastal Management, University of Newcastle, U.K.

Brown, AC., and McLachlan, A. 1990. Ecology of sandy shores. Elsevier, Amsterdam. 238 pages.

Chapman, G.A. and Newell, G.E. 1947. The role of the body fluids in relation to movement in soft–bodied invertebrates. I. The burrowing of *Arenicola*. Proceedings of the Royal Society, London B 134: 431–455.

Clifton, H.E. 1992. Estuarine deposits. Pages 179–190 In: Scholle, P.A., and Spearing, D. (Editors) Sandstone Depositional Environments. American Association of Petroleum Geologists, Tulsa.

Cookson, P., and Lepiece, A.G. 1997. Classification of oils supporting mangroves in the Sultan Qaboos Qurm Nature Reserve (Oman). Journal of King Abdulazziz University: Marine Sciences 8: 149–168.

Constable, C.G. and Stiffe, A.W. 1883. The Persian Gulf Pilot: The Gulf of Oman and the Makran Coast. Page 83, The Hydrographic Office, Admiralty, London.

Day, P.R. 1965. Particle fractionation and particle–size analysis. πages 545–567 In: Black, C.A., Evans, D.D., White, J.L., Ensminger, L.E. and Clark, F.E. (Editors) Methods of Soil Analysis Part 1. American Society of Agronomy, Madison.

Fouda, M.M. and Al–Muharrami, M. 1996. Significance of mangroves in the arid environment of the Sultanate of Oman. SQU Journal for Scientific Research Agricultural Sciences 1: 41–49.

Klute, A. 1965. Laboratory Measurement of Hydraulic Conductivity of Saturated. Pages 210–221 In: Black, C.A., Evans, D.D., White, J.L., Ensminger, L.E. and Clark, F.E. (Editors) Methods of Soil Analysis Part 1. American Society of Agronomy, Madison.

Landon, J.R. (Editor) 1984. Pages 75–77 In: Booker Tropical Soil Manual. Longman, U.S.A. 450 pages.

Chapter 32

THE GLOBALLY THREATENED CORNCRAKE *CREX CREX* (EGYPT)

Andrew Grieve[1] and Waheed Salama[2]
[1]*Ornithological Society of the Middle East, Caucasus and Central Asia, 3 Miriam Close, Second Avenue, Caister–on–Sea, Great Yaarmouth, Norfolk, UK, email ag@osme.org;*
[2]*Egyptian Environmental Affairs Agency, El Arish, North Sinai, Egypt*

1. INTRODUCTION

The Corncrake *Crex crex* is a globally threatened and vulnerable species due to a long–term and steep decline of its breeding numbers and range (BirdLife International 2000). It is listed in Appendix II of the Bonn Convention on the Conservation of Migratory Species of Wild Animals, to which many countries are signatories including Egypt. The Corncrake's breeding range extends over much of northern and central Europe and east to Russia and central Siberia. It migrates south in autumn, especially through Egypt and Arabia, to winter in sub–Saharan Africa (Cramp and Simmons 1980). The recent estimate of the total world population is between 1.6 and 3 million singing males (BirdLife International 2001). Corncrakes have been declining in Europe since the last century, with declines of 20–50 percent in many countries in the last 10 years (BirdLife International 2001).

2. LONG–TERM DECLINES

The long–term declines reported in many countries in Europe are now known to be mainly the result of low breeding success, due in particular to very heavy chick mortality. The decline of the Corncrake coincided with the introduction of mechanized mowing and a change to silage production rather than traditional hay cutting throughout Europe. Silage is cut earlier in the year when Corncrakes have small chicks in the hay fields. A secondary reason for declines has been the introduction of herbicide and chemical fertilizers that have reduced vegetation variety and reduced the number of invertebrates available for food to chicks and adults. A further loss of suitable nesting habitat has been caused an increase in sheep grazing with a consequent loss of tall vegetation (Green and Stowe 1993).

The wintering areas in southern Africa still appear to be suitable to sustain Corncrakes and at this time there is no indication of habitat change or any major threat to Corncrakes in winter (Stowe and Becker 1992).

The only other recognized danger to the survival of Corncrake has been identified during the autumn when large numbers of birds are trapped whilst on migration through Egypt (and to a lesser extent in other countries in this region) on their way to wintering grounds in Africa (Baha el Din 1991).

3. THE CORNCRAKE PROJECT IN EGYPT

The Corncrake project in Egypt between 1994 and 1997 comprised the following activities: (1)to carry out field surveys to calculate the number of Corncrakes trapped in autumn on the Mediterranean coast of Egypt; (2)to try to reduce the number of Corncrakes killed in Egypt during the autumn migration period by talking to hunters, local people, decision makers and distributing information to support the releasing of Corncrakes; (3)to carry out a study program on migrant Corncrakes to provide biometric data from trapped birds and feather samples for DNA analysis; (4)to provide training for personnel of the Egyptian Environmental Affairs Agency in survey techniques, public awareness campaigns and management of protected areas for migrant Corncrake; and (5)to assist with the development of a visitor center and educational program at Zaranik Protected Area, north Sinai, Egypt, to raise public awareness for the Corncrake.

Literature searches have revealed that Egypt appeared to be the most important country for concentrations of Corncrake during the autumn migration and accordingly this project utilized the status of Corncrake as a globally threatened species to promote sustainable biodiversity in the arid Mediterranean coastal belt of Egypt (Goodman and Meininger 1989).

4. SURVEY OF THE TRAPPING OF CORNCRAKE ON AUTUMN MIGRATION IN EGYPT

Sample 10 km long sections were surveyed for traps, guns and nets in a 1 km wide strip of land adjacent to the shore of the Mediterranean coast of Egypt in the autumn of 1994; the results are shown in Table 1.

Table 1. Numbers of traps and guns and metres of net in operation during surveys of randomly selected 10 km sections of the Mediterranean coast of Egypt in autumn 1994

Trap type	Coast	Delta	Sinai	Total
Trammel net	(1) 0 (2) 0	(1) 4,067 (2) 117,935	(1) 7,072 (2) 14,857	266,450
Mist net	(1) 22 (2) 1,158	(1) 0 (2) 0	(1) 0 (2) 0	1,158
Haleg trap	(1) 55 (2) 2,904	(1) 0 (2) 0	(1) 2 (2) 35	2,939
Eb net	(1) 1,147 (2) 60,797	(1) 1,381 (2) 40,068	(1) 1 (2) 16	100,881
Toraha	(1) 0 (2) 0	(1) 10 (2) 290	(1) 10 (2) 210	500
Romaka	(1) 13 (2) 669	(1) 0 (2) 0	(1) 0 (2) 0	669
Airgun	(1) 34 (2) 1,777	(1) 0 (2) 0	(1) 4 (2) 79	1,856
Shotgun	(1) 21 (2) 1102	(1) 2 (2) 68	(1) 0 (2) 3	1,173
(1)Traps, guns or net per 10 km section; (2)Total number of traps, guns or metres of net				

These data were used to estimate the number of traps, guns and nets in use along the whole of the coastal strip to 1 km inland. Further information was obtained by interviewing bird trappers and asking them how many Corncrakes they had trapped and this data was used to estimate the number of Corncrakes taken per trap, net and gun in the 1993 and 1994 seasons. The total number was then calculated to be as many as 11,000 and 17,000 respectively for the whole coast as shown in Table 2.

Table 2. Estimated numbers of Corncrakes (with bootstrap 95 percent confidence limits (CL) (Efron 1979) taken by all capture methods in autumn 1994 and autumn 1993.

	Coast	Delta	Sinai	All Areas
Trap type	95% CL	95% CL	95% CL	95% CL
Trammel net	–	16–206	965–4,257	1,055–4,332
Mist net	8–51	–	–	8–51
Haleg trap	205–704	–	1–6	205–700
Eb net	242–717	29–302	–	247–938
Toraha	–	399–5,061	833–3,563	130–6,27
Romaka	1352–2658	–	–	1,352–2,658
Airgun	2,554–4,441	–	30–442	2,761–4,707
Shotgun	242–498	–	0–3	242–900
All methods 1994	5,628–8,425	409–5,181	1,819–7,220	9,689–17,412
All methods 1993	4,782–8,739	225–2,050	326–2,092	6,572–11,641

Trapping methods and intensity varied with locality but the taking of Corncrakes was widespread as part of the by–catch of Quail. It was estimated that about 8,000 people were involved in trapping and shooting in the coastal strip with many more people trapping birds further inland as well. The results show that the "toraha," a small hand net, is the most lethal trapping method with an estimated total of as many as 6,827 Corncrakes taken by this method.

The average proportion of the world population of Corncrake taken on the coastal strip of Egypt per year, as evidenced by these surveys, was estimated to be in the range of 0.5 to 2 percent (Baha el Din et al. 1996). The total numbers trapped in the whole of Egypt will be considerably higher than this as trapping has been observed to take place up to 30 kms from the coast (Baha al Din 1991). It is also known from ringing recoveries that the world–wide population of Corncrakes does not migrate through Egypt. It is likely that this mortality may be impacting more severely on that part of the population that does pass through Egypt. It is remarkable that one small area, the Mediterranean coast of Egypt, should produce such a large measurable mortality over a period of 6–9 weeks of the autumn and thereby raising significant conservation concern for the future welfare of the species. It has since become known that a similar trapping intensity now takes place along the coast of Palestine and possibly in other countries (BirdLife International Palestine Officer, personal communication).

5. CONSERVATION MEASURES WITH TRAPPERS

During the autumns of 1994–1996 a concerted effort was made to reduce the number of Corncrakes killed along the Mediterranean coast of Egypt. The survey had discovered that for the majority of trappers, hunting was not their main occupation. Most trapping was carried out in spare time from jobs in farming, fishing and the commercial sector of towns and cities. Trappers were encouraged to avoid trapping and shooting Corncrakes and to release those they had caught. After every interview each trapper was presented with a booklet, children were given posters whilst stickers were widely distributed and displayed in prominent positions. These materials were in Arabic and were sponsored by the UK Royal Society for the Protection of Birds and BirdLife International. The booklet, written by Mindy and Sherif Baha el Din, BirdLife representatives for Egypt, describes the Corncrake the conservation and importance of Egypt on its migration. Most trappers showed interest and understanding of the situation and expressed a willingness to cooperate. The majority promised to release all the Corncrakes they caught in the future. The extent to which these promises have been kept is uncertain but some trappers visited again after the interview and in succeeding years claimed to have released or avoided catching Corncrakes.

6. DECISION MAKERS AND OTHER STAKEHOLDERS

A series of meetings was organized in Sinai as part of the public awareness program. In each year the Governor of Sinai was met in the Regional Government Offices in El Arish (a change of Governor occurred in

1995) and the importance of protecting Corncrakes on migration at sites such as Zaranik Protected Area in north Sinai was fully explored. This assisted in maintaining the Protected Area as a non–hunting area, the only section of the north Sinai coast with no trammel nets.

Additional meetings were arranged with district governments where local elected representatives and Bedouin sheikhs were present. The importance of Corncrakes and the need to maintain the Zaranik Protected Area was explained and generally a good response to the presentations was evident. Sustainable development of a visitor and training center, research station and bird observatory for Zaranik was welcomed and the prospects for local employment and contributions to the local economy by international and local visitors was explored.

Visits were also arranged for decision makers and important local people to the new visitor center at Zaranik when it was opened and subsequently and many guided visits have been undertaken around the protected area when the problems for migrants and in particular the Corncrake have been detailed.

7. PUBLIC AWARENESS AND CAPACITY BUILDING

The bird observatory was already present at Zaranik Protected Area at the start of this project and it has been used subsequently by up to 100 international visitors each year. In 1997, new visitor center was constructed by the Egyptian Environmental Affairs Agency. A team of 5 staff are now employed at Zaranik Protected Area to deliver protection for the area and to carry out a public awareness program.

8. THE FUTURE IS WITH THE YOUNG

The public awareness program aimed at providing an experience to visiting children from schools, students from colleges and universities and the general public at Zaranik has been funded by the World Wide Fund for Nature. The center manager or his assistant give all visitors a lecture in the lecture hall, visits are made to the displays and then the groups are taken around the protected area to see the wildlife and the coastal sabhka and lagoons. The Corncrake and the problems with over–hunting are discussed in the lecture and during the site visit. This is an important ongoing public awareness program that will slowly provide an awareness of the need to protect species and habitats to contribute towards global biodiversity.

9. RESEARCH AND TRAINING

Linked to the establishment of a bird observatory and research center, studies into the migration of Corncrakes through Zaranik have been undertaken in the autumn. A local Bedouin hunter was employed to catch Corncrakes with his "toraha" and 92 were trapped at Zaranik between 1994 and 1996, showing how easy it is to trap with this method. Biometric data were taken of the trapped birds to compare with data from other parts of their range and weight studies initiated. Interestingly, birds were found to range in weight from light to heavy birds indicating that the migration strategy is one of resting on migration along the Egyptian coast rather than a need to land due to lack of fitness to carry on.

Feathers were also collected from trapped birds before they were released back into the wild so that DNA analysis could be carried out. New techniques with radio isotopes may show the origin of these birds and from this information additional knowledge will be obtained on which parts of the breeding range migrants come from that pass through Egypt.

Further bird ringing and migration studies have been carried at Zaranik by German and Belgium ornithologists and the development of the bird observatory is expected to attract many more foreign ornithologists to carry out such studies. Zaranik is also to be used as a training center for ornithologists and conservationists from Egypt and the Middle East which will further enhance the role of the site and highlight the problems associated with hunting and in particular the trapping of Corncrakes.

10. NEXT STAGES AND FURTHER DEVELOPMENTS

The Corncrake project and the various elements that have developed from it will need to be continued for some time, particularly the public awareness and education program. There needs to be better legislation to protect Corncrakes. In many countries certain forms of hunting are restricted such as the use of pole traps in the UK and banning of the use of the "toraha" would dramatically reduce the number of Corncrakes trapped in Egypt. The Corncrake is a protected species in Egypt but its protection is not enforced by the law enforcement agencies.

More protected areas need to be established in coastal areas of Egypt to protect Corncrakes and other species breeding and resting on migration to sustain biodiversity. Further migration studies could then be carried out on the Corncrakes and other taxon at such sites as Zaranik to better understand their migration strategies. This would provide valuable scientific research possibilities for many people and training for future researchers and conservationists from Egypt and further afield. These developments will also contribute to the development of environmentally friendly green tourism, which is more sustainable than many other forms of tourism now developing along the Egyptian coasts. This will contribute to sustaining biodiversity for future generations and ultimately the quality of life for humans into the foreseeable future.

11. LESSONS FOR BEST PRACTICE

Protecting Corncrakes in Egypt has provided some valuable lessons showing possible best practices for conservation of biodiversity projects. Not all objectives were achieved and parts of the project are ongoing and will be so for many years. The objectives that were set for the project were simple and considered achievable and were given time limits. These are important first assessments when considering best practice.

The collection of scientific data demonstrated the extent of the problem and provided the basis for further assessment in future years. Ongoing monitoring was not fully achieved but can be accomplished in future years.

The project had adequate resource support during the period 1994–1997 including the level of manpower required to carry it out. Capacity building was also promoted by the government agencies but ongoing resources were probably not sufficient to sustain the level of the project after the current level of funding expired.

This project illustrates international cooperation that leads to an exchange of skills and the raising of skill levels by personnel involved in the project. The development of a training center has not been fully utilized but may be developed in future years.

Public awareness for the plight of Corncrakes and issues relating to the trapping of birds were raised with many audiences during the project and this is a positive aspect of the project. This involves local communities along the coast and the hunters actually carrying out the trapping. Integrating the development of a protected area with the local population is another good practice. The employing of local people to manage the protected area and the carrying out of the various aspects of the Corncrake project is another positive aspect of the project.

The development of the Zaranik Protected Area has also been environmentally friendly without destroying the fragile semi–arid habitats along this part of the Egyptian coast such as saline lagoons, sabhka and coastal dunes. These provide other important economical and biological functions such as sea defense, fishing, grazing and salt extraction together with the provision of a range of habitats for a variety of animals, birds, insects and plants.

Using the Corncrake as a flagship species to demonstrate the importance of coastal migration areas is also good practice and an alternative to using the habitat and ecosystem approach to conservation and sustaining biodiversity in the region. Monitoring the fortunes of the Corncrake on migration through Egypt may be one way to show whether Egypt is sustaining the biodiversity of its coastal areas.

12. ACKNOWLEDGEMENTS

We are grateful for the assistance of personnel from the Egyptian Environmental Affairs Agency who assisted with the carrying out this project and for many valuable contributions and suggestions by Mindy and Sherif Baha el Din, BirdLife International Representatives in Egypt. Dr. Rhys Green from the Royal Society for the Protection of Birds carried out the statistical analysis.

13. REFERENCES

Baha el Din, S. 1991. The Catching of Corncrakes *Crex crex* and Other Birds in Northern Egypt. Unpublished report. Royal Society for the Protection of Birds. Sandy, Bedfordshire, UK..

Baha el Din, S., and Salama, W. 1991. The Catching of Birds in North Sinai, Autumn 1990. (ICBP Study Report No. 43). International Council for Bird Preservation. Cambridge, UK.

Baha el Din, S.M., Salama, W., Grieve, A and Green, R.E. 1996. Trapping and shooting of Corncrakes *Crex crex* on the Mediterranean coast of Egypt. Bird Conservation International 6: 213–227.

(BirdLife International). 2000. Threatened Birds of the World. BirdLife International. Cambridge, UK. 365 pages.

Cramp, S., and Simmons, K.E.L. (eds.). 1980. The Birds of the Western Palearctic Vol. 6. Oxford University Press. Oxford, UK. 615 pages.

(BirdLife International) 2001. BirdLife's online World Bird Database Version 1.0. BirdLife International, Cambridge, UK. http://www.birdlife.net.

Efron, B. 1979. Bootstrap methods: Another look at the jack–knife. Annual Statistics 7: 1–26.

Goodman, S.M., and Meininger, P.L. 1989. The Birds of Egypt. Oxford University Press, Oxford, UK. 456 pages.

Green, R.E., and Stowe, T.J. 1993. The decline of the Corncrake in Britain and Ireland in relation to habitat change. Journal of Applied Ecology 30: 689–695.

Stowe, T.J., and Becker, D. 1992. Status and conservation of the Corncrake *Crex crex* outside the breeding grounds. Tauraco 2: 1–23.

Chapter 33

CONSERVATION OF FIG (FICUS CARICA L.) AND POMEGRANATE (PUNICA GRANATUM L.) VARIETIES IN TUNISIA

Messaoud Mars
Ecole Supérieure d'Horticulture et d'Elevage 4042 Chott–Mariem, Sousse, Tunisia; email messaoud.mars@laposte.net

1. INTRODUCTION

Arid lands cover about two thirds of Tunisia and are subjected to severe climatic conditions. The arid parts of the country are characterized by a large diversity of phytogenetic resources. The Matmata mountain chain is considered as a transition zone with high genetic potential. Many spontaneous and cultivated plants of this area can be well adapted to both saharan and humid regions. In this area, fruit trees, cereals and legumes are the main cultivated plants under rain–fed conditions. Farming practices are mainly based on Tabias/Jessours system giving a landscape specific to the region. Olive and fig are by far the most important fruit crops followed by almond and grape (Zammouri 1999).

The oases represent vast genetic reserves as many crops were developed there since ancient times. Several horticultural crops are grown under irrigation. Date and pomegranate are the most important crops. Their association in a three–level cultivation system along with herbaceous vegetables and forages is characteristic of the oases (Ghorbel et al. 1996)

Local fruit cultivars are developed by farmers. They are well adapted to local agroecological conditions and have been cultivated for centuries. Growers contributed efficiently to the increase and the maintenance of genetic diversity as well as to the development of valuable technical knowledge. However, this heritage is increasingly neglected and threatened due to: (1)social and economic pressures leading to the emigration of farmers looking for new financial prospects; (2)changes in the farming systems in rural areas following the introduction of mechanization; (3)introduction of new fruit species and varieties and propagation of few performant local cultivars in commercial nurseries; and (4)outbreak of diseases and water and soil salinization in some areas, especially oases (Mars et al. 1998, Mars and Marrakchi 1998)

Thus the problem of genetic erosion is real, whereas information available about local fruit germplasm is limited. The conservation of fruit genetic resources in Tunisia had been undertaken by national institutions, mainly the Institut National de la Recherche Agronomique de Tunisie (INRAT) (Valdeyron and Crossa–Raynaud 1950). Efforts were focused on olive and date palm, the major fruit crops. A specialized institute, Institut de l'Olivier and a specialized research center, Centre de Recherches Phoenicicoles were created for these species. Many other fruits were neglected.

In order to preserve arid land fruit germplasm many actions were undertaken since early 1990s by the Institut des Régions Arides (IRA), Médenine, in close collaboration with Ecole Supérieure d'Horticulture (ESH), Sousse, and Faculté des Sciences (FST), Tunis (Mars et al. 1994).

2. PROJECT OBJECTIVES

The overall objectives are: (1)inventory, identification, and conservation of arid land fruit genetic resources that can be used to stabilize fragile ecosystems while making these more productive through the development of new uses for fruits and subproducts; (2)sanitation and propagation of selected material for replanting; (3)safeguarding popular knowledge related to the dietary, pharmaceutical and industrial use of local plant material; (4)in situ and on–farm conservation of traditional varieties; and (5)enhancement of public awareness and collaboration between specialized research institutions, NGOs and local the population for conservation of plant genetic resources.

3. TARGET SPECIES

Target species for our study were fig and pomegranate. These are well adapted to arid and semiarid environments and are grown in rain–fed as well as irrigated fruit orchards. Both species have important agroecological and socioeconomic roles.

Fig tree (*Ficus carica* L., Moraceae) is common in Tunisia (Hodgson 1931). In the south, fig is grown in small rain–fed plantations, very often associated with other fruit species or as scattered trees in home gardens. Fruits are mainly used fresh. Some parts are sun dried and small quantities are used for the production of jam and alcoholic beverages. Some subproducts are used for animal feeding. Varieties are local and numerous. Fruits produced in traditional plantations are not adequate for the present market demand. Few varieties are being propagated commercially and used in new plantations. Others are neglected, particularly those that need pollination or caprification (Mars 1995).

Pomegranate (*Punica granatum* L., Punicaceae) is considered an excellent fruit tree in arid and desert areas and is grown for valorizing marginal lands and saline waters without special requirements; it is also considered the principal fruit of the oases. Traditional plantations are irrigated in limited acreage. Local varieties are numerous. Their denominations are strictly local in origin, mainly from the place of cultivation or the color of the fruit. Pomegranates are consumed fresh or processed into juice, syrup or jam. They also are used traditionally as medicine. Fruits normally are sold in local markets and small quantities (about one percent) are exported (Mars 1995).

Fig and pomegranate have earlier been considered as secondary fruit trees. In recent years, their cultivation has increased considerably. Total acreage today is about 12,000 ha and 19,000 ha respectively for pomegranate and fig plantations (Ministère de l'Agriculture 2000).

4. METHODOLOGY AND WORK PLAN

4.1 Prospection

Prospection took place mainly in old plantations during the fruiting season (summer/autumn) with the help of technicians of development agencies and farmers. Local cultivars encountered in each prospected area were registered on the basis of pomological characteristics and local denominations (Mars et al. 1994, Mars and Marrakchi 1998). While prospecting, open discussions were conducted with farmers about project activities and issues, cultivation techniques, and variety characteristics and uses.

4.2 Propagation

Collection of plant material representative of the observed diversity took place in winter. Labeled hardwood cuttings were taken during dormancy and planted in a nursery for rooting. The following season, rooted cuttings were transplanted in the orchards where the ex situ collected material is maintained as living plants. The same techniques were used for in situ and on–farm replanting. For some pomegranate and caprifig cultivars, preliminary trials were carried out to determine the optimal technique of vegetative propagation (Mars and Marrakchi 1998).

4.3 Evaluation

Assessment of fig and pomegranate cultivars was made primarily on the basis of pomological and agronomic characteristics (Mars and Sayadi 1992, Mars and Gaaliche 1993, Ben Salah et al. 1995, Mars et al. 1998). Fruits were first sampled on trees in situ for preliminary characterization. Once the collections were made, all accessions were assessed under the same conditions. Other parameters related to shoots, leaves and flowers were used for further characterization of accessions and to determine the overall degree of polymorphism within local germplasm.

4.4 In Situ Conservation

In situ on–farm replanting and conservation of local varieties was carried out in the framework of research and development activities implemented with the help of a local NGO "Association des Jeunes de Zammour" (AJZ) and with the aid of development agencies relevant to different ministries. Farmers were selected on the basis of their receptiveness, awareness and activities related with genetic resources; the diversity of plant material in their plots was also considered. Target varieties were planted directly or propagated in the nursery and then transplanted in the selected plot.

5. RESULTS

Almost all cultivation zones of pomegranate and many traditional fig cultivation zones were prospected (Mars et al. 1994, Mars et al. 1998, Mars and Marrakchi 1998, Arayedh 2000). Ex situ collection of fig established at IRA (Médenine) contains 54 accessions representing 39 local denominations: 21 of Smyrna type, 9 of Common type, 5 of San Pedro type and 4 caprifigs. Another small collection containing 20 accessions is located at ESH (Sousse) in Central Tunisia. Ex situ collection of pomegranate is established at Gabès (South East Tunisia) and contains 63 accessions covering 20 local denominations. Another collection containing half the accessions is located at ESH (Mars and Marrakchi 1998). Distribution maps of local major pomegranate and fig varieties have been updated (Mars and Gaaliche 1993).

Propagation methods and nursery techniques have been improved. For example, satisfactory results were obtained both at nursery and after transplantation. Howeverr, use of basal hormonal treatment was necessary to enhance rooting ability of cuttings in some pomegranate and caprifig cultivars (Mars and Marrakchi 1998).

Accessions of fig (20) and pomegranate (30) are already well described on the basis of pomological and agronomic characteristics with a horticultural classification (Mars and Gaaliche 1993, Ben Salah et al. 1995, Mars 1995, Mars et al. 1998, Mars and Marrakchi 1998, Mars and Marrakchi 1999).

Genetic diversity of local fig and pomegranate germplasm was estimated using morphometric studies that revealed considerable infra–specific variability among accessions. Some areas presented particular richness of ecotypes (e.g., oases for pomegranate). Many individual cultivars and multiclone varieties were identified. Some misidentifications and taxonomic discrepancies were detected (Mars and Marrakchi 1999).

Although many cultivar names referred to the external fruit skin color, particularly in fig, this characteristic was not so useful for cultivar discrimination. Self–devised variety descriptors used at the beginning were well improved. A descriptor list for pomegranate has already been finalized, reviewed by experts in the Mediterranean countries and presented to International Plant Genetic Resources Institute (IPGRI), Centre International des Hautes Etudes Agronomiques Méditerranéennes (CIHEAM) for publication (Mars et al. 1997).

Traditional techniques of fruit handling, fruits and other subproducts in the farms were registered. Several traditional products were displayed at local, national and international exhibitions in the Mediterranean region. New potential uses such as juice production and essential oil extraction from sour pomegranate variety seeds were suggested. Best clones were identified and preselected. Several are already propagated and replanted in small plantations and home gardens.

Attempts of in situ on–farm conservation involving farmers themselves were undertaken (Neffati and Mars 1999). For fig, the Valley of Zammour (west of Médenine) was selected as the target area. It is situated in the middle part of Matmata mountains and is considered as an agroecosystem where different activities are developed by villagers.

An inventory of fruit species and varieties was carried out in the selected area. Rare and endangered cultivars were identified (Zammouri 1999). Best units (Jessours) that may be considered for the establishment of a "Network of genetic resources conservation" were identified and many rooted cuttings of selected local fig cultivars were replanted there.

Fundamental aspects related to floral biology and reproduction were studied (Mars and Marrakchi 2000). Major constraints affecting productivity and fruit quality were determined (Haj Ahmed 2000, Arayedh 2000). Some solutions were suggested.

These experiments are now extended to another region, Douiret, that is considered as a traditional area for fig and olive cultivation. All activities including plant material assessment, rehabilitation and replanting are implemented with the help of a local NGO "Association pour la Sauvegarde de la Nature et la Protection de l'Environnement à Douiret" (ASNAPED), United Nations and Development Program (UNDP) and Agencia Espanola de Cooperacion Internacional (AECI).

Research results were presented and discussed in several local, national, regional and international scientific meetings and workshops. Field days and demonstration sessions were organized with technicians and farmers. Project team members participated in several local festivals and exhibitions. Various academic groups frequently studied the plant material collections. Thus, the public awareness of the importance of local germplasm in sustainable development has been improved. Technicians and growers are aware of the relative lack of adaptation of introduced varieties. Growers associations are increasingly interested in the selection and improvement of local high value clones for use in new plantations.

6. COLLABORATION AND PARTNERSHIPS

Main research activities are carried out at IRA in collaboration with the Universities (ESH and FST) and development agencies relevant to the Ministries of Agriculture and Environment and Land Use Planning and Local Communities. It was envisaged at the beginning of the project that, for such activities, it is necessary to strengthen relationships between all partners namely the research institutions, development agencies, local organizations, local communities, authorities and growers. Authorities at local and national levels were gradually involved through the dissemination of technical and scientific documents generated by the project.

Collaboration is extended to regional levels and an important network "Collaborative Working Group on Underutilized Fruit Crops" (CIHEAM) was established in 1994. International symposia on fig and on pomegranate respectively were organized in 1997 and 1998. Experiences and results were shared among participants. Plant material exchange and collaborative research programs were discussed (Mars 1999). IRA is the focal point for pomegranate within the CIHEAM working group and the FAO's "Mediterranean Selected Fruits Inter–country Network" (MESFIN).

Members of the project team were involved in some activities of REMERGE "Réseau Méditerranéen des Ressources Génétiques" as well as the regional project "Germplasm for Arid Lands" conducted in the framework of "Initiative of Collaboration to Control Natural Resources Degradation" (World Bank/ICARDA). Recently, IPGRI–CWANA has initiated collaboration with some central Asian countries.

7. LESSONS LEARNED

Rural areas and home gardens remain the main source and beneficiary of horticultural crops diversity. Farmers are the producers and the end users of plant diversity and knowledge of food and agriculture. These factors should be taken in account while planning projects. Diversity of fruit genetic resources is huge. It is impossible to collect all diversity. Since the main objective is to conserve the evolutionary and dynamic potential of genetic systems, it is necessary to emphasize more on in situ conservation of gene pools. Ex situ collections of perennial woody plants cannot exceed a limited number of accessions because they present many technical problems, particularly in arid zones with scarce water resources.

It is relatively easy to demonstrate the importance of local genetic resources to growers. Due to external pressures, it is still hard to convince them to conserve cultivars with low economic value. Conservation of native land–races could be easier through the development of novel uses to promote them as performant cultivars. However, this is not a guarantee unless it is integrated into a larger sustainable agriculture development system. Such activities cannot be efficiently initiated and implemented without adequate financial support.

It is necessary to develop diverse and complementary plant genetic resources and conservation strategies adapted to different situations. It is also necessary to strengthen collaboration with research institutions in other countries having similar conditions in order to develop new approaches and long–term studies on biodiversity issues.

8. GLOBAL SIGNIFICANCE OF CASE STUDY

The technical and socioeconomic impact of this project will increase agricultural production, stabilize rural populations and slow down desertification. It will contribute to the establishment of sustainable agriculture through rural development programs. Results and lessons learned can be shared with specialists and users in other countries. Identified material could be exchanged after the establishment of safety guidelines for its movement. Ecotypes developed in this transition area between Sahara and humid regions of Africa should be of high interest because of their adaptive potential for global warming and climatic change.

9. FUTURE PROSPECTS

Future plans include the development of new methods (markers) for cultivar identification and genetic studies, establishment of a database for plant material and local knowledge, an inventory of diseases infecting fig varieties and sanitation of plant material, expansion of fig in situ replanting experience, development of different uses of fig and pomegranate fruits and subproducts and strengthening of collaboration and interchange of experiences, results and plant material.

10. REFERENCES

Arayedh, W. 2000. Contribution à l'étude de la caprification et de la diversité des caprifiguiers au sud–est de la Tunisie. Projet de Fin d'Etudes du Cycle Ingénieur, Ecole Supérieure d'Horticulture de Chott–Mariem, Juin, 2000. 34 pages.

Ben Salah, M., Ancilotti, M. and Loumerem, M. 1995. Etude pomologique de six variétés de figuier (Ficus carica L.) typiques de Beni Khédache. Plant Genetic Resources Newsletter 104: 16–20.

Evreinoff, V.A. 1949. Le grenadier. Fruits d'Outre–Mer. 4: 161–170.

Ghorbel, A., Kchouk, M.L., Fnayou–BenSalem, A., Mars, M. and Rhouma, A. 1996. Morphoqualitative descriptions in ex situ collections of date palm, fig and pomegranate. Pages 125 –140 In: INIA Proceedings of the first MESFIN Plant Genetic. Resourses Meeting, Tenerife, Spain, 2–4 Oct. 1995. 334 pages.

Haj Ahmed, N. 2000. Contribution à l'étude des problèmes d'éclatement et de pourriture des fruits du grenadier dans les oasis littorales. Mémoire de Fin d'Etudes du Cycle Ingénieur, Ecole Supérieure d'Horticulture de Chott–Mariem, Novembre, 2000. 43 pages.

Hodgson, R.W. 1931. Le Figuier. Pages 96–100 In:.La culture fruitière en Tunisie, son état actuel, ses possibilités et son amélioration. Rapport de mission d'études fruitières en Tunisie. Ed. Soc. Anonyme de l'Imprimerie Rapide de Tunis. 131 pages.

Mars, M. and Sayadi, S. 1992. Etude comparative de la qualité des fruits de cinq variétés de grenadier (Punica granatum L.). Revue des Régions Arides 4: 45–57.

Mars M. and Gaaliche, F. 1993. Les variétés de grenadier en Tunisie. Ed. GOVPF/Alpha S.A. 32 pages.

Mars, M., Carraut, A., Marrakchi, M., Gouiaa, M. and Gaaliche, F. 1994. Resources génétiques fruitières en Tunisie (poirier, oranger, figuier, grenadier). Plant Genetic Resources Newsletter 100: 14 –17.

Mars, M. 1995. La culture du grenadier (Punica granatum L.) et du figuier (Ficus carica L.. en Tunisie. Cahier Options Méditerranéennes 13: 85–95.

Mars, M., Melgarejo, P., Amoros, A. and Martinez, R. 1997. Pomegranate Descriptors. CIHEAM Collaborative Working Group on Underutilized Fruit Crops in the Mediterranean region, April 1997. CIHEAM–IAMZ (Spain). 20 pages.

Mars, M. and Marrakchi, M. 1998. Conservation et valorisation des ressources génétiques du grenadier (Punica granatum L.) en Tunisie. Plant Genetic Resources Newsletter 114: 35–39.

Mars, M., Marrakchi, M. and Chebli, T. 1998. Multivariate analysis of fig (Ficus carica L.) germplasm in southern Tunisia. Acta Horticulturae 480: 75–81.

Mars, M. and Marrakchi, M. 1999. Diversity among pomegranate (Punica granatum L.) germplasm in Tunisia. Genetic Resources and Crop Evolution 46: 461–467.

Mars, M. 1999. Pomegranate in the Mediterranean region: Needs of collaborative research programs. Pages 99–104 In: Tuscu, O. and Kuden, A.B. (Editors) Proceedings of the Third General Meeting of MESFIN, Adana, Turkey, June 10–12, 1999.

Mars, M. and Marrakchi, M. 2000. Variabilité intra–arbre chez le grenadier (Punica granatum L.): application à l'échantillonnage des fruits. Fruits 55: 347–355.

Ministère de l'Agriculure, 2000. Résultats de l'enquête sur le suivi de la campagne agricole 1998–1999. S/D Statistiques Agricoles–DG/PDIA, Tunis. Février 2000. 60 pages.

Valdeyron, G. and Crossa–Raynaud, P. 1950. Les fruits de Tunisie. Annales du S.B.A.T. 23: 1–124.

Zammouri J., 1999. Inventaire des espèces et des variétés fruitières de la Vallée de Zammour. IRA Médenine. 13 pages.

Chapter 34

A SOCIOLOGICAL PERSPECTIVE ON IN SITU CONSERVATION OF ON–FARM CROP GENETIC DIVERSITY IN MOROCCO

F. Nassif[1] and A. Birouk[2]
[1]Institute National de la Recherche Agronomique (INRA), P.O. Box 589, Settat, Morocco, email fnassif@iam.net.ma; [2]Institut Agronomique et Veteinaire Hassan II (IAV–Hassan II), Rabat, Morocco

1. INTRODUCTION

This case study presents the sociological work of the Moroccan component of a global project on "strengthening the scientific basis of in situ conservation of agricultural biodiversity" coordinated by the International Plant Genetic Resources Institute (IPGRI) and implemented in nine countries. The main objectives of the global project are: (1)to support the development of a framework of knowledge on farmer decision making processes that influence in situ conservation of agricultural biodiversity; (2)to strengthen national institutions for the planning and implementation of conservation programmes for agricultural biodiversity; and (3)to broaden the use of agricultural biodiversity and the participation in its conservation by farming communities and other groups.

The project's objectives and methodologies were the result of several discussions between IPGRI and national partners; they were agreed on at the first participants meeting held in Rome in July 1995. Two years later, the second participants meeting provided the opportunity to national partners, IPGRI staff and technical experts to discuss and agree on the types of needed information, focal areas and methodological tools to achieve agreed upon objectives. More importantly, within the global project framework there was an emphasis on integrating key socioeconomic variables such as gender, ethnicity, and socioeconomic status in data gathering efforts as well as in decision making.

The participation of Moroccan scientists from the two leading national research and teaching institutes in the Rome meetings and in the global project preparation over the period 1995–1998 was instrumental in helping the national project team to develop the necessary national and international partnerships for project implementation. In addition, Moroccan researchers increasingly gained knowledge and understanding of a new research area where methodologies and models were yet to be developed. In fact, the preparation of the country component was undertaken as a joint venture between IPGRI and the national team (Jarvis and Hodgkin 1998).

Effective implementation of Morocco's component started in May 1998 with the signature of the memorandum of understanding between IPGRI and directors of the Agronomic and Veterinary Institute Hassan II (IAV–Hassan II) and the National Agricultural Research Institute (INRA). Target crops include durum wheat, barley, faba bean and alfalfa. Selected sites include: (1)Bouhrazen and Lahrouna farming villages in the province of Azilal (Atlas Mountains), (2)several villages in the High Ziz Valley in the Errachidia province (Oasis area), and (3)several villages in the province of Taounate (pre Rif mountains).

The main goal of Morocco's component was to develop the scientific basis for setting up a national strategy of in situ conservation of genetic diversity of cropped plants. Specifically, the three main objectives of the project were: (1)strengthening the national capacity in the area of in situ conservation of cropped plants, (2)understanding the factors and processes affecting crop genetic diversity, and (3)identifying adding value options for target crop landraces and associated local knowledge.

Preparation and planning are crucial elements to project success. Two planning workshops were held in Morocco before the project was implemented. The organization of these meetings was facilitated through the National Plant Genetic Resources Committee (NPGRC), a consultative and coordination body for research activities on plant genetic resources. The first five–day workshop was held at the end of February 1998. Participants included IPGRI global and regional staff, representatives of Morocco's the funding agencies, several national and local nongovernmental organizations, and scientists. The workshop focused on stakeholders' analysis, problems analysis, and objectives identification. A major output of the workshop was a consolidated document with agreed upon objectives. Another important output was the participatory process itself through which most participants became acquainted with the project and contributed to the definition of its objectives on the basis of the country specificity. The participation of the funding agency of the country component in the workshop was useful since it contributed to the definition of the project agenda. Participants to the first workshop agreed also on the composition of project teams with the identification of the members of the Project Advising Committee (PAC) and the coordinators for the three sites. In addition to the project coordinators from IAV Hassan II and INRA, PAC included scientists in plant genetics, agroecology and sociology, and representatives of local extension institutions. Key tasks of the committee were to provide technical back up, to approve plans and reports, and to insure links with the coordination of the Global Project at IPGRI and the NPGRC.

The second workshop was held in June 1998 to develop a detailed project work plan. Workshop participants came from research and teaching institutions, provincial directorates of agriculture, and nongovernmental organizations (NGO), and the IPGRI global project coordinator. This workshop specified project outputs and indicators, needed activities and their time frame, target sites and people, and estimated budget and staff requirements (Birouk and Mellas 2000). The project's long preparatory phase both at global and national levels was extremely useful because it facilitated the smooth implementation of the project, especially insofar as helping to raise participants' awareness on the necessity to work in multidisciplinary and multiinstitutional teams.

2. ISSUES ADDRESSED BY THE CASE STUDY

The three main issues addressed by the in situ project in Morocco are: (1)What are the environmental, biological and human factors which affect crop genetic diversity? (2)How crop diversity expressed in landraces or local cultivars of four crops is maintained in farmers fields? (3)How can in situ on–farm conservation be not only a required condition for sustainable crop genetic diversity but also an economically attractive option for farmers whose livelihoods depend on these crops (see, e.g., Jarvis and Hodgkin 2000)?

In addressing the complex issues required to understand crop diversity, scientific efforts were coupled with both abroad and in–country training opportunities as well as awareness raising activities. One of the main issues repeatedly raised throughout all the research teams meetings was the concern with standardization of adopted methodologies or at least a common agreement on key themes, data collection tools, and where they ought to be applied. In fact, during the first months of project implementation, the scientific teams of plant scientists, sociologists and agroecologists held several meetings. The agenda was to define priority areas and discuss methodologies, and harmonize the work to be done in the various sites. In the field, research teams have worked along disciplinary and multidisciplinary lines as well. Research activities focused on: (1)the characterization of the farming systems and farmer management of target crops, (2)the identification of partner farmers, (3)the acquisition of baseline information, and (4)the understanding of women and farmer perceptions of crop landraces.

3. DEVELOPING AN APPROACH

The four elements of Morocco's project approach were participation of all concerned parties, gender balance, multidisciplinary work, and multiinstitutional collaboration. One might argue that these elements are common to most projects, or at least they are commonly claimed by most projects. Indeed, the challenge faced by the project team and probably also true in other projects was not merely conceptual but operational. The task was how to effectively conduct and implement multi–institutional, multidisciplinary, and participatory and gender sensitive research in an area where full experience in these areas is lacking.

The project team was aware of the advantages of multi–institutional collaboration. In fact, IPGRI Director and Directors of IAV Hassan II and INRA signed the Letter of Agreement for Morocco's participation in the project. Consequently, the project advisory committee and research teams included scientists from both institutions as well as from other national, regional and local institutions. This was done on the basis of the two preparatory meetings as well as visits to the sites and contacts with key development people. The result was the development of partnership relations with various institutions at the site level such as the Provincial Directorates of Agriculture (DPA) in Azilal and Taounate Provinces and the Office Régional de la Mise en Valeur Agricole du Tafilalet (ORMVA–TF) at Errachidia Province and their representatives at the local level. Also, existing local nongovernmental organizations (NGO) have been encouraged to participate in project activities. Examples of these NGOs are the Local Development Association (ADL) in the case of the Bouhrazen locality and ADRAR in the case of the High Ziz Valley site. The inclusion of a member of the PAC and a representative from local extension institutions also helped foster coordination. As the project activities progressed, multi–institutional representation and involvement became a permanent feature of project management and decision making.

Another key element that was emphasized during project development was the multidisciplinary composition of research teams and effective implementation of project activities by multidisciplinary field teams.

Research teams included cereals geneticists and breeders, alfalfa geneticists, faba bean geneticists, sociologists, biotechnology specialists, soil scientists, and agroecologists. Some of these scientists are also members of the site working teams.

One of the most important aspects of the in situ project in Morocco is the participatory and integrated character of the approach in overall project planning and implementation. Not only has the project succeeded in building strong multi–institutional and multidisciplinary teams but also in effectively conducting project research and other activities in a multidisciplinary manner with socioeconomic and biological scientists working together with farmers and other local groups. The project team was convinced that the use of a participatory approach was necessary to achieve project objectives (see, e.g., Jarvis and Hodgkin 2000). Consequently, special efforts were taken to establish a common understanding of the project goals and objectives and to establish dialogue with local partners, especially with farmers and local extension agents. Although the advantages and benefits of participatory work are many, they often are limited by constraints of money, time, and energy.

Since its inception, the project "strengthening the scientific basis of in situ conservation of cultivated crops" was gender sensitive. One of the underlying principles of the project was the recognition that in situ on–farm conservation of agricultural biodiversity is practically grounded in the work of both women and men farmers. In Morocco, attention was given to the gender dimension of the project early in its life cycle and was emphasized during the two major planning workshops held in Rabat between February and June 1998. Survey questionnaires were designed to be filled out by male or female farmers, and data points on household composition, labor supply and agricultural work for major farm operation were gender disaggregated (Nassif 2000a). Women's perceptions of target crop landraces and their perspectives on in situ on–farm conservation of crop diversity were addressed with groups of women in project sites. Understanding women's roles and contributions to maintaining crop landraces is considered an integral part of the general understanding of in situ conservation of crop genetic diversity. Currently, one of the concerns of the project team is how to promote the recognition and preservation of women's local knowledge on uses of durum wheat, barley, alfalfa, and faba bean landraces as it relates to the genetic diversity of the crops. With respect to gender, participation and national capacity building, the project research teams include both male and female scientists. In–country training opportunities and inter–sites tours and visits have also been gender sensitive.

4. METHODOLOGY

The project's methodology revolved around formal survey methods with participatory techniques. As indicated earlier, a formal survey was conducted in the beginning of project implementation with representative samples of farm–households in selected villages. The questionnaire was developed by the project team and was administered to about 60 heads of farm–households in each of the three sites. Although attention was paid to include women farmers in village samples, there were not many farmers who grew target crops in selected communities. Consequently, a total of 173 male farmers participated in the survey (Table 1). The questionnaire emphasized target crop production, cropping systems and practices, different varieties of crops utilized, comparison between local and improved varieties, sources of seed, and end use of crop production.

Since the questionnaires were completed by male farmers, focus groups discussions were conducted with groups of women from the same farming communities. Discussions focused on the role of women in crop production and use, their participation in key farming decisions, their knowledge of used varieties, and desirable characteristics of preferred varieties. The objectives of the work with women were to: (1)examine any issues insufficiently covered in the questionnaire, and (2)understand women's views and perspectives regarding the use and conservation of crop landraces of selected crops. A total of 89 women participated in these group discussions (Table 1).

Training workshops, field visits, and other participatory techniques were also used to capture farmers' perceptions and knowledge systems with respect to cropped plants and to identify ways of strengthening existing practices towards better conservation of crop genetic diversity and provision of greater benefits to farmers and their families.

Table 1. Number of farmers and women in surveyed focus groups by province and village, Morocco

Province	Village	Surveyed farmers	Women
	Tamagourte	16	13
Errachidia	Tabia	15	12
	Ait Ben Yahia	13	12
	Other	16	
Azilal	Lahrouna	17	19
	Bouhrazen	42	19
Taounate	Timedite		9
	Ljir/Chaynou	54	5

5. SELECTED FINDINGS

Major findings focus on: (1)farm–household and farmer characteristics; (2)prevailing farm systems and the place of target crops in these systems; (3)farmers' generalized use and high appreciation of target crop landraces; (4)farmers' reasons for maintaining landraces; (5)farmers seed management and sources of annual seed; and (6)the important contributions of women to crop production, processing and utilization and their knowledge about studied crop landraces. Characteristics of farmers are shown in Table 2.

Table 2. Farmer and farm household characteristics by province, Morocco

Characteristic	Errachidia Province	Azilal Province		Taounate Province
	High Ziz Valley villages	Lahrouna	Bouhrazen	Ljir Chaynou
Sample size	60	17	42	54
Age of household head	53	50	55	48
Farmer level of education (years in school)	3	4	0.5	–
Average household size	8	6.5	6.8	8
Share of cropping to total farm income	83	70	70	87
Share of non–cropping to total farm income	17	28	29	13
Percent of the land operated by owner	87	83	99	85
Average number of parcels	11	8	6	5
Average farm size (ha)	1.0	5.7	7.2	3.4
Average number of parcels	11	8	6	5
Cultural/ethnic groups	Berber Arab	Berber Arab	Berber Arab	Berber Arab

The average age of farmers varies between 48 years for sampled farmers in the Taounate province and 55 years in the Bouhrazen site (Azilal province). With respect to farmers' level of education measured as the number of years in school, most surveyed farmers have never been to school and the average number of years of schooling is low. In addition, the two main ethnic groups in selected sites and nationwide are the Berbers and Arabs. Other important features which characterize farm operations in study sites are the small size of holdings and the extent of land fragmentation, especially in areas where crops are grown under irrigation conditions such as Lahrouna and villages of the High Ziz Valley (Nassif 2000b). As indicated in Table 2, the average farm operation in the High Ziz Valley is about one hectare with more than 10 parcels on average. Target crops in the cropping system and farmer use of landraces are shown in Table 3.

Table 3. Farmers' use of target crop landraces and sources of seed by crop and site or village, Morocco

Site/village	Ljir/Chaynou			High Ziz Valley		Bouhrazn	Lahrouna
Crop	Barley	Durum wheat	Faba bean	Durum wheat	Alfalfa	Barley	Alfalfa
Area in target crop as percent of total cultivated area	44	24	11	50	31	67	24
Percent farmers using landraces	100	100	100	98	100	100	100
Percent farmers using their own seed	81	76	92	73	67	88	67

Barley covers two thirds of farmers' cultivated area in Bouhrazen while durum wheat is grown on half of farmers cropped area in the High Ziz Valley. The use of crop landraces is extremely high in all of the different study sites and for the different target crops. However, the reasons for farmers' maintenance and appreciation of these crop landraces vary from one crop to another and from one site to another. In general, landraces of durum wheat, barley, alfalfa and faba bean are valued by farmers for their adaptation and their responsiveness to farmers' diverse needs and production objectives. As indicated in Table 4, these crops are essentially used to feed household members, feed the livestock, provide seeds for next year crops and when there is surplus it is sold in the local market.

Table 4. End uses of target crops and farmers reasons for using landraces by crop in Morocco

Crop	Main end–use	Main reasons for using landraces
Alfalfa	Animal feed Sales of seed	High grain yield High seed price Adaptation
Faba bean	Human consumption Animal consumption	Availability Adaptation Quality Tradition
Durum wheat	Human consumption	Availability Adaptation Grain Quality Yield
Barley	Human consumption Animal consumption	Availability Multiple use Adaptation Grain quality Grain and straw yield

Data on farmers' seed management included: (1)sources of annual seed, (2)seed selection practices, (3)seed storage, and (4)seed quality. Farmers' own production is the main source of annual seeds for barley, durum wheat, faba bean and alfalfa across all studied sites (Nassif 2000c) (Table 3). The second most important source of seed supply is the local market. The highest value is associated with faba bean in the Taounate site where the quasi–majority of farmers seem to draw their faba bean seeds from previous on–farm production, and the lowest value is found for alfalfa with two thirds of the cases of farmers in the Lahrouna High and Ziz Villages depending on the production for their own needs.

Farmers' seed selection practices vary by crop. In general, durum wheat followed by alfalfa are the crops receiving farmers' greatest care while barley tends to receive the least care. As stated by farmers, parcels are carefully selected and closely watched when destined to seed production of durum wheat. Criteria used for parcel selection include preceding crop and soil type; once sown by durum wheat parcels usually receive fertilizer or manure and are kept weed free. Selection of best spikes represents farmers' most common selection practice and seed cleaning is the only means used to prepare seed before planting. Unlike durum wheat, specific parcels are seldom chosen for barley production. Rather, barley is often grown on marginal fields and very little or no selection is done. Seed conditioning is never practiced for both durum wheat and barley. According to farmers' statements, seed renewal intervals for durum wheat is between three to four years in the High Ziz Valley while barley seeds are never renewed in Bouhrazen.

The analysis of seed storage separate from the rest of harvested grain storage (unless the grain was initially produced as seed) is complicated because of the size of farmland, the diversity of the cropping system, the structural production constraints faced by farmers, and the amount of harvested grain of any particular crop. In general, storage systems vary by site and by crop. For example, traditional wooden boxes are commonly used in the High Ziz Valley whereas fiber or plastic bags are widely used for cereals in the other sites. Because of the economic value of alfalfa seeds and their limited availability, women store alfalfa seeds in fiber or metallic containers which are carefully protected from contamination and spoilage by oil and humidity.

Interestingly, one of the most important issues raised by surveyed farmers and women was seed quality seen as landrace seed purity in the case of alfalfa and durum wheat. Farmers differentiate between their own seed and commercial seed, and farmers and women expressed their concerns about seed mixing and complained about merchants who mix low quality seed generally with farmers' seed.

Both the formal survey and focus group discussions revealed the substantial contribution of women to crop production, particularly in the case of alfalfa and faba bean. Women's responsibility for livestock and food preparation in the villages of the High Atlas mountain agroecosystem make them very knowledgeable about crop production and utilization. However, there are differences among sites with respect to women's participation in marketing and farming decisions. For example, women in Lahrouna village are very informed and involved in marketing of agricultural products. Women in the oasis system of the high Ziz Valley are experts about desirable quality traits of durum wheat for human consumption. In Taounate of the Rif mountain site, women are heavily involved in grain marketing of selected crops in local markets. Women know the landraces grown on their farms and in their communities. Importantly, women know more than male farmers with respect to the relationship between on–farm use and quality and food products from concerned crops. Equally important is the fact that women's views were consistent with those collected previously through the formal survey conducted with farm–household heads.

For all the crops, the existing genetic diversity is quite broad at each site and between sites. The statistical analyses are now underway for every crop in order to identify the most distinctive traits used by plant breeders and compare them with the most pertinent traits identified by farmers according to their perceptions of their own landraces. For example, in the case of alfalfa the results of the trials at the Errachidia and Tassaout experimental stations showed that green and dry matters of local populations were higher than those of check varieties. Certain local populations collected from High Ziz and Lahrouna sites outperformed the improved varieties for winter forage production. The agromorphological characterization of on–farm of alfalfa landraces in the High Ziz and Lahrouna sites confirmed clear differences between the two sites. Lahrouna farmers were able to exploit their landraces all year around with more than eight cuts per year, while High Ziz valley farmers were able to obtain only four cuts because of winter cold and summer drought. Differences in forage quality and disease susceptibility were also noted between the two sites and between the populations of each site. It is interesting to point out that obtained results in the case of alfalfa confirmed that the local populations have an important agronomic potential that should be conserved and utilized. From farmers' perspectives, these results confirm their appreciation and interest in landraces adaptation potential. In fact, many farmers emphasize that adaptation is one of the reasons for their maintenance of local cultivars.

6. LESSONS LEARNED

The three years of the in situ on–farm conservation project in Morocco has been an important learning experience for all project participants, farmers, scientists, development workers, NGO members, and many others. The study of in situ on–farm conservation and its implications for the sustainable use of crop genetic resources embodies scientific and institutional challenges. Some of the lessons learned are summarized below.

6.1 Scientific Lessons

The most important lesson learned is the solid grounds upon which farmers have founded their continued use and maintenance of crop landraces. Adaptation, quality, price premium, and availability were clearly used by farmers to explain their variety choices and decisions. Scientific research must capitalize on these features in order to give landraces the valuable place they deserve; e.g., breeding programs can certainly benefit from landraces adaptation potential and hardiness and their genetic heterogeneity.

The second lesson is that in situ conservation of agricultural biodiversity on–farm is a complex and multifaceted process requiring long term scientific research, commitment and competence. Requirements include adopting the right approach, raising the relevant research questions, adapting and developing the adequate methodologies to answer them and setting up the appropriate mechanisms for translating the scientific findings to strategic principles and development actions for implementation at the national, regional and local levels.

A third lesson is that multidisciplinary research and strong working teams comprised of agroecological scientists, plant genetic resources specialists, and social scientists are necessary for conducting research on in situ conservation of crop genetic diversity on–farm. A wide array of disciplines and specialties are needed to design and implement integrated research programs on conservation and sustainable use of plant genetic resources. Otherwise, less effective deterministic and reductionistic analyses are likely. Multidisciplinary work is often faced with methodological challenges of standardization, appropriate scaling and sampling, qualitative and quantitative approaches, data integration and correspondences and a number of other scientific issues.

A fourth lesson is that approaches to the study of in situ conservation on–farm need to be participatory and gender based. Farmers and women's contributions would not have been understood nor clearly acknowledged using conventional approaches and methodologies. Women's responsibility for livestock, grain storage, processing and food preparation result in their substantial involvement and use of agricultural crops such is the case of target crops in Morocco. Women do all the processing for household consumption in the case of barley, durum wheat, and faba bean. Women's responsibility of livestock underlies their daily use and associated acquired knowledge of alfalfa considered which is considered the most important forage crop in the country. Therefore, the involvement of women in discussions and decision making on conservation and sustainable use of crop biodiversity should be increased.

A fifth lesson pertains to the correspondence between farmers and women's knowledge systems and experiment–based findings; in fact, the two systems are complementary. Adaptation potential of landraces is highly appreciated by farmers and under existing conditions farmers use landraces because they represent the best options from their point of view. Not only do landraces satisfy farmers' crop production objectives and immediate needs but they seem to harbor a great potential for adaptation to existing environmental and socioeconomic conditions. However, key questions are: Would farmers continue to grow these crop landraces if better alternatives are made available? Under what conditions will farmers continue or abandon using landraces? The answers to these questions will contribute to the understanding of the conditions under which the use and conservation of crop genetic diversity on–farm can be sustained.

6.2 Public Policy and Management Lessons

The most important policy and management lesson learned is that policy issues and constraints are necessary components and need to receive more attention in the future. Without favorable policies for in situ conservation and farmers' rights, the most elaborate scientific research will not result in sustainable in situ conservation. This challenge is particularly great in Morocco where there is a double gap between the country's international obligations and national policy on one hand and between the country's formal policy and necessary institutional conditions for in situ on–farm conservation of cropped plants on the other.

Morocco is a signatory of the Convention on Biological Diversity (CBD) and has established a National Plant Genetic Resources Committee. Morocco has also adopted the Global Action Plan of 1996 on Plant Genetic Resources (PGR). The county is a member of the FAO Commission on Genetic Resources for Food and Agriculture. Recently, Morocco has adhered to the International Treaty on Plant Genetic Resources for Food and Agriculture that was adopted by the FAO Conference on November 3, 2001. However, there is a gap between these accomplishments and the fact that currently there is no national strategy on biodiversity. A national gene bank is lacking although a regional seed gene bank facility is being established at INRA Settat with the objective of conserving cereals and forage crops from semiarid areas of the north central region of the country. Consequently, germplasm collections of cultivated species are dispersed between the country's various research and teaching institutions and International Research Centers, namely ICARDA.

The second gap is the prevailing assumption of conflict between the country's formal policy and in situ conservation. As argued throughout this study, farmers' use of landraces of selected crops is an essential ingredient for in situ conservation objectives. However, the government's official policy is based on the promotion and the diffusion of improved and newly released varieties. Several meetings have been held with decision makers and development workers to explain the in situ project approach and objectives and to foster discussions on the importance of landraces and their role in the conservation of crop genetic diversity. Talking about landraces was often perceived as contradictory with the official position regarding the promotion of improved varieties. The apparent contradiction between using landraces and adopting improved varieties has been lessened as the project has progressed and as activities have been carried out in various sites with the participation of farmers, local extension and local development associations.

Many institutions of the Ministry of Agriculture have been continuously informed about the project progress. The most recent event in terms of informing policy makers and raising public awareness was the national seminar on in situ conservation of crop genetic diversity held in Rabat during January 2002. Primary results have been presented and discussed with national and international scientific colleagues. Policy and decision makers from relevant departments of the Ministry of Agriculture and the Ministry of Environment participated in the seminar. Nevertheless, structural difficulties still remain. The issue of supportive policy to conservation is exacerbated by the lack of funds for translating policy changes when they occur into concrete development actions. The challenge in a country like Morocco is how to channel some of the already scarce public funds to tackle the very urgent aspects of biodiversity conservation in general and in situ conservation in particular. From the policy makers point of view, there are always more urgent priorities.

6.3 Partnerships, Cooperation and Capacity Building

In Morocco, like in many other developing countries, institutional collaboration and partnerships are not only desirable but are necessary especially under conditions of scarcity of public funding and human resources. The institutions involved in this case study's in situ project activities represent the main leading agricultural research and teaching institutions in Morocco.

One of the lessons learned regarding partnerships and capacity building was the necessity to establish appropriate mechanisms that allow not only the development of partnerships between institutions but also that facilitate their smooth and efficient functioning. Agreement protocols with clearly specified rules on the contributions (financial, manpower, facilities, etc.) and the distribution of benefits generally offer a framework for collaborative work and partnership. Networking is another mechanism that can facilitate multi–institutional collaboration. However, networking was not always easy because individual members tend to use their institutional affiliation to obtain self–serving benefits or they are unable to find a balance between individual career and institutional capacity building. In addition, some institutions may not always succeed in creating favorable conditions for their members to work together for common goals and shared outcomes.

Diffusion of information and exchange of views between the members of project teams is an important task of the project coordination. In the present case study, meetings of the project advising committee were held whenever necessary and all members were invited to participate. Project documents and reports were often made available and distributed. Connection through electronic mail was made possible for those who have not been connected before the project. In fact, it was one of the most used means of communication among the project participants. At the sites where electronic mail was not possible, local extension services and collaborating nongovernmental organizations were provided with fax machines.

Efforts to establish the right synergy between all participants in the project were not always successful because they were either insufficient, not properly channeled, or because of external factors. Sometimes, a simple change in personnel be it a researcher, an extension agent or an NGO member tended to drastically affect the work of other project participants. Sometimes, various people did not understand the rules of collaboration in the same way. One of the attempts to handle these difficulties was to formalize the relationship between the project and its partners. For example, letters of agreement were established between the project and the local NGO's at Bouhrazen and Rich sites.

The wide geographic coverage and variation in sites and in crops was not only a source of learning and richness but also was a source of difficulty. The smoothness in achieving planned activities in each of the sites was dependent on institutional, environmental, and sociocultural site–specific factors. Availability and willingness of local institutions and farmers to engage in project activities vary from one site to another. For institutional collaboration to be fruitful, participating institutions should respect agreed–upon rules and contributions and secure a sufficient degree of continuity in assigned staff to a particular project. However, these types of institutional arrangements are necessary but insufficient because it is also important that participating individuals should be able to work in multi–institutional teams. Without individual strong commitment to teamwork and shared goals, institutionalized frameworks remain mere formal exercises in the field.

Throughout the three years of the project, the project team was very aware of the need to build the national capacity in areas where it was relatively weak. Building the teams and ensuring some sort of continuity and equitably distributed opportunities were key concerns of project members. Consequently, a series of in–country training workshops have been organized and implemented in various project sites. The two major themes of these workshops were participatory approaches and seed technology. Collaborating local extension and nongovernmental organizations have been active participants in all training activities. Partner farmers, however, have participated only in those where the emphasis was on restitution of results, discussion of local varieties, and where other relevant information was available. In addition, several young scientists have received short–term training abroad.

Developing the right partnerships and the proper institutional collaboration arrangements is never an easy task particularly when individuals and institutions are accustomed to working in their own institutional boundaries and mandates. Provincial and local institutions, governmental and nongovernmental bodies are expected to create favorable conditions for insuring the sustainability of the work on in situ conservation on farm. Although not formally institutionalized, partnerships with farmers based on trust and respect are of utmost importance. Internationally, IPGRI headquarters in Rome and at the Regional Office for Central and West Asia and North Africa (CWANA, Aleppo) remain key collaborators.

7. CONCLUSION

A major conclusion of the study is the urgent need to pay attention to the policy dimension of in situ conservation research. Farmers' current use of landraces must be recognized and supported through policies, research and extension that are favorable to in situ conservation. It is necessary to develop the present initiative on in situ conservation of landraces into a more global perspective on the sustainable use and conservation of agricultural biodiversity in Morocco. Such a perspective needs to go beyond farmers to the government responsibility in preserving agricultural biodiversity for future generations.

8. ACKNOWLEDGEMENTS

The work was carried out with joint funding from the International Plant Genetic Resources Institute (IPGRI), the German Federal Ministry for Economic Cooperation and Development and the German Agency for Technical Cooperation (BMZ/GTZ), the Agronomy and Veterinary Institute Hassan II (IAV Hassan II) and the National Agricultural Research Institute (INRA).

9. REFERENCES

Birouk, A., and Mellas, H. 2000. Methods used to create a framework for implementation and management of in situ conservation on farm: Morocco's country component. Pages 12–14 In: Jarvis, D., Sthapit, B., and Sears, L. (Eds.). Conserving Agricultural Biodiversity In

Situ: A Scientific Basis for Sustainable Agriculture. Proceedings of a Workshop, 5–12 July 1999. IPGRI, Pokhara, Nepal.

Jarvis, D., and Hodgkin, T. (Eds.). 1998. Strengthening the Scientific Basis of In Situ Conservation of Agricultural Biodiversity On–Farm. Options for Data Collecting and Analysis. Proceedings of a Workshop to Develop Tools and Procedures for In Situ Conservation On–Farm, 25–29 August 1997. IPGRI, Rome.

Jarvis, D., and Hodgkin, T. 2000. Farmer decision making and genetic diversity: Linking multidisciplinary research to implementation on–farm. Pages 261–278 In: Brush, S. (Ed.) Genes in the Field: On–Farm Conservation of Crop Diversity. Lewis Publishers, Boca Raton, FL, USA.

Nassif, F. 2000a. Socioeconomic data collection and analysis: The case of Morocco. Pages 51–53 In: Jarvis, D., Sthapit, B., and Sears, L. (Eds.) Conserving Agricultural Biodiversity In Situ: A Scientific Basis for Sustainable Agriculture. Proceedings of a Workshop, 5–12 July 1999. IPGRI, Pokhara, Nepal.

Nassif, F. 2000b. Gender: Methods used for increased access, participation and decision making: The case of Morocco. Pages 76–77 In: Jarvis, D., Sthapit, B., and Sears, L. (Eds.) Conserving Agricultural Biodiversity In Situ: A Scientific Basis for Sustainable Agriculture. Proceedings of a Workshop, 5–12 July 1999. IPGRI, Pokhara, Nepal.

Nassif, F. 2000c. Seed supply systems: The case of Morocco. Pages 157–158 In: Jarvis, D., Sthapit, B., and Sears, L. (Eds.) Conserving Agricultural Biodiversity In Situ: A Scientific Basis for Sustainable Agriculture. Proceedings of a Workshop, 5–12 July 1999. IPGRI, Pokhara, Nepal.

Chapter 35

BIODIVERSITY CONSERVATION AND SUSTAINABLE DEVELOPMENT: A CASE STUDY OF OMAN'S NATIONAL BIODIVERSITY STRATEGY AND ACTION PLAN

Reginald Victor
Department of Biology and Center for Environmental Studies and Research (CESAR), Sultan Qaboos University, P.O. Box 36, PC 123, Sultanate of Oman; email victor@omantel.net.om

1. INTRODUCTION

Biodiversity and sustainable development are terms that have been used in the arena of environmentalism with sheer abandon in the past decade. As a consequence the definitions of these terms have been overextended to cover matters far beyond their intended scope. One of the major difficulties causing this confusion is the inability to explain the link between biodiversity and sustainable development in unambiguous terms. Good examples are the National Biodiversity Strategies and Action Plans produced by the nations signatory to the Convention on Biodiversity (CBD). The worst–case scenarios in some of these strategies and action plans indicate very remote, if at all any links between biodiversity and sustainable development.

This paper presents a case study of a fragile mountain ecosystem in the Sultanate of Oman with reference to National Biodiversity Strategy and Action Plan (NBSAP)– Oman 2001. As a preamble, a critical overview of biodiversity conservation and sustainable development is provided. The purpose of this overview is to compare the past and present day thinking on biodiversity conservation and sustainable development and recommend a balanced perspective while formulating important instruments such as national strategies and action plans. This overview emphasizes the need for a balanced integrated approach, so that often oversubscribed development initiatives in strategies and action plans do not smother the primary objective of biodiversity conservation.

2. OVERVIEW

2.1 Definitions

For the purposes of this paper, the most commonly used definition of biodiversity, the variability among living organisms from all sources including terrestrial, marine and other aquatic ecosystems and the ecological complexes of which they are part, including diversity within species, between species and ecosystems is accepted. Similarly, sustainable development is defined, as development that meets the needs of the present without compromising the ability of future generations to meet their own needs.

2.2 Perceptions on Conservation and Development

Prior to 1970, nature conservation predominantly referred to the protection of ecosystems (e.g., rainforests, coral reefs) and/or communities and species of special interests (e.g., large mammals, birds). The habitat complexities, biogeochemical interdependencies and the interrelationships between biological components were well understood and concerns for overall biodiversity was implicit in nature conservation programs. Conservation strategies involved development, but this development was in favor of nature like the establishment of national parks, implementation of nature protection laws and even culling programs. This approach was at variance with the development of the omnipotent component of the ecosystem, the human beings. Nature conservation goals were at loggerheads with human needs for land, food, shelter and natural products to name a few.

The reversal of thoughts on nature conservation is echoed in the call made in 1970 by Budowski, the then Director General for the International Union for Conservation of Nature and Natural Resources (IUCN, now known as the World Conservation Union) to consider conservation as "an indispensable ingredient in development planning." Recent views note, rather accusatorily that biodiversity and conservation in the past have been purely treated as scientific and technical issues requiring specialized skills, whereas the present focus should be on institutional development and governance (Mugabe 1998). It is therefore necessary to briefly examine the reasons for the link between conservation and development.

The global view supported at present by environmental organizations at most levels believes that linking conservation with development is the only way to protect nature. The CBD, a legal instrument ratified by more than 165 countries does not promote conservation only. It also seeks the sustainable use of biodiversity components and the fair and equitable sharing of benefits obtained from biological resources; these two goals are clearly associated with development. Oates (1999) presented a controversial view that linking conservation and development is not clearly the only way to protect nature, but it is a result of a series of historical events, influence of conservation strategists, political compromises and financial expediency.

The argument favoring the link between conservation and development could be traced back to Nicholson (1970) who stated that planning for conservation should permit "beneficial natural processes to operate to the fullest extent compatible with whatever justifiable human demands may need to be satisfied." This is a critical statement that puts human needs ahead of nature conservation. What are justifiable human demands? What is the limit for satisfaction? What are the criteria determining compatibility between human demands and natural processes? These are context specific open–ended questions for which there are a multitude of answers and none of them are simple and directly solution orientated.

The counter argument favors the old traditional thinking that nature must be protected for its own intrinsic value and not because of the use to which it might be put by people (Grzimek and Grzimek 1960). Proponents of sustainable development dismiss this argument as outdated. Oates (1999) in his book "Myth and Reality in the Rainforest: How Conservation Strategies are Failing in West Africa" provided a powerful discussion debunking the unity of conservation and economic development. Conserving nature and biodiversity for its intrinsic value is strongly supported by renowned conservationists like George B. Schaller, Michael E. Soulé, Arne Naess and Thomas Struhsaker. Environmental activism defending plants and animals seldom supports the link between conservation and sustainable development unless development is clearly perceived as a secondary objective.

Any one who has had the opportunity to work on nature conservation projects in poor developing countries in the last decade would readily admit that sustainable development as defined here is not the ultimate vehicle to conserve either nature or biodiversity. Even more confusing is the realization that in these areas there is no consensus on what sustainable development is. The so–called sustainable development projects (e.g., fish farming as an alternative to protect against the overexploitation of inland fisheries; bush–meat farming to relieve hunting pressure) fail due to inadequacies in governance including the pilfering of funds, nepotism, corruption and kickbacks. Thus sustainable development is yet to be proven as a globally acceptable concept. The global public is becoming fast disillusioned with the sale of sustainable development as a mechanism to resolve issues that address human needs impacting biodiversity conservation. This skepticism is less publicized when compared to the vigor with which sustainable development is internationally promoted. Environmental professionals have been discussing the need for holistic approaches to sustainable development for some years (Fyfe et al. 1993). These in fact are not very different from the original objectives of the Brundtland Commission except for the recognition of new ideas on methods of development, economic systems, global economy, geopolitical conditions and advanced technologies. Otherwise, these approaches are utopian, idealistic and unpractical.

2.3 Road to Rio (1970–1992)

The perceptions discussed above require a review of important milestones on the road to Rio de Janeiro where the fabled Earth Summit was held in 1992.

"The Environmental Revolution" by Nicholson (1970) advocated the wise use and sustained yield exploitation of nature. These views, as expected received a strong support from WWF (World Wide Fund for Nature, then known as World Wildlife Fund) and the IUCN. " Only one earth," popularly known as the Stockholm Report by Ward and Dubos (1972) focused on pollution, population growth and development. It called for cooperative global planning, but had very little on conservation (Oates 1999). In 1974, the IUCN announced the formulation of World Conservation Strategy and its second draft was released in 1978 emphasizing sustainable development and conservation for utilitarian purposes. World Conservation Strategy was formally launched in 1980. Munro, the Director General of IUCN called it "a response to the need to make development sustainable." World Conservation Strategy was seen as an evolving plan that would be revised as new knowledge was gained and values changed (Oates 1999). This resulted in "Caring for the Earth: A Strategy for Sustainable Living" jointly published by IUCN, WWF and UNEP (United Nations Environment Program) in 1991. The culmination of these events was at the Earth Summit in Rio de Janeiro in 1992 where a revised strategy more strongly orientated towards human well–being was released. The concept of sustainable living, cross–sector planning by national governments and the need for communities to create secure sustainable societies were advocated.

This review clearly shows that views on nature conservation by protection for its intrinsic value has not only transformed, but also has formed an inseparable link with economic development. As Oates (1999) pointed out that conservation had "fallen in love" with development and

WWF, an organization that once advocated conservation for the intrinsic value of nature ironically played a role in promoting "the unlikely marriage" of conservation and development.

However, the "marriage of conservation and development" has lasted long enough and what are the results? There is an enormous amount of literature on the sustainable management of biodiversity (Mugabe and Clark 1998). Despite these efforts, none of the signatory nations to the CBD have been able to show that there is a significant improvement in biodiversity conservation. There have been improvements in human development, but the biodiversity loss continues. Some aspects of biodiversity such as agrobiodiversity may be relevant to human development, but may not be beneficial to natural biodiversity, if not detrimental. As Robinson (1993) pointed out sustainable use of an area neither supports nor reflects the biodiversity conservation of that area. Sustainable exploitation of fishing resources in an area may be beneficial to the fishermen, but it would not improve biodiversity as would a moratorium on commercial fishing for a few years allowing recovery. Sustainable use of cleared forests to plant oil–palm and rubber may provide a higher material quality of life to forest inhabitants, but the loss of forest and its biodiversity cannot be supplemented by sustainable development. Thus, "Caring for Earth" addresses human well–being more than the well–being of biodiversity.

2.4 Road from Rio to Johannesburg (1992–2002)

A decade of knowledge gained from "Caring for Earth" clearly indicates that views on conservation have further changed in support of development. The road from Rio to Johannesburg is not leading to a revised World Conservation Strategy; it is leading to a World Summit on Sustainable Development (WSSD). African Ministerial Statement to the WSSD (Anonymous 2001a) is a good example that reaffirms "poverty eradication" is an indispensable requirement for sustainable development and reiterates its commitment to address "all three components of sustainable development: namely, economic growth and development, social development and environmental protection as interdependent and mutually reinforcing pillars–in a balanced way in conformity with the fundamental principle of common but differentiated responsibilities." Environmental protection indirectly encompasses biodiversity conservation, but problems center on "the balanced way" in which it will be addressed against the other pillars, economic growth and development and social development.

3. BIODIVERSITY IN THE ARABIAN REGION

With the above overview as a backdrop, the status of biodiversity in the Arabian region requires scrutiny. As in other parts of the world, the present biodiversity scenario in Arabia is dominated by sustainable development. A summary of this issue was provided in the draft conclusions of Oman International Forum on Environment and Sustainable Development held in Muscat and attended by Ministers and Heads of Delegations from Arab Gulf Cooperation Council Countries in December 2001. Under the fourth and last agenda item addressing major environmental problems in the region, the protection of biodiversity is penultimate in the order of importance (Anonymous 2001b). The issues listed were, (1)increasing the percentage of protected areas by 50 percent by 2005, (2)increasing forest cover, especially in mountain areas, by 50 percent by 2005, (3)focus on coral reefs and mangrove systems as major sites for generating tourism–based revenues and the protection of biodiversity in arid regions, (4)prevention of sewage release into rivers and regional seas, and (5)establishment of adequate reception facilities in ports and harbors to receive oily wastes, ballast water and other wastes from oil tankers and other ships. It was suggested, "FAO/UNEP/UNESCO/IUCN, in addition UNCCD and CBD Secretariats could lead the way towards implementation and follow up." It was believed that the combined effect of these objectives "will have multiple benefits including protection of human health, conservation and development of freshwater resources, protection of fisheries and biodiversity in the regional seas." Issues of disparate importance are strung together in this list with unrealistic expectations. All participating states in the above forum are signatory to the CBD and therefore should have National Strategies and Action Plans for the protection and management of biodiversity.

4. OMAN CONTEXT

In the Sultanate of Oman, as in other developed and developing countries of the world, sustainable development has been receiving considerable attention in several initiatives. One of the tenets of Oman's Vision 2020, a major initiative to be accomplished in the next 20 years is the optimal use of natural resources of which biodiversity is allegedly a part.

The vehicle for protecting biodiversity in Oman is the NBSAP–Oman 2001. The former Ministry of Regional Municipalities and Environment (MRME, now including the Ministry of Water Resources as MRMEWR) with the help of all government ministries, Sultan Qaboos University, United Nations Development Program (UNDP), Global Environmental Facility

(GEF) and IUCN produced the NBSAP–Oman 2001 in complete agreement with the modern conservation thinking. Strategic goals and guiding principles of NBSAP–Oman are orientated towards sustainable development. However, in keeping with Oman's vision 2020, the optimal use of biodiversity requires conservation that should result in improvement of biodiversity for current and future generations.

The implementation of NBSAP–Oman is now in progress. Strategies and action plans do not succeed if the use and conservation of biodiversity do not adopt best practices during implementation. Best practices are those that consider biodiversity as important without compromising development. If development takes priority, biodiversity will continue to deteriorate against the best intentions of strategies and action plans.

5. CASE STUDY

To consider biodiversity conservation in the context of NBSAP–Oman, Jebel Akhdar (Green Mountain) region, a fragile arid mountain ecosystem in the Western Hajar range of northern Oman is chosen as a case study.

5.1 Environment and Biodiversity

The Sultanate of Oman is located in the southeastern corner of the Arabian Peninsula (Lat. 16° 38'–26° 52'N; Long. 51° 52'–59° 50'E). There are two mountain ranges, one in the north and the other in the south of Oman. The northern mountains, Al Hajar, are divided into eastern and western Hajar by the Semail gap. Jebel Akhdar is the largest structural domain in the western Hajar with a maximum elevation of 3009 masl. (Gardner and Fisher 1996, Llewllyn–Smith and Launay 2001).

Specific studies contributing to the knowledge on Jebel Akhdar are many. The basic geology of this mountain is known and its climate has been described as semi–arid to arid (Glennie 1995, Hanna 1995, Gardner and Fisher 1996). The hydrogeology of the northern mountains has also been investigated in detail and water chemistry data are available for surface and groundwater in the region (Stanger 1986). The juniper woodlands of the area have been relatively well studied and the basic vegetation structure including phytogeography has been described for some areas (Ghazanfar 1991, Gardner and Fisher 1994, 1996; Fisher and Gardner 1995). Victor and Al–Ujaily (1998, 1999) investigated the water qualities of mountain reservoirs and related management issues. In addition, studies on selected biota such as zooplankton and Charophyta have been conducted (Victor and Victor 2002, Hussain et al. in press). Reports on the flora and fauna of the mountains are also available in several isolated publications (Harrison 1977) and unpublished manuscripts.

Biodiversity of the Jebel Akhdar region is unique. It has considerable endemism with similarities to the fauna and flora of the northwestern oriental region comprising Iran, Afghanistan and Baluchistan (Mandaville 1977, Arnold 1987, Ghazanfar 1998, Gardner and Fisher 1996, Gardner 1999). It is believed that 60 percent of Oman's 1204 species of vascular plants are found in these mountains, of which 16 are endemic. Similarly, three national and five regional endemic species of lizards are also known from this area (Arnold and Gallagher 1977, Gardner 1999). Gallagher (1977) listed 71 bird species in Jebel Akhdar area including 28 residents and 41 migrants. In addition, five species of birds with restricted distribution in the Middle East are also found here (Evans 1994).

Jebel Akhdar also is the home for several interesting species that are not found elsewhere in the lower plains. Some of these species are also considered endangered. Among trees, *Juniperus excelsa* and *Olea europea* are good examples. *Dionysia mira*, a primula associated with olive–juniper woodlands is an endemic restricted to a small area in the northern mountains. *Daphnia longispina*, a temperate water flea found in the reservoirs of Jebel Akhdar has never been reported in the Gulf region (Victor and Victor 2002). This area supports a population of the lappet–faced vulture, *Torgos tracheliotus*, a species listed as vulnerable in the 2000 IUCN Red List (Llewellyn–Smith and Launay 2001).

Despite the brief review provided here on the environment and biodiversity characteristics of Jebel Akhdar, the flora, fauna and the ecology of the area remain poorly studied. Investigations so far conducted are isolated, concerned with a few specific groups or habitats, and severely short–term. The World Wide Fund for Nature (WWF) considers this area as an important terrestrial ecoregion (Llewellyn–Smith and Launay 2001).

Jebel Akhdar area is a fragile ecosystem. Its fragility is signified by the delicate interrelationships between the physical environment, the flora, fauna and most of all man. Man, to a large extent has control over the physical environment, the flora and the fauna. For example, the disappearance of the Arabian leopard, *Panthera pardus nimr* and the scarcity of Arabian wolf, *Canis lupus*, and striped hyaena, *Hyaena hyaena* in the mountains are the result of possible persecution by man (Spalton and Willis 1999, Fisher 1999). Similarly, man is equally responsible for the presence of goats and feral donkeys that severely impact the vegetation.

5.2 Jebel Akhdar in NBSAP Framework

Biodiversity issues of Jebel Akhdar are addressed here under the themes listed in NBSAP–Oman 2001. Under each theme problems are identified and areas requiring focus or mitigation are discussed.

5.2.1 Theme 1: Conservation of Natural Resources

At present there are no protected areas in Jebel Akhdar. A decade ago, certain areas of Jebel Akhdar like Saiq Plateau were inaccessible to the general public except for the residents of the area. Permits when issued only allowed day visits. Camping was prohibited. This is because of the military installations and training facilities in the area. Although the permit system is still in place, access is easy. Camping is permitted and a new hotel opened on Saiq Plateau in 2000 facilitating access to tourists. Implementation of NBSAP should involve the designation of protected areas as nature reserves. The focus should be on juniper and olive woodlands. Research in progress in these woodlands shows severe degradation and very limited regeneration (Victor and Al Jamei unpublished data). These woodlands are unique to the region and therefore require protection. Major threats are expansion of human settlements, local visitors cutting wood for camp fire and cooking, collection of olive foliage as forage for livestock, olive branches lopped off to make ceremonial walking sticks and at least in one instance an agricultural farm.

Evaluation of endangered flora and fauna requires immediate attention. Lists of endangered species for the area need to be generated for various groups. Habitat loss is severely impacting flora. *Caralluma* spp. and *Dynosia mira* require protection. There are two herbaria at the National Museum and Sultan Qaboos University. These institutions with the help of MRMEWR should monitor the plant collecting activity in the area. Among fauna, the Grey wolf, *Canis lupus arabs* and Arabian tahr *Hemitragus jaykari* are important. The population size and conservation status of these two species should be assessed.

5.2.2 Themes 2 and 3: Terrestrial and Freshwater Flora and Fauna

Despite limited surveys on flora and fauna of the area, species lists for many groups are incomplete. Taxonomic studies in the area should be encouraged, but under a strict organizational hierarchy to be established by the MRMEWR or National Museum of Natural History. National Museum however should be a repository for type and representative specimens and Sultan Qaboos University should be a facilitator to procure and monitor the activities of both international and local experts. Studies should also focus on the evaluation of conservation status and production of local red lists for the mountain.

5.2.3 Theme 4: Marine Resources

Not relevant to the study area.

5.2.4 Theme 5: Agricultural Resources

Agrobiodiversity of the area is limited and well known. Minor crops like garlic and green vegetables are grown in small scale, not even enough for local consumption. The focus should be on three important issues, fruit production, Jebel Akhdar rose and saffron farming.

The major fruit crops in the area are pomegranate (*Punica granatum*), apricot and peach (*Prunus* spp.). These are seasonal fruits. Pomegranate production is high enough to reach markets in the foothills. Apricot and peaches are consumed locally. Commercial scale farming of these fruits and walnut (*Juglans regia*) does not seem feasible. Main limitations are the availability of arable land and irrigation water.

Saiq valley, the major fruit growing area has been affected by the contamination of soil and water by a diesel spill that occurred about 10 years ago. The effect of this spill on fruit crops were not detected for the first few years, but the farmers now complain that the fruits appear normal externally, but are black inside and are inedible. The diesel contamination has also caused a significant accumulation of heavy metals in soil (Victor et al. unpublished data). Remediation measures so far taken have not been very effective. Ecological restoration of this area is an immediate priority. Otherwise, the value of fruit production to the future generations of this area will be lost irretrievably.

Jebel Akhdar is the only place in Oman where rose flowers are grown in farming quantities and not as ornamentals in gardens. Rosewater making is a traditional industry that seems to have thrived on the Saiq Plateau for hundreds of years. Rosewater is primarily used to flavor coffee and sweets, especially *halwa* (made from wheat flour and sugar). Some claim that it is also used as a perfume despite smelling smoky, like wood–distillate. Jebel Akhdar rose (*Rosa* sp.) is highly seasonal. Flowering lasts from last week of March to first week of May with peak flowering in April. The largest area of rose cultivation is on the terraces of two villages, Al Ayn and Shuraijah that have perennial spring flow harnessed by a *falaj* (i.e., man–made channel; *aflaj* plural).

Bottled rosewater is sold locally and is more expensive than perfumed commercial rosewater imported from Pakistan and India. Commercial production of rosewater will not be feasible simply due to the low quantities of rose produced, only in one or two months of a year. The rose plants are of unique biodiversity value and the flowering season attracts local and foreign tourists. The traditional technique of making rosewater is of cultural importance. These traditional methods are slowly disappearing from the modern Oman society and therefore, require conservation for their heritage value.

Saffron (*Crocus sativus)* has been grown in Jebel Akhdar in small quantities in home gardens of villages since ancient times. Flowers are harvested, dried and deep–red thread–like stigmas are dried and used for flavoring food and also as medicine. It is an expensive condiment cultivated in India, Spain, France and Italy. At present, saffron is grown intensively in one farm in Jebel Akhdar at about 2300 masl. A few hectares of olive–juniper woodland were cleared to establish the farm. Crocus patches are drip irrigated. This is a new venture in the area and the cost–benefit of this operation is not known. Some environmental costs are the value of cleared woodland, the water use, effect of pesticides and fertilizers, especially the export rates via surface run–off and waste disposal. The benefits are revenues earned from saffron production that will be of support to local communities and employment of local labor. Resource economics of saffron farming in the area should be worthy of investigation.

5.2.5 Theme 6: Energy Resources

Not relevant to the study area.

5.2.6 Theme 7: Mineral Resources

Not relevant to the study area.

5.2.7 Theme 8: Industry, Technology and Services

Tourism is the only industry that has been important to the region since the year 2000. The presence of a hotel on the Saiq Plateau encourages tour operators to bring foreign tourists to the area. The foreign tourist activities are limited partially because the tour operators are not fully aware of the tourism potential of the area. Expatriate residents in Oman to a large extent and other foreign residents in the nearby Gulf countries utilize the area extensively for recreation. Local visitors to the mountains have also increased considerably during this period. Unorganized tourism in the area will damage the natural, social and cultural environment. The concepts and practicalities of ecotourism in the Jebel Akhdar should be investigated as a matter of priority. This does not necessarily mean the infrastructure development to support tourism, but also to assess the impacts of tourism on a fragile ecosystem. Administrative measures to minimize impacts should be planned. Controlled, but informative ecotourism may be an option.

5.2.8 Theme 9: Urban Environment

Jebel Akhdar region has experienced unparalleled expansion since 1993. Some villages in Saiq Plateau have doubled and tripled in size. Traditional dwellings are being replaced by modern housing and new houses are added. In 1989, there were only graded roads and tracks linking villages. Some villages were in remote locations connected by footpaths and donkey–tracks or by helicopter during emergencies. Now most roads on the Saiq Plateau are tarmac and all remote villages are accessible by graded roads. In 1993, the supply of electricity was very limited in the area. Power generation was through oil–generators. There were very few reliable telephone links. Since then there has been considerable improvement in infrastructure covering power supply and communication. Health care and education sectors have been revamped. Commerce is also developing steadily.

Urban development if not checked and guided can be detrimental to biodiversity conservation in this fragile ecosystem. For example, a major road running through between 2000–2300 masl has fragmented woodlands and the habitat of the rare *Dynosia mira*. Road and building construction activities increase erosion via surface runoff causing slope destabilization. Solid and liquid waste disposal is likely to become a problem in the not too distant future. NBSAP should focus on the impacts of economic development on the biodiversity of the area. Regulatory mechanisms governing physical urban development should be stringent. Environmental impact assessments must be mandatory for all development projects. Designation of buffer zones and protected areas should limit urban development to the existing villages. The establishment of new settlements and expansion old settlements beyond designated boundaries should be prohibited. Commerce in the area should be sustainable. Easy access to the region will bring in cheaper food products (e.g. fruits, vegetables, meat) that would easily out–compete the local produce. The market for local products should be protected in appropriate seasons.

5.2.9 Theme 10: Water Resources

Jebel Akhdar has freshwater that could support small communities. Historically, the human colonization of this mountain must have been governed by the availability and distribution freshwater. Natural springs and spring–streams are perennial (e.g., Wadi Bani Habib, Al Ayn, Azeezi, Masirat Ruwijh). In Oman, the flowing water is harnessed by man–made aflaj and a very efficient system of water distribution for irrigation has been practiced for over 1500–2700 years BP (Wilkinson 1974). There are designated aflaj keepers in villages whose sole job is to maintain the channels and distribute water to various farm plots by diverting water from the main channel for specific time periods.

Jebel Akhdar receives more rainfall than desert plains. The mean annual rainfall is about 300–400 mm. Rain water is drained by *wedian* (i.e., seasonal rivers; *wadi* singular). During periods of high rainfall, the wedian flood and are thus responsible for flash floods in lower altitudes. Small traditional dams intercepted the floodwaters and water was stored in small reservoirs for human use. The now defunct Ministry of Water Resources planned to construct 64 dams in the Jebel Akhdar area and by 1994 completed 24 retention dams. These new reservoirs were considerably larger and stored more water. Water quality and management of these reservoirs has been discussed by Victor and Al–Ujaily (1998, 1999). All these reservoirs are eutrophic and the water is not fit for drinking. The water is, however used as drinking water for livestock and for other domestic uses.

Groundwater extraction in Saiq and in the foothills provides the drinking water distributed to villages by government tankers under the jurisdiction of the Ministry of Water and Electricity.

Rapid development in the Jebel Akhdar region is exerting pressure on water resources. Biodiversity directly depends on the sustainable use of water resources. NBSAP should concentrate on two basic issues: the water use efficiency and the water quality. Water in the reservoirs is under–utilized and efficient exploitation requires improvement in water quality. Poor water quality caused by eutrophication is due to the enrichment of water by animal feces. Fecal contamination of water also has health implications. Overexploitation of groundwater exceeding the water recharge potential of the area will severely impact both natural and agrobiodiversity. Water quality affected by pollution (e.g., Falaj Azeezi by diesel and fecal contamination of Al Ayn) should be remedied.

5.2.10 Theme 11: Environmental Emergencies

Jebel Akhdar region is not a high–risk area for environmental emergencies, especially those affecting biodiversity in a large scale. There are no industries and there are no mining operations in the area. Villages suffer from flood caused emergencies during times of high rainfall that is mostly unpredictable. Flood damages roads and communication systems. The presence of the Ministry of Defense in the area ensures restoration of services usually within 24 hours.

Some pollution related issues due to diesel and fecal contaminations were presented under the themes, agricultural resources and water resources. The Ministry of Defense Battle training camp on Saiq Plateau operates a firing range that uses a variety of weapon grade explosives. Surface run–off from the area during rains is likely to carry nondegradable and biologically refractory chemicals into the wedian. Groundwater contamination is also likely. Severe consequences of chemical contamination from military testing grounds have been documented elsewhere (Tashiro 2001). All these pollution related issues may not be perceived currently as environmental emergencies, but monitoring is necessary to prevent unexpected disasters.

5.2.11 Theme 12: Public Participation, NGOs and the Private Sector

Public participation in decision making is limited. It is believed that villagers are consulted on key environmental issues such as reservoir construction (Victor and Al–Ujaily 1999). The Ministry of Defense has a Jebel Liaison Office on Saiq Plateau that acts as a link between the people of the area and the Ministry administration. Most Ministries have offices on the Saiq Plateau and concerns of the people relevant to the ministry in question are heard (e.g., dialogue between the Ministry of Water Resources and Saiq village about the diesel contamination of Falaj Azeezi). NGOs like the Historical Society of Oman and Horticultural Association are interested in the biodiversity of the Jebel Akhdar region, but their activities are restricted to occasional visits, which are of heuristic value. The involvement of private sector in Jebel Akhdar region is for development activities that are antithetical to biodiversity conservation.

NBSAP should focus on utilizing all three components of the theme. Public participation in the area will offer the benefit of traditional wisdom and sustainable practices used in the area since ancient times. For example, issues like water use efficiency, resource sharing and overgrazing were problems traditionally handled by mountain communities. These traditional practices should be compiled and evaluated. NGOs are very useful in providing volunteers for assistance with biodiversity research, and environmental education at grass–root levels. They are also useful to stimulate cross–border international interest and secure international funding. Development oriented private sector should be monitored for compliance with environmental regulations, given incentives for best practices and encouraged to provide free services for restoration activities.

5.2.12 Theme 13: Societal Values

Public awareness of the importance of Jebel Akhdar as an important ecoregion is poor. Environmental education is in its rudimentary stages. Research concerning the region is patchy and limited. Trained personnel are few. Rangers monitoring conservation affairs are allegedly present in the area; if so, their activities are not visible. The rapid and unchecked development in the area indicates that environmental impact assessments are poorly implemented.

NBSAP should focus on rising environmental and biodiversity awareness in local people and visitors through educational methodology. Establishment of a visitors' center, publication of information brochures and guided ecological and geological tours should be considered. A long–term, integrated environmental research initiative is needed for the region. Recruitment of additional rangers and increasing the visibility of existing rangers to monitor protected habitats (e.g., woodlands) and endangered fauna and flora is essential. A ranger–training program should be initiated and local people should be employed and trained. Laws for biodiversity protection should be promulgated and implemented.

5.2.13 Theme 14: Quality of Life

The development of the Jebel Akhdar region in the last five years has clearly improved the quality of life for the local people. Improvement in transport, communication, education, health care and commerce has been beneficial. Many people who originated from the mountains and migrated to the cities for livelihood are returning to build or rebuild family homes with modern amenities. The physical growths of some villages are phenomenal. Unfortunately, most development activities work against biodiversity conservation (e.g., land clearing activities; unsustainable water extraction).

NBSAP should advocate policies and practices that promote sustainable development, but this is likely to be a difficult challenge.

5.2.14 Theme 15: Spiritual Values

This is an unusual theme for a national biodiversity strategy. It is also controversial because it is non–materialistic. However, this theme reminds us of rationales for conserving biodiversity for its own sake and not for utilitarian reasons of development, sustainable or otherwise. Communities in the Jebel Akhdar region have always followed the Islamic way of life that advocates concerns for the environment and its creatures. This way of life supports the view of "Deep Ecology" that nature should be protected for its own intrinsic value and because of the "psychological value to humans of integrating their lives more closely with those of other species" (Oates 1999).

Development brings in new values of life. Values of materialistic significance are often contradictory to those of spiritual value. NBSAP should therefore be concerned that cultural and spiritual erosion associated with development does not promote the indifference towards biodiversity protection.

5.2.15 Synthesis

NBSAP–Oman with reference to Jebel Akhdar region could be organized as four biodiversity dominant themes (Themes 1–3, 5) and seven development orientated themes (Themes 8–14). Themes on biodiversity conservation, flora and fauna (1–3) form the core issue, while agrobiodiversity (Theme 4) has a utilitarian value requiring sustainable development and management. All other themes are utilitarian and if not managed will have severe impact on the biodiversity of the area. The theme on spiritual values focuses on the intrinsic values of biodiversity and implies the need for man to live in harmony with other species that share this earth with him. In the order of priority, this theme usually is given less priority by most people whereas it should receive top priority. Human commitment to the conservation of biodiversity should be first and foremost, without which attempts via sustainable development will not preserve biodiversity. If people come first, biodiversity will have to take a backseat and this is why many conservation projects in developing countries have failed and are still failing.

5.2.16 Threats and Management

It is imperative that NBSAP should recognize and prioritize threats to biodiversity in the Jebel Akhdar region as follows: (1)overgrazing by goats and feral donkeys, (2)introduction of intensive farming as opposed to traditional farming, (3)impacts caused by poor water use efficiency, (4)construction activities, (5)pollution related issues and (6)recreational activities including tourism. Pressures exerted by these threats on biodiversity should be assessed and the present status should be evaluated. This knowledge then should be used to formulate responses to protect biodiversity with judicious allowances made for development, if and when necessary.

What biodiversity parameters are measurable to assess pressures and status of threats? Knowledge on the biodiversity of flora and fauna in the Jebel Akhdar region is incomplete and it is not likely to be complete in the near future due to lack of expertise, infrastructure and problems like paucity of funds for taxonomic work. However, habitat loss could be assessed as an indirect indicator to measure biodiversity loss with reasonable reliability. Pressures–States–Responses methodology applied to habitat conservation will protect biodiversity (Bayfield 2001). Habitat loss caused by each threat should be estimated and mitigation measures must be developed as best practices. In most cases, these best practices are likely to be those borrowed from successful experiences elsewhere. Two examples are: organized culling to relieve overgrazing pressures by feral animals and restoration of eutrophic reservoirs by controlling nutrient inputs.

How could NBSAP do this? For each theme in the strategy, realistic links between key issues and objectives should be made. Uncertainties relating to decision making should be recognized. List of options and priority actions should not just be wish lists. Priority matrices of actions should spell out allocation of responsibilities, accountability criteria and timeframe for achievement of objectives and value judgments should be precautionary in favor of biodiversity protection. Finally, themes in NBSAP orientated towards development issues should not be allowed to compromise conservation themes. On the contrary, development if necessary should be sacrificed to protect valuable biodiversity. Development objectives most often have viable alternatives, whereas biodiversity losses in most cases are irreplaceable.

6. CONCLUSIONS

This paper contains two distinct sections. The first section is an overview of biodiversity conservation and sustainable development and the second section is a case study with reference to NBSAP–Oman 2001.

The intention of the first section is to lead the reader from basic definitions of biodiversity and sustainable development through various perceptions on conservation and development leading to the Earth Summit in

Rio de Janeiro (1992) and the World Summit on Sustainable Development in Johannesburg (2002). The historical review of the evolution of thoughts on conservation and sustainable development including contradictory opinions is absolutely necessary for the environmental professionals charged with the administration of biodiversity affairs.

When a new proverbial bandwagon rolls along, there often is a human tendency to endorse it with or without fully understanding its objectives. This seems to be the case with biodiversity and sustainable development scenarios that are in vogue today. After listening to the interpretations, explanations and discussions of numerous scientists, environmentalists and administrators of local, regional and international organizations, one cannot help but marvel at their different conceptions, especially about sustainable development. Similarly, some advocates for sustainable development, so taken up with the utilitarian values of human requirements often leave conservation behind. The overview presented implies the need for an integrated approach to deal with conservation and development.

The second section is a summary of the status of biodiversity conservation and sustainable development in the Arabian region and in the Sultanate of Oman. NBSAP–Oman is a recognized legal instrument that should protect biodiversity. Strategies and action plans are conceptual documents that identify key issues under various themes, state the objectives and propose priority actions. However, these strategies should be analyzed on a case–by–case basis. The case study considered here is that of a fragile mountain ecosystem that is undergoing rapid development at present, in the context of NBSAP–Oman. The entire analysis was based not only on what is known about the region in literature, but also on primary research conducted by this author and his graduate students from 1992–2002. This study gives a blueprint for biodiversity protection and sustainable development in the Jebel Akhdar region under NBSAP framework. It recommends the use of Pressures–States–Responses methodology for habitat conservation and the use of habitat loss as an indicator to evaluate biodiversity loss. It also suggests the methods for developing priority matrices for mitigation measures through best practices. This case study approach will be helpful for the implementation of NBSAP in Oman.

7. REFERENCES

Anonymous 2001a. African Ministerial Statement to the World Summit on Sustainable Development. African Preparatory Conference for the World Summit on Sustainable Development, 15–18 October 2001, Nairobi. 12 pages.

Anonymous 2001b. Draft conclusion of the Oman International Forum on the Environment and Sustainable Development, 22–23 December 2001, Muscat, Sultanate of Oman. 4 pages.

Arnold, E.N. 1987. Zoogeography of the Reptiles and Amphibians of Arabia. Pages 245–256 In: Krupp, F., Scheider, W. and Kinzelbach, R. (Eds.) Proceedings of the Symposium on the Fauna and Zoogeography of the Middle East, vol. Reihe A (Naturwissenschaften) 28. Dr. Ludwig Reichert Verlag, Wisbaden.

Arnold, E.N., and Gallagher, M.D. 1977. Reptiles and Amphibians from the Mountains of Northern Oman with Special Reference to Jebel Akhdar Region. Pages 59–80 In: Harrison, D.L. (Ed.) The Scientific Results of the Oman Flora and Fauna Survey 1975. The Journal of Oman Studies special Report No.1. Ministry of Information and Culture, Muscat, Sultanate of Oman.

Bayfield, N. 2001. Mountain Resources and Conservation. Pages 7–38 In: Warren, A. and French, J.R. (Eds.) Habitat Conservation. Managing the Physical Environment. John Wiley & Sons, Ltd. New York.

Evans, M.I. 1994. Important Bird Areas in the Middle East. Birdlife International, Cambridge, U.K. 320 pages.

Fisher, M. 1999. The Conservation Status of the Terrestrial Mammals of Oman: A Preliminary Red List. Pages 109–127 In: Fisher, M., Ghazanfar, S.A. and Spalton, J.A. (Eds.) The Natural History of Oman: A Festschrift for Michael Gallagher. Backhuys Publishers, Leiden.

Fisher, M., and Gardner, A.S. 1995. The status and ecology of a *Juniperus excelsa* subsp. *polycarpos* woodland in the northern mountains of Oman. Vegetatio 119: 33–51.

Fyfe, W.S., McEwen, J., and Sadar, M.H. 1993. Sustainable Development: The need for Holistic Approaches. Pages 1–23 In: Anonymous (Ed.) Alfred Wegener Stiftung (Hrsg), Die Benutze Erde–Ökosysteme, Rohstoffgewinnung, Herausforderungen 1993. Ernst and Sohn Verlag für Architektur und Technische Wissenschaften GmbH, Berlin.

Gallagher, M. 1977. Birds of Jabal Akhdar. Pages 27–58 In: Harrison, D.L. (Ed.) The Scientific Results of the Oman Flora and Fauna Survey 1975. The Journal of Oman Studies special Report No.1. Ministry of Information and Culture, Muscat, Sultanate of Oman.

Gardner, A.S.1999. The Reptiles and Amphibians of Oman: Herpetological History and Current Status. Pages 65–88 In: Fisher, M., Ghazanfar, S.A. and Spalton, J.A. (Eds.)The Natural History of Oman: A Festschrift for Michael Gallagher. Backhuys Publishers, Leiden.

Gardner, A.S., and Fisher, M. 1994. How the forest lost its trees: Just so stories about juniper in Arabia. Journal of Arid Environments 26: 199–201.

Gardner, A.S., and Fisher, M. 1996. The distribution and status of the montane juniper woodlands of Oman. Journal of Biogeography 18: 299–309.

Ghazanfar, S.A. 1991. Vegetation structure and phytogeography of Jabal Shams, an arid mountain in Oman. Journal of Biogeography 18: 299–309.

Ghazanfar, S.A. 1998. Vegetation of the Plains. Pages 175–190 In: Ghazanfar, S.A. and Fisher, M. (Eds.) Vegetation of the Arabian Peninsula. Geobotany 25. Kluwer Academic Publishers, Dordrecht.

Glennie, K. 1995. The Geology of the Oman Mountains; an Outline of Their Origin. Scientific Press, Beaconsfield. 92 pages.

Grzimek, B., and Grzimek, M. 1960. Serengeti Shall Not Die. Hamish Hamilton, London. 344 pages.

Hanna, S. 1995. Field Guide to the Geology of Oman. The Historical Association of Oman, Muscat. 178 pages.

Harrison, D.L. (Ed.) 1977. The Scientific Results of the Oman Flora and Fauna Survey 1975. The Journal of Oman Studies special Report No.1. Ministry of Information and Culture, Muscat, Sultanate of Oman. 267 pages.

Hussain, M.I., Victor R. and Khoja, T.M. In press. Charophyta of the Sultanate of Oman, Southern Arabia. Nova Hedwigia.

Llwellyn–Smith, R. and Launay, F. 2001. Al Hajar montane woodlands (AT0801). http://www.worldwildlife.org/wildworld/profiles/terrestrial/at/at0801_full.html. 6 pages.

Mandaville, J.P. Jr. 1977. Plants. Pages 229–267 In: Harrison, D.L. (Ed.) The Scientific Results of the Oman Flora and Fauna Survey 1975. The Journal of Oman Studies special Report No.1. Ministry of Information and Culture, Muscat, Sultanate of Oman.

Mugabe, J. 1998. Introduction. Pages 1–29 in Mugabe, J. and Clark, N. (Eds.) Managing Biodiversity. National Systems of Conservation and Innovation in Africa. ACTS Press, Nairobi.

Mugabe, J., and Clarke, N. (Eds.) 1998. Managing Biodiversity. National Systems of Conservation and Innovation in Africa. ACTS Press, Nairobi. 279 pages.

Nicholson, E.M. 1970. The Environmental Revolution: A Guide for New Masters of the World. Hodder and Stoughton, London. 366 pages.

Oates, J.F. 1999. Myth and Reality in the Rain Forest. How Conservation Strategies are Failing in West Africa. University of California Press, Berkeley, Los Angeles and London. 310 pages.

Robinson, J.G. 1993. Limits to caring: Sustainable living and the loss of biodiversity. Conservation Biology 7: 20–28.

Spalton, A., and Willis, D. 1999. The Status of the Arabian leopard in Oman: First Results of the Arabian Leopard Survey. Pages 147–60 In: Fisher, M. Ghazanfar, S.A. and Spalton. J.A. (Eds.) The Natural History of Oman: A Festschrift for Michael Gallagher. Backhuys Publishers, Leiden.

Stanger, G. 1986. The Hydrogeology of the Oman Mountains. Volume 1, Main text, Ph.D. Thesis, The Open University, U.K. 355 pages.

Tashiro, A. 2001. Discounted Casualties. The Human Cost of Depleted Uranium. The Chugoku Shimbun, Transnet, Hiroshima. 144 pages.

Victor, R., and Al–Ujaily, S.R. 1998. Reservoir Limnology in Arid Northern Mountains of Oman: Problems and Prospects. Internationale Revue gesamten Hydrobiologie 83: 73–82.

Victor, R. and Al–Ujaily, S.R. 1999. Water Quality and Management Strategies of Mountain Reservoirs in Arid Northern Oman. Pages 307–350 In: Goosen, M.F.A. and Shayya (Eds.) Water Management, Purification and Conservation in Arid Climates. Technomic Publishing Co., Inc., Lancaster and Basel.

Victor, R., and Victor, J.R. 2002. Zooplankton of an arid zone mountain reservoir in the Sultanate of Oman, Arabia. Verhaltnisse Internationale Vereingung Limnologie 28: 507–512.

Ward, B., and Dubos, R. 1972. Only One Earth: The Care and Maintenance of a Small Planet. W.W. Norton, New York. 225 pages.

Wilkinson, J.C. 1974. Water and Tribal Settlement in Southeast Arabia: A Study of the Aflaj of Oman. Clarendon Press, Oxford. 276 pages.

Chapter 36

ACHIEVING INSTITUTIONAL COOPERATION FOR IMPLEMENTATION OF SUSTAINABLE DEVELOPMENT PLANS AND STRATEGIES

Donald A. Brown

Senior Counsel for Sustainable Development, Department of Environmental Protection, Commonwealth of Pennsylvania, USA; Director, Pennsylvania Consortium for Interdisciplinary Environmental Policy, Commonwealth of Pennsylvania, USA; emaill brownd@state.pa.us

1. INTRODUCTION

Thick sustainable development plans and strategies are gathering dust around the world, soon to be forgotten, and likely to be thrown away with the next serious housecleaning. This is so lamentably because many sustainable development plans and strategies frequently fail to be implemented, become quickly outdated, and are reinvented by the next group with responsibility to develop a plan or strategy. This next group is often unfamiliar with previous plans and strategies and sometimes spends enormous amounts of energy fighting the same fights and drawing similar conclusions about what needs to be done to move a place to a sustainable future only to have the new plan become another in a series of documents created by wishful thinking. For instance, since the Earth Summit in 1992 in Rio de Janeiro alone there are hundreds of decisions that have been made by the United Nations Commission on Sustainable Development (UNCSD) and the United Nations Environment Program (UNEP) in their yearly meetings that have failed to be taken seriously by the parties that participated in negotiating these decisions (Brown 1998). As these issues are reconsidered in future negotiations, negotiators often fight the same fights and reach similar compromises only to have the new decision become the next in a series of decisions to be ignored. As another example, in the United States there are innumerable land use plans that have never been implemented.

There are many reasons for these failures. They include the party preparing the plan or strategy: (1)has no political will to implement it, (2)has no authority to implement it, (3)has no money to implement it, or (4)has failed to obtain the cooperation of institutions necessary to implement it.

If the party preparing a plan or strategy has no political will or authority to implement a sustainable development plan, not much can be done. If implementation of the plan or strategy depends on financial resources that don't exist, nothing will happen until the money is found. In this paper, I identify barriers to and methods for overcoming the last of these common reasons for failure to implement plans or strategies, namely the need to obtain cooperation between institutions necessary to implement a sustainable development plan or strategy.

2. BARRIERS TO INSTITUTIONAL COOPERATION

To develop a plan for institutional cooperation to implement sustainable development plans or strategies, it is helpful to have a better and realistic understanding of why institutional cooperation is often so difficult and challenging. The following discussion identifies barriers to institutional cooperation to implement sustainable development plans that are often overlooked or are not well understood. Although this list describes a daunting number of barriers to obtaining institutional cooperation, it has been prepared in the belief that better understanding of these barriers will lead to greater success in implementation of sustainable development plans and strategies. Occasionally, some plans and strategies change the status quo by overcoming these barriers. A better understanding of these barriers to implementation of sustainable development plans and strategies is also a key to the institutional reform needed to implement sustainable development programs.

2.1 Failure to Fix Responsibility for Interdisciplinary Strategic Planning

Institutions can only cooperate effectively to solve sustainable development problems if at least one of the institutions is responsible for overall strategic planning for the problem. That is so because to develop effective strategies to solve many sustainable development problems requires analysis of sustainable development trends, determination of the causes of negative trends, and the selection of responses that are capable of reversing negative trends. To perform these tasks often requires scientific, technical, legal, and sometimes social analyses. In other words, sustainable development problem solving often requires rigorous interdisciplinary strategic planning. Yet, many institutions working on sustainable development issues have never made rigorous interdisciplinary strategic planning a central focus of institutional operations or don't have the mandate to do so (Meyers 1993). If institutions needed to develop or implement a plan or strategy don't engage in serious strategic planning or don't have the ability to implement needed strategic actions that have been identified by others, then the plan or strategy will not be implemented or will be ineffective. There are several reasons why government agencies haven't made interdisciplinary strategic planning a focus of their institutional mission are as follows.

2.1.1 Institutional Priorities Not Set On Strategic Planning Basis

Some institutions' sustainable development program priorities are not determined through an interdisciplinary strategic planning process. There are several examples of this. The program of work for some international institutions such as UNEP and the UNCSD are proscribed by their international governing body. These bodies are comprised of representatives from nations who often have greatly different views on how limited funds should be spent. For instance, the UNEP governing body is often split along regional and developed versus developing nations. The sustainable development problems of priority in the developed nations tend to be different then the priorities in the developing nations lines. The industrialized nations, for instance, want UNEP to focus on global environmental issues, while developing nations want assistance with domestic problems such as sewage treatment and drinking water. Although UNEP staff draws up a proposed program of work before UNEP's budget negotiations begin that includes some strategic thinking, UNEP's authorized program of work is actually determined during negotiations over UNEP's budget that is developed through intense negotiations among nations who are competing to direct UNEP's limited financial assets to serve individual national interests rather than to meet strategic environmental needs. In fact, many of the negotiators fighting over UNEP's budget have little understanding of overall strategic global sustainable development needs. As a result, not only are grossly inadequate resources provided to run UNEP programs (UNEP as of a few years ago had over 100 programs and less than $50 million budget), but negotiators representing nations frequently give UNEP new mandates to meet national interests rather than fund priorities that have been determined in sustainable development strategic planning. And so, UNEP's program of work is determined through political negotiations rather than interdisciplinary strategic planning. Because of intense political pressure, UNEP's budgets are often are set without serious reflection upon what is actually needed to solve sustainable development problems. As a result UNEP's programs are often, as the saying goes, a "mile wide but an inch deep." That is, most of UNEP's many programs, through no fault of UNEP, are run with resources insufficient to do meaningful strategic planning.

Another reason why institutions that develop sustainable development plans and strategies fail to focus on strategically determined priorities is that many environmental institutions' programs are set by legislative mandates rather than through interdisciplinary strategic planning. For instance, most environmental agencies in the United States have usually been preoccupied and consumed with implementing complex environmental statutes on air, water, and waste pollution. These statutes often overtax the resources of implementing agencies with such responsibilities as adopting regulations, issuing permits or approvals under the statutes, taking enforcement actions against those who do not comply with the law, and defending government decisions in court proceedings. As a result these agencies are often primarily staffed by engineers that review permits and write regulations and lawyers responsible for enforcement and defend the agency in legal proceedings. Very little resources are available for strategic planning and the work of the employees that staff these agencies tend to be programmed by their narrow professional backgrounds. Because United States' environmental agencies are consumed with implementing specific laws, they are often structured on a media basis with separate departments for air, water, and solid waste among others. These departments mostly work on implementing the individual statutes that created specific environmental programs. The statutes and regulatory structures are sufficiently complex that personnel working in different departments have little understanding of the regulatory logic of other department's programs. As strange as it may seem, an environmental professional working on water pollution often has no understanding of how the air pollution program works within the same agency. Because it takes several years to understand the regulatory logic of environmental programs because of their complexity, political appointees that are frequently brought in to administer the program leave the program to be replaced by the next political appointee just as he or she is beginning to understand how the program works. Budget proposals are often derived from an analysis of how many people or how much money is needed to implement the program mandated by the legislature that created them rather than on desired resources needed to meet strategically determined sustainable development problem solving needs. In addition, these agencies often assume that their mission is to implement only those programs that have been created by legislative bodies and as a result little energy is expended in serious sustainable development strategic planning. In fact, as strange as it seems, heads of environmental agencies often think it is none of their business to identify new causes of problems; such activity would amount to environmental advocacy, and a bureaucracy's function is to narrowly carry out the specific mandate of the legislature.

2.1.2 Institutional Fragmentation

Although sustainable development problem solving requires interdisciplinary analyses, many institutions have narrow jurisdictional responsibilities. For instance, some of the most serious water quality problems in many parts of the United States stem from non–point sources of pollution. To understand the causes of non–point caused water pollution, one must understand land use trends and activities as well as pollutant deposition to water bodies from air–borne contaminants. Yet the regulatory activity of water pollution programs is focused almost exclusively on point sources of water pollution. For example, in Pennsylvania over 85 percent of the streams not meeting water quality standards fail to do so because of non–point pollution loading yet most of the Pennsylvania Bureau of Water Pollution Control's resources are focused on point sources of pollution because that is what the authorizing legislation is most focused on. To control pollution from non–point pollution in Pennsylvania it will be necessary to control land use activities and air–borne contaminants, yet the Pennsylvania Bureau of Water Pollution Control, the state institution responsible for water pollution, has no authority to work on land use and air pollution problems. Air pollution is the responsibility of a Bureau of Air Quality that is focused on ambient air quality concerns and that correspondingly has no authority to work on water quality issues. Land use is a local government responsibility and therefore not a state concern.

Not only are environmental institutions fragmented in terms of the scope of their responsibility, they are often also staffed by a limited number of types of expertise. For instance, water pollution programs in many parts of the United States have largely been staffed by engineers because the main function of these programs has been to review water pollution control projects from point sources. Yet to understand the causes of non–point pollution expertise in local land use law and regional planning is essential. However, usually neither the technical nor legal employees of most environmental agencies in the United States have knowledge about land use planning and law. And so although sustainable development problems often raise deep interdisciplinary questions, most environmental institutions look at these problems through a narrow technical or legal lens. Workers are often trained to know something about one aspect of sustainable development problems such as biology but know little about others such as economics, international institutions, or law, not to mention numerous other technical disciplines. As amazing as it is, in my experience environmental agencies rarely look to ecologists for policy guidance and in fact have few, if any, ecologists on staff, because most environmental regulatory programs are based on human health endpoints and decisions are based upon an engineer's review of whether models indicate human health derived standards will be exceeded by a proposed project. As a result, environmental agencies to this very day rarely think about ecological risk of human activities and haven't developed the resources to think about ecological damage.

Universities rarely train students in broad policy issues and experience is obtained on the job in institutions that were designed for an era when linkages between institutions were not seen to be an important priority. As a result, persons working on one international treaty such as the United Nations Convention on Climate Change (CCC) have little working knowledge of other international treaties such as the United Nations Convention on Biological Diversity (CBD) despite the fact that there are linkages between climate stresses and biodiversity losses and a need to consider these linkages in formulating policy (Institute for Sustainable Development 2001).

When institutions have only limited and fragmented responsibility for aspects of sustainable development problems, then nobody has responsibility for overall interdisciplinary strategic thinking. Yet to solve emerging sustainable development problems, large amounts of cooperation will be necessary among institutions that rarely cooperate because they are consumed with existing but narrow responsibilities. In fact, many institutions have narrow mandates about what they can do (Meyers 1993). Additionally, cooperative efforts to coordinate strategies is difficult to accomplish if no one has overall strategic planning responsibility for the problem.

2.1.3 Failure to Consider What is Needed for Implementation of Environmental Plans and Strategies

Often strategic plans or strategies are prepared by organizations that have no legal authority to implement the plans, programs, or strategies. Where cooperation is needed among many institutions to implement a plan or strategy, plans or strategies are often prepared without agreement by all the persons who must implement the plan, program or strategy. As a result, sustainable development plans and strategies often never amount to much more than wishful thinking and planning documents are soon stored away and forgotten.

Those who have the authority to implement environmental plans and strategies are often not included in plan development. For instance, in the United States land use plans are often prepared by county government in states where land use authority exists at the municipal level. These county plans are often ignored by the local governments who did not participate in their development.

The United States after the 1992 Rio Earth Summit relied on the President's Council for Sustainable Development (PCSD) to develop programs for the United States. Yet, PCSD was only an advisory group to the United States and had no authority to implement its recommendations. As a result may of its worthwhile recommendations were ignored. Other governments created new government bodies after Rio to develop national sustainable development strategy. These governments have been more successful in getting recommendations implemented because the new institution creating the recommendations had authority to implement proposed strategies. For this reason, those preparing environmental plans or strategies must seek to include those with authority to implement the plan or strategy in the development of the plan or strategy.

In many cases, a plan's or strategy's implementation will require adoption and implementation by multiple parties. In these cases it is particularly important that all parties that need to cooperate to have a plan or strategy implemented participate in plan development. Where no clear authority to implement a plan, program or strategy exists, plans should lead to the adoption of legislation to create needed authority. In these cases early involvement of legislative bodies may be crucial. In any event, at minimum every environmental plan or strategy should identify who has the authority to implement the plan or strategy or what legislation is necessary to implement the plan. Yet many plans are prepared by scientists who are not trained to think about implementation issues.

2.1.4 Failure to Match the Scope of Investigation with the Scale of the Problem

Environmental problems transcend political jurisdictions. Yet decision makers who authorize the development of a plan or strategy only have authority over decisions within their jurisdiction. For this reason, sustainable development plans and strategies often fail to be developed that match the scale of the problem. In order to solve sustainable development problems it is often necessary to understand and control human activities that will threaten entire ecosystems of concern. To do this requires the cooperation all parties with authority to implement a plan or strategy within ecosystem. It also requires study of how human activities may affect the entire ecosystem. Yet sustainable development plans and strategies are often prepared for only part of an ecosystem because the government agency preparing the plan or strategy has jurisdiction over only part of the ecosystem. For instance, municipalities have sometimes adopted ordinances to protect ground water only to see their ground water degraded by activities in other municipalities that share groundwater resources yet that have not adopted similar protection strategies.

Similarly, watershed management protection programs can only be implemented if all agencies with authority in the watershed to implement the watershed management plan commit to do so. This requires ecosystem–wide cooperation among all relevant institutions at all stages of a strategic planning process. From the very beginning, the scale of investigation used to formulate environmental plans and strategies should match the ecological scale of the problem to the maximum extent possible.

2.1.5 Failure to Consider Interdisciplinary Implementation Questions

Sustainable development problems raise scientific, economic, legal, social, and ethical questions (Lemons and Brown 1995). Yet, much of the research performed on environmental and sustainable development problems is focused in on important but narrow scientific issues. For instance, a biological analysis of a threatened aquatic ecosystem might be performed, yet data on trends of human activities that are stressing the ecosystem are not studied. As a result, environmental decisionmakers rarely have enough information to make accurate analyses of causes of negative environmental trends. To have a good understanding of the causes of negative environmental trends often requires additional study after a determination of negative environmental trends has been made. For instance, biological monitoring might establish that an aquatic ecosystem is in decline. Yet even if it is known that the case of this decline is non–point sources of pollution, establishing the root causes of the nonpoint pollution might require significant additional investigation. That is, investigation might be needed to determine: (1)what are the major sources of this pollution, an investigation that might require additional monitoring data; (2)who has the authority over the activities that are causing the pollution, an investigation that might require legal analysis; and (3)what are the barriers preventing those with authority to control activities that are degrading the ecosystem from exercising that authority, an investigation that might require economic, sociological, or political analysis.

For this reason, the selection of a government response in a plan or strategy to negative environmental trends that has the potential to correct the negative trends must be based upon interdisciplinary root cause analyses of negative trends. Yet many scientifically based environmental plans and strategies often fail to sufficiently consider interdisciplinary implementation issues.

2.1.6 Failure to Build a Sustainable Development Information Base for Plans and Strategies

Cooperation among multiple parties to solve environmental problems requires reliance on information that is accessible to multiple parties working at different spatial and temporal scales and which information can be supplemented in the future to determine positive or negative environmental trends. Yet, often environmental information is collected without understanding the need to use it at different spatial and temporal scales.

For instance, the UNCSD created a list of sustainable development indicators that are being used by many nations in a way that allows international comparison of success, yet the United States developed its own methods and indicators that cannot be aggregated at global scales because of incompatibility with the United Nations' global scale indicators. Nor has the United States encouraged the development of indicators by governments at state, regional, and local scales that would allow aggregation at the national scale and comparisons among governments at the same scale. In addition, it is quite common for studies to be performed of biological conditions of ecosystems without consideration of the need to reproduce the same study in the future to show positive or negative trends. In order for biological information to be usable for long–term trend analysis its acquisition methods must be reproducible in the future. Because of the failure to consider long–term implementation uses of biological information, biological studies are often only valuable for the particular time and place at which the information was gathered.

2.1.7 Failure to Consider Prior Plans, Decisions, or Information in Decisionmaking

Many international bodies meet and discuss problems with little understanding of the prior legal or institutional decisionmaking that has considered the same or similar issues. For instance, for many years various international institutions have considered conservation of fresh water, including UNCSD, UNEP, WHO, UNDP, the Global Water Forum and others that have put global fresh water conservation on their agenda. As each new institution has considered global water issues they have often begun deliberations without an understanding of prior decisions on the very same subjects under consideration. Much of the negotiation time therefore has focused on issues on which plans, decisions, or recommendations have been previously adopted. Rather than building on prior decisions, environmental and sustainable development planning projects often waste huge amounts of time reinventing decisions that have already been made while failing to focus on practical action steps that could make a difference in the status quo. Deliberations and negotiations on sustainable development should minimize to the extent possible issues and matters that have already been considered.

Many environmental investigations are carried out without an understanding of how the project might fit into national or regional strategy. As a result, the investigation is not useable in implementing the strategy. For instance, the CBD requires each nation to develop national strategies to protect or allow for sustainable use of biodiversity. The national strategy that has been adopted to implement the CBD might prioritize biologically rich regions as requiring most attention, yet few scientists working in institutions doing biodiversity research may be aware of national strategies to protect biodiversity.

2.1.8 Institutional Rivalry, Overlap, and Issues of Vertical and Horizontal Institutional Scale

Government institutions are frequently created whose mandate overlaps with other institutions. For instance, at the international level there is large overlap between UNEP, UNCSD, UNDP, WHO, and the World Bank. These institutions often compete with each other rather than cooperate to solve sustainable development problems. As a result much duplication of work is performed rather than filling strategically important policy gaps through sharing scarce resources. In addition to horizontal duplication, vertical duplication and overlap of functions is also quite common. That is, international, national, regional, and local governments often fail to work on those parts of a problem that are appropriate for their scale. In fact, harmonization of work at the appropriate scale is rare. Municipal governments may collect information, for instance, that is not usable at a county or national scale although it would be of interest if it was collected with a view toward multi–scale use. For this reason, work performed at one scale is often not useable at other scales.

Another related problem is the failure of higher scale program development to include lower scale institutions in program design, development, and implementation. For instance, a national government might develop a sustainable development indicator without any knowledge of how information on the same subject is collected at regional or municipal scale. Much environmental plan and strategy failure stems from this failure to have all relevant institutions participate in problem analysis, strategy formation, and implementation.

3. METHODS TO ACHIEVE INSTITUTIONAL COOPERATION

Because of these barriers to institutional cooperation in environmental problem solving the following steps focusing on creating inventory information should be taken by any institution responsible for sustainable development plans or strategies: (1)identify other institutions whose missions and scope of duties and responsibilities can contribute to solution of the problem; (2)identify and obtain data and information that has already been collected by other institutions relevant to the problems under concern; (3)fill in gaps in information where possible; (4)identify and obtain relevant legal commitments in treaties, national, local laws, or decisions of relevant sustainable development institutions; (5)obtain guidance to implement treaties, national laws, or commitments; and (6)determine what people will be affected by or otherwise have a stakeholder interest in the project.

3.1 Obtain Agreement to Cooperate and Participate by Relevant Institutions and Potentially Affected Peoples and Interested Stakeholders

Having determined which institutions have a role in developing or implementing a plan or strategy under consideration, each institutions should attempt to get the full participation and cooperation of all relevant institutions, affected people, and interested stakeholders to: (1)assign responsibility for interdisciplinary strategic planning; (2)eliminate overlap and duplication among institutions to the maximum extent possible by dividing tasks among institutions; (3)agree on common methods of quantitative measurement to the extent possible so that aggregation at appropriate scales can take place; (4)obtain agreement on which institutions at which scales will perform specific tasks; (5)determine which prior existing plans, agreements, commitments or legal obligations shall be built upon rather than duplicated; and (6)obtain agreement from relevant institutions and stakeholders on elements of the strategy or plan as it is developed and implemented.

3.2 Integrate Environmental, Economic Development, and Social Policy at All Scales

One of the most frequent reasons why sustainable development projects and strategies fail is because environmental protection, economic development and social justice activities and policy making have not been integrated at all scales. If there is no party charged with institutional integration of environment, economic, and social issues, steps should be taken to see that those responsible for economic development, environmental problems, and social justice policy commit to develop and implement the project or strategy.

3.3 Assure that Legal Authority and Commitments Exist to Implement Sustainable Development Strategies

Many plans and strategies developed to solve sustainable development problems have failed to change the status quo because plans or strategies were prepared without consideration of whether authority and commitment to implement the plan or strategy exists. Where no legal authority to implement the plan and strategy exists, those preparing the plan or strategy should seek cooperation from the legislature to obtain the authority to implement the plan or strategy. Where authority to implement the plan or strategy already exists, persons preparing the plan or strategy should seek to obtain commitment from institutions with authority to implement the plan or strategy.

3.4 Build Quantitative Information on Long Term Trends of Sustainable Development Success

Sustainable development program success requires that policy makers monitor environmental, economic, and social trends of concern, human activities that are causing negative trends, and human responses to correct negative trends. Often, scientific information is collected without understanding either the needs to use that information over the long–term or needs to aggregate data from various scales. Cooperation between institutions is necessary to build baseline information that is useable by all parties at all scales and to build on the information over time to monitor success and to identify actions needed to reverse negative trends. Therefore, parties preparing sustainable development plans and strategies should seek to create data collection procedures and an information baseline that will be useable for implementation of plan and strategy.

3.5 Conduct Training Activities

It is important that persons participating in the project or strategy development or implementation understand the objectives of the programs, how the program fits into global, national, regional, and local scale activities, information and data development needs and methods, and other information relevant to program or strategy success. To accomplish this goal, training programs need to be conducted particularly of those that will develop the plan or strategy. For instance, projects at local scales to develop programs to protect biodiversity should understand how the local problem fits into national strategies to protect biodiversity. For this reason, persons preparing a sustainable development plan or strategy need to be trained on how the project might fit into regional and national obligations.

3.6 Secure Interdisciplinary Expertise to Develop and Implement Strategies

It is important to have participation of scientific, economic, legal, ethical, engineering, public policy expertise involved with all phases of plan or strategy development and implementation. These disciplines must be deeply integrated into the program development because sustainable development program success requires that governments no longer split the world into disciplinary components.

4. REFERENCES

Brown, D. 1998. Making the United Nations Commission On Sustainable Development Work, Earth Negotiations Bulletin. Winnipeg, Manitoba.

International Institute for Sustainable Development. 2001. Summary Report of the Informal Regional Consultations of Interlinkages, Synergies and Coordination Among Multilateral Environmental Agreements 48: 1–11.

Lemons, J. and D. Brown. (eds.) 1995. Sustainable Development: Science, Ethics, and Public Policy. Kluwer Academic Publishers, Dordrecht, The Netherlands. 281 pages.

Myers, N. 1993, The Question of Linkages in Environment and Development. BioScience 43: 302–310.

Subject Index

MIX
Papier aus verantwortungsvollen Quellen
Paper from responsible sources
FSC® C105338

If you have any concerns about our products,
you can contact us on
ProductSafety@springernature.com

In case Publisher is established outside the EU,
the EU authorized representative is:
Springer Nature Customer Service Center GmbH
Europaplatz 3, 69115 Heidelberg, Germany

Printed by Libri Plureos GmbH
in Hamburg, Germany